Advances in Fire Suppression

Editors

Song Lu
Changcheng Liu
Guohui Li
Paweł Wolny

MDPI

Basel • Beijing • Wuhan • Barcelona • Belgrade • Novi Sad • Cluj • Manchester

Editors
Song Lu
University of Science and Technology of China
Hefei
China

Changcheng Liu
North University of China
Taiyuan
China

Guohui Li
Tianjin Fire Science and Technology Research Institute of MEM
Tianjin
China

Paweł Wolny
Łódź University of Technology
Łódź
Poland

Editorial Office
MDPI AG
Grosspeteranlage 5
4052 Basel, Switzerland

This is a reprint of articles from the Special Issue published online in the open access journal *Fire* (ISSN 2571-6255) (available at: https://www.mdpi.com/journal/fire/special_issues/Fire_Suppression).

For citation purposes, cite each article independently as indicated on the article page online and as indicated below:

Lastname, A.A.; Lastname, B.B. Article Title. *Journal Name* **Year**, *Volume Number*, Page Range.

ISBN 978-3-7258-2401-4 (Hbk)
ISBN 978-3-7258-2402-1 (PDF)
doi.org/10.3390/books978-3-7258-2402-1

Advances in Fire Suppression

Contents

Preface

As the Guest Editors of this Special Issue, we are honored to present a collection of scholarly articles that delve into the intricate world of fire safety and prevention. This reprint encompasses a comprehensive array of topics, from predictive modeling using machine learning to innovative firefighting techniques and the synthesis of eco-friendly flame retardants. Our aim is to provide a platform that bridges the gap between theoretical research and practical applications, offering insights that can inform and enhance the strategies employed in mitigating fire hazards.

The motivation behind this compilation stems from the pressing need for advancements in fire safety technology and the desire to share knowledge that can save lives and protect property. We have carefully selected these articles for their contributions to the field, each offering a unique perspective on the challenges and opportunities in fire science.

This reprint is addressed to a diverse audience, including researchers, engineers, policymakers, and public safety professionals, who share a common interest in advancing the frontiers of fire safety. We extend our heartfelt thanks to the authors for their invaluable contributions and to the peer reviewers as well as editorial staff for their diligent work in ensuring the quality of this publication.

We are also grateful for the support and assistance from various institutions and individuals who have helped bring this Special Issue to fruition. Their collaboration has been instrumental in creating a reprint that we hope will serve as a valuable resource for years to come.

In particular, we would like to express our deep appreciation for the financial support provided by the National Natural Science Foundation of China (grant nos. 12202410 and 51906238), the Project Funded by the China Postdoctoral Science Foundation (grant nos. 2023T160734 and 2023M733935), the Natural Science Foundation of Hunan Province (grant no. 2023JJ40726), the Changsha Municipal Natural Science Foundation (grant no. kq2208277), the Research Project Supported by the Shanxi Scholarship Council of China (grant no. 2022-139), the Fundamental Research Program of Shanxi Province (grant nos. 20210302123017 and 202303021211145), the Fund Program for the Scientific Activities of Selected Returned Overseas Professionals in Shanxi Province (grant no. 20220012), and the Opening Foundation of Key Laboratory in North University of China (grant no. DXMBJJ2023-03).

Song Lu, Changcheng Liu, Guohui Li, and Paweł Wolny
Editors

fire

Editorial

Brief Introduction on Advances in Fire Suppression

Changcheng Liu [1,2,3], Guohui Li [4,*] and Song Lu [5,*]

1 School of Environment and Safety Engineering, North University of China, Taiyuan 030051, China; ccliu@nuc.edu.cn
2 Institute of Advanced Energy Materials and Systems, North University of China, Taiyuan 030051, China
3 Shanxi Key Laboratory of Efficient Hydrogen Storage & Production Technology and Application, Taiyuan 030051, China
4 Tianjin Fire Science and Technology Research Institute of MEM, Tianjin 300381, China
5 State Key Laboratory of Fire Science, University of Science and Technology of China, Hefei 230022, China
* Correspondence: liguohui@tfri.com.cn (G.L.); lusong@ustc.edu.cn (S.L.)

Citation: Liu, C.; Li, G.; Lu, S. Brief Introduction on Advances in Fire Suppression. Fire 2024, 7, 309. https://doi.org/10.3390/fire7090309

Received: 26 June 2024
Accepted: 27 August 2024
Published: 30 August 2024

1. Introduction

As society rapidly evolves and urbanization accelerates, fire safety has become an increasingly pressing concern. Fire-extinguishing technology serves as a vital tool for the prevention and control of fires, capable of averting casualties and mitigating severe damage to the environment. The performance of extinguishing agents, a central component of fire-extinguishing technology, is pivotal to both the effectiveness of fire suppression and its environmental repercussions. Notably, technologies utilizing extinguishing agents such as water, dry powder, gas, and foam have demonstrated significant utility across a variety of application scenarios [1–4]. In light of the recent advancements in efficient fire-extinguishing technologies, access to timely research findings and profound insights is imperative for expanding the knowledge base in this domain. This Special Issue compiles 11 papers that tackle emerging issues and challenges in various fire suppression technologies, offering the necessary insights for effective fire mitigation.

2. Descriptions

The research scope of this Special Issue encompasses the development of fire-extinguishing agents, the design of fire-extinguishing systems, the transport dynamics of agents, dispersion kinetics, and simulation studies. These areas of research lay a robust foundation for the evolution of fire-extinguishing technology and its practical application. As environmental consciousness grows and demands for fire-extinguishing efficiency escalate, the innovation of novel extinguishing agents has emerged as a prevalent field of study. Notably, water-based fire-extinguishing agents, which leverage water as their primary component, have seen enhancements in their fire-suppression capabilities through the incorporation of chemical additives or the modification of water's physical properties. Among these, fine water mist systems, recognized for their cleanliness and efficacy, have garnered considerable interest [1]. According to the national standard of China for "Water mist extinguishing equipment" [5], fine water mist is characterized by droplets with a flow-weighted cumulative volume distribution of less than 200 μm ($D_{v0.50}$) and 400 μm ($D_{v0.99}$) under the minimum design operating pressure. Systems are categorized based on operating pressure into high pressure ($p \geq 3.50$ MPa), medium pressure (1.20 MPa $\leq p <$ 3.50 MPa), and low pressure ($p <$ 1.20 MPa), where P denotes the pressure of the medium within the distribution network. The fire-extinguishing mechanisms are illustrated in Figure 1.

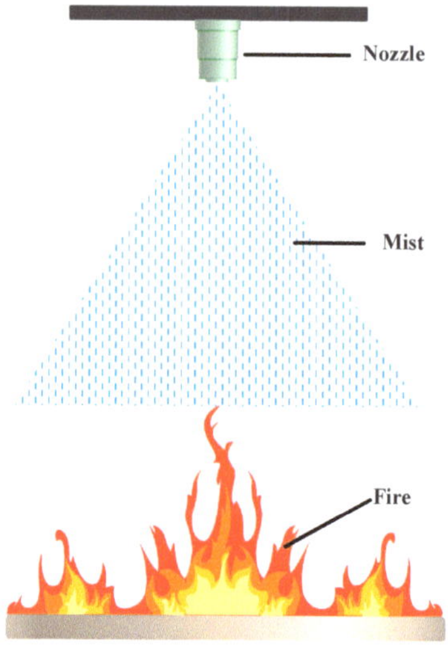

Figure 1. Water mist extinguishing mechanism.

Fine-water-mist firefighting technology markedly enhances both the efficiency and versatility of firefighting applications through the incorporation of physical and chemical additives. Physical additives, such as surfactants, bolster cooling and suffocation effects, work by diminishing the surface tension of water, augmenting the diffusivity and wettability of the water mist, and promoting the amalgamation of water with fuel. Surfactants may be hydrocarbon-based or fluorinated [6]. While fluorinated surfactants offer superior hydrophobicity and the capacity to create aerosolized water films, they may also exert adverse environmental effects due to their persistence and potential for bioaccumulation in natural ecosystems [7–9]. Chemical additives, including alkalis, transition metals, and their hydroxides and salts, contribute to fire suppression by engaging in chemical reactions and physical processes. For instance, sodium and potassium compounds slow flame propagation and create a foam blanket to smother the fire source [10], whereas metal hydroxides quench combustion through neutralization reactions and by sequestering free radicals [11]. Furthermore, ultrafine water mist, engineered via high-pressure atomization, exhibits exceptional heat transfer efficiency and swift evaporation traits. This technology can precipitously lower the temperature of the fire source, offering a novel and efficient direction for environmentally friendly firefighting techniques.

Gaseous extinguishing agents are a class of fire suppression media that inhibit or extinguish fires through physical or chemical actions, playing a significant role in the field of fire extinguishing. Perfluorohexanone, as a new generation of gaseous fire-extinguishing agents, has garnered widespread attention in the international firefighting community due to its highly efficient fire suppression capabilities and excellent environmental characteristics [3]. This agent features an exceptionally low extinguishing concentration of 4.2%, rapid extinguishing speed, and the ability to effectively prevent fire re-ignition, with an extinguishing efficiency that significantly surpasses that of the traditional halon 1301 agent [12]. Perfluorohexanone has an Ozone Depletion Potential (ODP) value of zero, indicating no harm to the atmospheric ozone layer, and it possesses a relatively low Global Warming Potential (GWP), with an atmospheric lifetime of merely 0.014 years, thereby exerting a negligible impact on global climate change [13]. During the fire-extinguishing process, perfluorohexanone does not generate harmful corrosive gases and exhibits good compatibility with metals and polymeric materials, along with excellent electrical insulation properties. It leaves minimal residue post-extinguishing, reducing environmental pollution and post-fire cleanup efforts. Its excellent long-term storage stability also minimizes the maintenance costs associated with fire suppression systems. Amidst the global focus on environmental protection and sustainable development, the market demand and application prospects for perfluorohexanone are expanding. It is anticipated to play an increasingly prominent role in the realm of fire-extinguishing agents, contributing significantly to the protection of human life, property safety, and the environment.

Foam extinguishing agents are employed to extinguish fires through their mixture with water, resulting in the formation of foam. These agents are categorized into the following two primary types: protein-based and synthetic. Protein-based foam agents incorporate a range of reinforcing agents into their animal protein base liquid. Notably, fluoroprotein types enhance fire-extinguishing efficiency by incorporating fluorocarbon compounds [14]. On the other hand, synthetic foam agents exhibit a spectrum of characteristics. For instance, aqueous film-forming foam (AFFF) can create a protective film on the surface of combustible materials, as depicted in Figure 2. This protective film bolsters the fire-extinguishing process and the material's resistance to re-ignition [15].

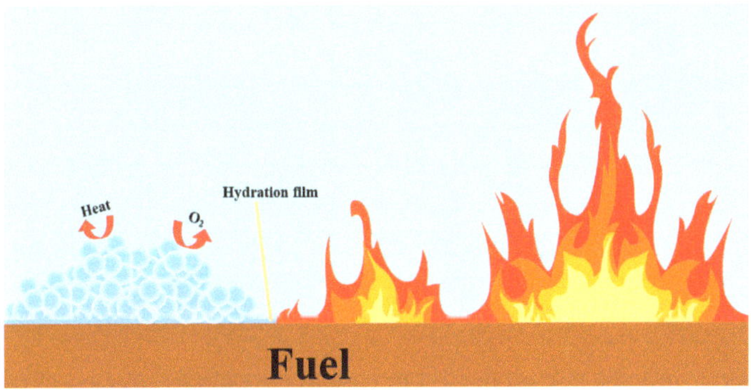

Figure 2. Foam cooling schematic.

Dry powder extinguishing agents are a class of efficient, economical, and environmentally benign fire suppressants that integrate the functions of chemical extinguishing with physical suppression. Their fire-extinguishing efficacy can be several times greater than that of halon, leading to their widespread adoption across various applications [2]. A derivative of dry powder, dry water extinguishers, leverage the combined benefits of dry powder and water mist technologies, making them exceptionally suitable for extinguishing lithium-ion battery fires [16]. In their preparation, water and hydrophobic silica are blended via high-speed agitation, yielding a core–shell structured powder extinguishing agent (as illustrated in Figure 3). The extinguishing mechanisms of this agent encompass physical cooling, chemical inhibition, thermal radiation blocking, and free radical trapping [17]. During the firefighting process, the substantial volume of water discharged by the dry water extinguisher rapidly evaporates, drawing heat away from the fire source. Concurrently, the hydrophobic silica particles produced create a barrier over the fuel surface, sequestering the fuel from external heat exchange and oxygen, thereby effectively interrupting the combustion chain reaction. Moreover, dry water extinguishers exert minimal environmental impact post-use, aligning with contemporary requirements for green and efficient firefighting technologies.

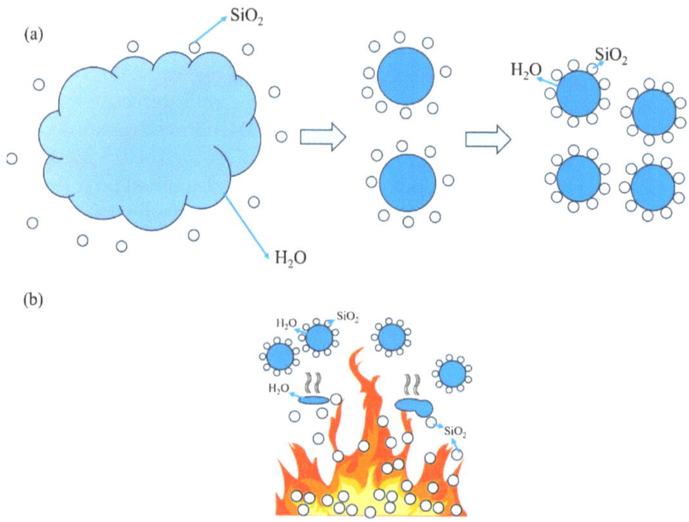

Figure 3. (**a**) Formation mechanism of the dry water extinguishing agent; (**b**) mechanism of dry water extinguishing [17].

Recent scholarly research underscores the pivotal role of piping design in the transport and dispersion dynamics of extinguishing agents, which is essential for enhancing fire prevention and suppression strategies. An optimized piping design enhances the flow characteristics and spatial dispersion efficiency of extinguishing agents by considering critical fluid dynamics parameters, including flow velocity, pressure loss, and turbulence [18]. By meticulously designing the pipe diameter, length, elbows, and branches, the system's pressure loss can be minimized, thereby augmenting the delivery efficiency of the extinguishing agents. The utilization of simulation technology further refines piping design, facilitating the virtual simulation of extinguishing agent flow. This enables pre-assessment and refinement of the design scheme in a controlled environment, ensuring the high efficiency and reliability of the fire suppression system prior to its actual deployment [19].

3. Future Research Direction

The evolution of big data and artificial intelligence technologies is propelling fire suppression research to unprecedented heights. The integration of sophisticated AI algorithms with fire suppression technologies is poised to expand into broader application domains. Concurrently, the exploration of multimodal fire suppression technologies that integrate physical and chemical suppression mechanisms is underway to accommodate a wider array of fire scenarios. Furthermore, innovation in pipeline design, coupled with the application of simulation and modeling technologies, should be further advanced. These advancements should be integrated with transport and dispersion models within actual fire contexts to optimize the distribution efficiency of extinguishing agents. This optimization is essential to ensure the high efficiency and effectiveness of firefighting operations.

Author Contributions: Investigation, C.L.; Writing—original draft preparation, C.L.; Writing—review and editing, G.L.; Project administration, S.L. All authors have read and agreed to the published version of the manuscript.

Funding: This study has been sponsored by National Natural Science Foundation of China (12202410 and 12472370), Project funded by China Postdoctoral Science Foundation (2023T160734 and 2023M733935), Supported by Fundamental Research Program of Shanxi Province (20210302123017 and 202303021211145), Fund Program for the Scientific Activities of Selected Returned Overseas Professionals in Shanxi Province (20220012), Research Project Supported by Shanxi Scholarship Council of China (2022-139), Natural Science Foundation of Hunan Province (2023JJ40726), Changsha Municipal Natural Science Foundation (kq2208277), Supported by the Opening Foundation of Key Laboratory in North University of China (DXMBJJ2023-03). Also, the authors thank Wei Yue from Shiyanjia Lab (https://www.shiyanjia.com) for the DSC analysis.

Conflicts of Interest: The authors declare no conflicts of interest.

References

1. Hamzehpour, A.; Verda, V.; Borchiellini, R. Simulation Study on Suppressing Shielded Fires by Water Mist Systems. *Fire* **2023**, *6*, 129. [CrossRef]
2. Lu, G.; Zhao, J.; Zhou, Y.; Fu, Y.; Lu, S.; Zhang, H. Study on Flowability Enhancement and Performance Testing of Ultrafine Dry Powder Fire Extinguishing Agents Based on Application Requirements. *Fire* **2024**, *7*, 146. [CrossRef]
3. Ni, X.; Chen, Y.; Huang, Q.; Zhao, C.; Li, S.; Huang, J.; Wang, J. An Experimental Study on the Transportation Characteristics of Perfluoro(2-methyl-3-pentanone) in a Straight Pipe. *Fire* **2023**, *6*, 156. [CrossRef]
4. Kang, W.; Yan, L.; Ding, F.; Guo, X.; Xu, Z. Experimental study on fire-extinguishing efficiency of protein foam in diesel pool fire. *Case Stud. Therm. Eng.* **2019**, *16*, 100557. [CrossRef]
5. *GA 1149*; People's Republic of China Public Security Industry Standards. Technical Report; Standardization Administration of China: Beijing, China, 2014.
6. Robinet, A.; Chetehouna, K. A Review of Additives for Water Mist Fire Suppression Systems. *Fire Technol.* **2024**, *60*, 2923–2961. [CrossRef]
7. Yun, X.; Lewis, A.J.; Stevens-King, G.; Sales, C.M.; Spooner, D.E.; Kurz, M.J.; Suri, R.; McKenzie, E.R. Bioaccumulation of per- and polyfluoroalkyl substances by freshwater benthic macroinvertebrates: Impact of species and sediment organic carbon content. *Sci. Total Environ.* **2023**, *866*, 161208. [CrossRef] [PubMed]
8. Burkhard, L.P. Evaluation of Published Bioconcentration Factor (BCF) and Bioaccumulation Factor (BAF) Data for Per- and Polyfluoroalkyl Substances Across Aquatic Species. *Environ. Toxicol. Chem.* **2021**, *40*, 1530–1543. [CrossRef] [PubMed]
9. Alexandrino, D.A.M.; Mucha, A.P.; Almeida, C.M.R.; Carvalho, M.F. Atlas of the microbial degradation of fluorinated pesticides. *Crit. Rev. Biotechnol.* **2022**, *42*, 991–1009. [CrossRef] [PubMed]
10. Wei, S.; Yu, M.; Pei, B.; Xu, M.; Guo, J.; Hu, Z. Experimental and numerical study on the explosion suppression of hydrogen/dimethyl ether/methane/air mixtures by water mist containing $NaHCO_3$. *Fuel* **2022**, *328*, 125235. [CrossRef]
11. Bokka, S.; Ameta, P.; Achary, S.N.; Chowdhury, A. A simple and economical fire test setup for examining the fire retardancy/extinguishing ability of water additive fire-retardant materials on class A fires. *Fire Mater.* **2024**, *48*, 93–101. [CrossRef]
12. Liang, C.; Jin, K.; Liu, P.; Wang, C.; Xu, J.; Li, H.; Wang, Q. The Efficiency of Perfluorohexanone on Suppressing Lithium-Ion Battery Fire and Its Device Development. *Fire Technol.* **2023**, *59*, 1283–1301. [CrossRef]
13. He, Y.; Deng, J.; Yi, X.; Chen, W.; Xiao, Y.; Deng, Y.; Zhu, X.; Yin, L. Effect of fluorine-containing explosion suppressants on methane explosions. *J. Therm. Anal. Calorim.* **2024**, *149*, 3711–3722. [CrossRef]
14. Zhao, M.; Ni, X.; Zhang, S.; Cao, W.; Guan, Y.; Liang, C.; Wang, X.; Zhang, H. Improving the performance of fluoroprotein foam in extinguishing gasoline pool fires with addition of bromofluoropropene. *Fire Mater.* **2016**, *40*, 261–272. [CrossRef]
15. Liu, Z.; Li, C.; Yuan, Z.; He, W.; Zhang, L.; Cheng, Z. Synthesis of Y-Type Fluorinated Surfactant for Aqueous Film-Forming Foam Extinguishing Agent with High Performance. *Ind. Eng. Chem. Res.* **2024**, *63*, 12288–12296. [CrossRef]
16. Li, X.; Du, K.; Zhu, Y.; Zhou, Z.; Zhou, X. Dry water: Toward an ideal extinguishant for lithium-ion battery fire. *J. Energy Storage* **2024**, *80*, 110204. [CrossRef]
17. Wang, Y.-Y.; Zhu, F.-H.; Zhou, H.-L.; Jiang, J.-C.; Huang, A.-C. Development of a novel dry-water fire extinguishing agent containing additives. *J. Therm. Anal. Calorim.* **2023**. [CrossRef]
18. Li, Q.; Li, Z.; Chen, R.; Zhang, Z.; Ge, H.; Zhou, X.; Pan, R. Numerical Study on Effects of Pipeline Geometric Parameters on Release Characteristics of Gas Extinguishing Agent. *Symmetry* **2021**, *13*, 1766. [CrossRef]
19. Mohamed, M.A.; New, T.H.; Ng, B.F. Modelling of fire-suppressant injection into engine nacelle for various flight regimes. *Proc. Inst. Mech. Eng. Part G J. Aerosp. Eng.* **2023**, *237*, 3111–3125. [CrossRef]

fire

Article

Optimized Machine Learning Model for Fire Consequence Prediction

Wei Zhong [1], Shuangli Wang [1], Tan Wu [2], Xiaolei Gao [1,*] and Tianshui Liang [1]

[1] School of Mechanics and Safety Engineering, Zhengzhou University, Zhengzhou 450001, China
[2] School of Energy and Power Engineering, Zhengzhou University of Light Industry, Zhengzhou 450001, China
* Correspondence: gaoxiaolei@zzu.edu.cn

Abstract: This article focuses on using machine learning to predict the distance at which a chemical storage tank fire reaches a specified thermal radiation intensity. DNV's Process Hazard Analysis Software Tool (PHAST) is used to simulate different scenarios of tank leakage and to establish a database of tank accidents. Backpropagation (BP) neural networks, random forest models, and the optimized random forest model K-R are used for model training and consequence prediction. The regression performance of the models is evaluated using the mean squared error (MSE) and R^2. The results indicate that the K-R regression prediction model outperforms the other two machine learning algorithms, accurately predicting the distance at which the thermal radiation intensity is reached after a tank fire. Compared with the simulation results, the model demonstrates higher accuracy in predicting the distance of tank fire consequences, proving the effectiveness of machine learning algorithms in predicting the range of consequences of tank storage area fire events.

Keywords: pool fire; machine learning; prediction; PHAST

Citation: Zhong, W.; Wang, S.; Wu, T.; Gao, X.; Liang, T. Optimized Machine Learning Model for Fire Consequence Prediction. *Fire* **2024**, 7, 114. https://doi.org/10.3390/fire7040114

Academic Editors: Song Lu, Changcheng Liu, Guohui Li and Paweł Wolny

Received: 16 February 2024
Revised: 18 March 2024
Accepted: 27 March 2024
Published: 1 April 2024

1. Introduction

In the production process of chemical enterprises, due to their special nature, large quantities of flammable and explosive liquids and gases are stored in the storage tank area. If an oil tank leaks and is exposed to an ignition source, it may lead to a serious fire accident. Once a disaster occurs, personnel exposed to the radiant heat of the fire may suffer serious injuries. In the UK, a significant chemical plant tank leakage and fire incident took place on 11 December 2005 at the Buncefield oil depot, which is situated less than 50 km from London. The explosion and subsequent fire ravaged over 20 substantial oil storage tanks, resulting in injuries to 43 individuals but no fatalities. The direct economic damage amounted to GBP 250 million. This event triggered the most profound ecological crisis in Europe during peacetime, substantially affecting the ecological landscape of London and the entire nation of Great Britain [1].

By collecting and analyzing data on domestic and international chemical incidents from 10 December 2003 to 10 January 2024, we obtained the following findings: domestically, there were 1060 poisoning incidents, 797 fire incidents, 759 explosion incidents, 592 leakage incidents, and 1 other type of incident; internationally, there were 36 poisoning incidents, 285 fire incidents, and 638 explosion incidents, with leakage incidents being the most frequent at 933 occurrences. Based on these data, we conducted a comprehensive analysis of the proportions of each type of incident and visually displayed the proportion of each incident type in Figure 1. This paper conducts a study with a certain oil transfer station as the background, focusing on petroleum products such as gasoline and diesel. These chemicals, when leaked or burned, generate toxic gases or smoke, posing a threat to the surrounding environment and human health, with toxic accidents likely being a consequence of leakage and fire incidents. Among these, explosions are instantaneous reactions that are difficult to intervene in, so this paper does not involve the study of explosions. By combining the frequencies of accident types in Figure 1, this paper mainly focuses on tank leakage and fire accidents.

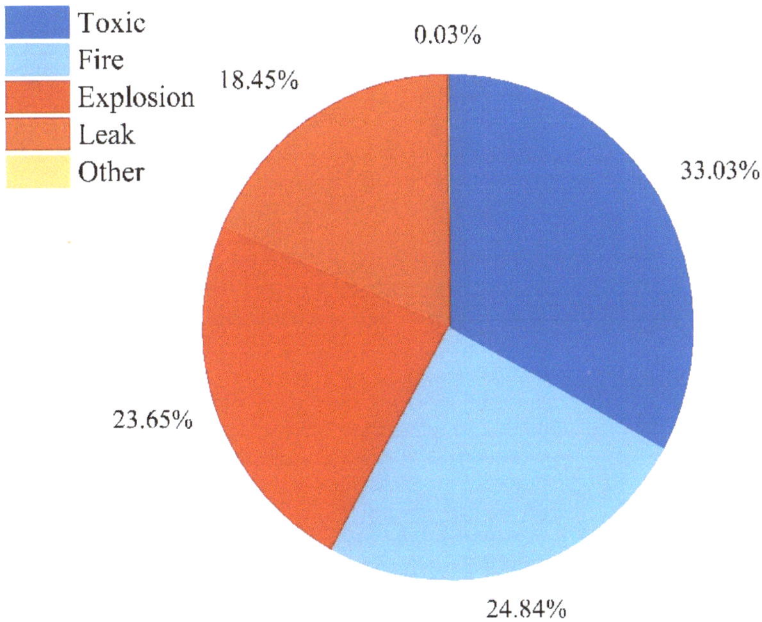

(a)

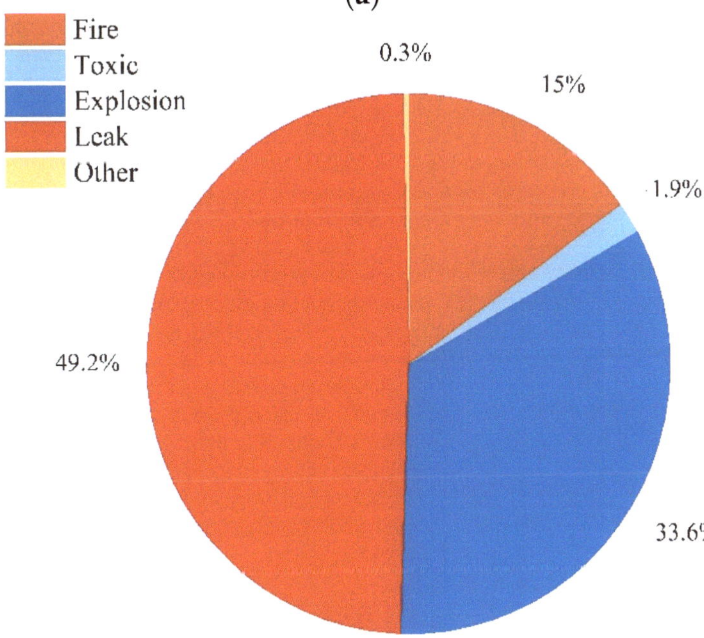

(b)

Figure 1. (**a**) Data chart of China's domestic chemical park accidents. (**b**) Data chart of foreign chemical park accidents.

In existing research, the consequences of leakage and fire accidents in chemical enterprises can be calculated using empirical models, computational fluid dynamics (CFD), and integrated models such as the Process Hazard Analysis Software Tool (PHAST 8.7). Although empirical models provide quick calculations, their accuracy is limited due to factors affecting the overall evaporation rate, such as surface roughness. CFD models like the Fire Dynamics Simulator (FDS) and FLACS can capture these effects, but they come with higher computational costs and longer execution times. The integrated model used in PHAST strikes a balance between accuracy and computational cost while still capturing the effects of surface roughness on propagation, pool evaporation, and mass evaporation—this has been validated by the fire scene experimental results obtained by Henk W.M [2]. Tan Wu et al. [3] used PHAST for a risk analysis and system safety integrity study of enclosed ground flame devices; their findings are useful for designing various types of material windbreaks and flame systems' strength and for determining the safety distances of surrounding personnel and facilities. Wang Kan et al. [4] studied the impact of leakages of different forms of high-pressure hydrogen storage containers on the scope of consequences after accidental hydrogen leaks or explosions, as well as the variations in environmental temperature differentials and wind conditions. Song Xiaoyan [5] employed the Unified Dispersion Model (UDM) in the PHAST program to conduct numerical simulations of liquefied petroleum gas leakage explosions, analyzing the impact of changes in the wind speed and leak orifice diameter on the consequence range after a leakage.

Although PHAST has significant advantages in process simulation and consequence analysis, with the continuous development of deep learning technology, we can also explore the combination of PHAST and machine learning to achieve more accurate predictions of the consequences in chemical industrial parks. This integrated approach will provide us with more comprehensive and detailed prediction results, helping to further optimize safety management and emergency response strategies. Additionally, machine learning is now widely used in the field of fire decision making. More and more experts and scholars are applying it to the auxiliary decision-making domain. For example, Manuel J. Barros-Daza et al. [6] proposed a data-driven method using artificial neural networks (ANNs) for classification, providing real-time optimal decision making for firefighters in underground coal mine fires. Bilal Umut Ayhan et al. [7] utilized a latent class clustering analysis (LCCA) and artificial neural network (ANN) to develop a new model to predict the outcome of construction accidents and determine the necessary preventive measures. Peng Hu and colleagues [8] performed an analysis and machine learning modeling of the highest roof temperature in the longitudinal ventilation tunnel fire, introducing a new model called the genetic algorithms backpropagation neural network (GABPNN) to predict the highest roof temperature in the ventilation tunnel. Aatif Ali Khan et al. [9] established an intelligent model based on machine learning to predict smoke movement in fires and enhance emergency response procedures through predicting key events for dynamic risk assessment. Mohd Rihan and co-authors [10] conducted an assessment of forest fire susceptibility areas in specific regions using geographic information technology, machine learning algorithms, and uncertainty analysis based on deep learning technology, achieving significant results. Sharma et al. [11] compared eight machine learning algorithms and concluded that the enhanced decision tree model is the most suitable for fire prediction. They also proposed an intelligent fire prediction system based on the site, considering meteorological data and images, and predicted early fires.

Machine learning has made significant progress in the field of fire safety, and we should not overlook the potential advantages of combining PHAST with machine learning. Recent studies have shown that Sun and his team [12] simulated chemical leakage scenarios using PHAST and built causal models of three types of fires using ANN to predict radiation effects on distance. Wang et al. [13] proposed a method for source estimation using ANN, particle swarm optimization (PSO), and simulated annealing (SA). In addition, Yuan et al. [14] established a database of organic compounds with low flammability limits using PHAST, analyzing and predicting the lower flammability limits of organic chemicals. Makhambet

Sarbayev et al. [15] proposed a method involving mapping a fault tree (FT) to an ANN and verified the significant progress of combining a numerical simulation software and machine learning to predict the consequences of fire accidents in the chemical industry through the analysis of system failures in the Tassero-Anacortes refinery accident. Jiao et al. [16] used PHAST to establish a toxic diffusion database and developed a model to accurately predict the downwind toxic diffusion distance based on this database. This study aims to accurately estimate the risk level of the oil tank area by utilizing simulated data from the PHAST 8.7 software based on previous research. A comprehensive oil tank leakage fire database was constructed, and machine learning technology was employed to predict the potential impact ranges of accidents. The proposed method seeks to enhance the effectiveness of chemical safety management and risk assessment. However, it is important to note that the created database solely relies on PHAST simulation results and lacks the integration of actual accident data or validation using empirical formulas, consequently limiting its applicability and accuracy. Subsequently, the database is utilized for conducting a predictive analysis of accident consequences through machine learning technology. It is crucial to emphasize that all input data exclusively originated from the PHAST software simulation results, signifying that the database employed for machine learning training is constrained by the scope of PHAST simulation. Hence, this study elucidates the model's limitations and the restricted application scope of the database, offering valuable insights for future research avenues and data integration.

This paper focuses on the oil storage tank area of a certain oil transfer station as the research object. The station area includes a diesel tank area, gasoline tank area, mixed oil tank area, and mixed oil processing tank area. The main materials involved are gasoline and diesel hydrocarbons. Consequently, the database is constructed based on the PHAST fire consequence simulation results of gasoline and diesel, and it is then used to train the predictive model.

2. Methodology

2.1. Workflow of the Model Design

The main steps of machine learning prediction for the consequence range of storage tank leakage are as follows:

(1) Identify the research subject and scenarios; simulate the consequences of tank leaks and fires using PHAST.
(2) Construct a database of fire consequences based on the simulation results from PHAST.
(3) Develop a quantitative prediction model for the range of consequences based on the database; these include a BP neural network, random forest regression, and K-R regression prediction models.
(4) Tune the prediction models to determine the optimal model.
(5) Evaluate the model's performance by calculating the mean squared error (MSE) and the R^2 coefficient of determination [17].
(6) Apply machine learning algorithms to predict the consequences of fire accidents caused by leaks from storage tanks and draw conclusions by comparing and analyzing actual cases with the predictive results.

The workflow diagram is shown in Figure 2:

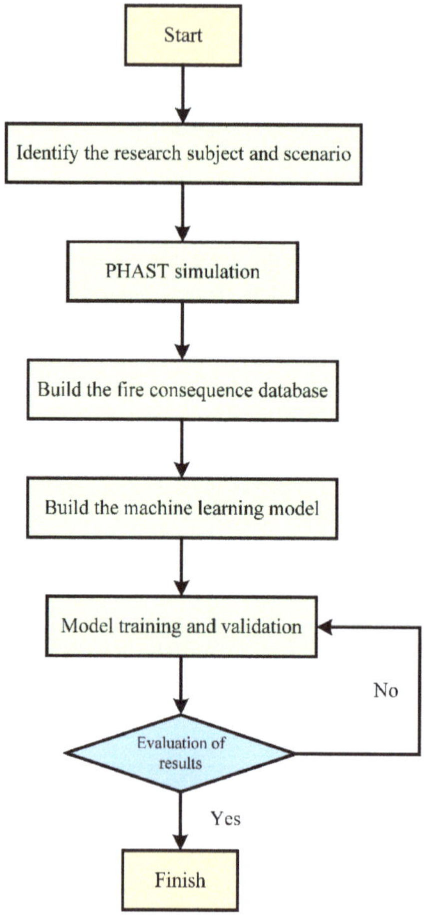

Figure 2. The workflow diagram.

2.2. PHAST Model

This study is based on the research of a certain oil depot's storage tanks. The specific distribution of storage tanks in the oil depot is shown in Figure 3 below. The tank area consists of four sections, labeled according to Figure 3, where "1" represents the diesel tank area, "2" represents the gasoline tank area, "3" represents the mixed oil tank area, and "4" represents the mixed oil processing area. The specific specifications of the tank area are shown in Table 1.

Table 1. Storage tank sizes in each tank area.

Area (m^3)	Specification Model	Quantity	Capacity (m^3)	Storage Medium
1	Φ60 × 19.92	4	50,000	diesel fuel
2	Φ38 × 19.8	8	20,000	gasoline
3	Φ22 × 14.852	1	5000	diesel fuel
3	Φ22 × 14.852	1	5000	gasoline
4	Φ8.920 × 12.565	2	500	gasoline
4	Φ8.920 × 12.565	2	500	diesel fuel

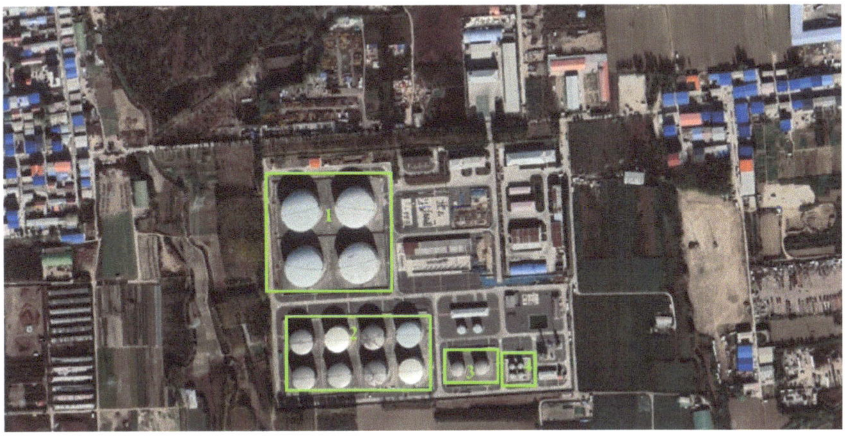

Figure 3. Tank area distribution.

To systematically analyze the impact of the wind speed and direction on the outcomes of fires resulting from tank leakages across varying orifice diameters, this study classifies ten distinct orifice sizes. The sizes range from small (5 mm) to medium (25 mm) and large (100 mm) orifices, extending to scenarios involving catastrophic ruptures. Additionally, this study investigates the influence of environmental temperature on tank leakage outcomes, considering a range of temperatures from −17.9 °C to 42.5 °C, including intermediate conditions at 10 °C, 20 °C, and 25 °C. Moreover, this study evaluates the impact of atmospheric stability on tank leakage outcomes by categorizing atmospheric conditions into ten stability classes labeled A through G. The analysis also explores the effects of preset leak orifice sizes on tank leakages containing diesel, gasoline, and mixed oil. The wind speeds vary from 1 to 16 m/s in 1 m/s increments, resulting in 16 unique meteorological scenarios. These methodological configurations are detailed in Table 2, outlining the foundational settings for the PHAST (Process Hazard Analysis Software Tool) model utilized in this investigation.

Table 2. PHAST model settings.

	Range	Interval	Total Category
Leakage pore size/(mm)	5 mm-Catastroptic rupture	5/10/15	10
stability	A-G	-	10
Wind velocity/(m/s)	1-16 m/s	1	16
Material	N-OCTANE N-OCTADECANE N-HEXADECANE	-	3
Capacity/(m^3)	500, 500, 5000, 20,000, 50,000	-	5

Taking the example of a 5 mm leak orifice in a diesel storage tank, the specific configuration parameters are shown in Table 3.

Table 3. The 5 mm leakage setting parameters of the storage tank.

Leakage Pore Size/mm	Tank Materials	Volume/m^3	Filling Level	Temperature/°C	Wind Velocity/(m/s)
5 mm	N-HEXADECANE	50,000	80%	25	1

Based on previous research results, the impact of tank leakage on the surrounding environment mainly depends on its thermal radiation intensity, which varies in its level of harm to the human body. The losses caused by different levels of thermal radiation intensity are shown in Table 4 [18]. According to the results of the literature review, the thermal radiation intensity level for this study is determined to be 4 kW/m^2.

Table 4. Losses caused by different incident intensities of thermal radiation.

Incident Intensity/(kW/m^2)	Damage to Equipment	Injury to Individuals
37.5	Complete damage to the operating equipment	1% death (10 s) 100% death (1 min)
25	The minimum energy required for wood combustion under flameless, prolonged radiation	Severe burns (10 s) 100% death (1 min)
12.5	The minimum energy required for wood combustion and plastic melting in the presence of flames	First-degree burn (10 s) 1% death (1 min)
4	-	Pain lasting for more than 20 s, not necessarily accompanied by blisters
1.6	-	Long-term radiation without any discomfort

2.3. Machine Learning

2.3.1. BP Neural Network

The machine learning algorithm [19] learns and recognizes patterns from historical data and uses these patterns for prediction and decision making. The main machine learning methods include supervised learning, unsupervised learning, and reinforcement learning. This article uses the supervised learning algorithm, employing the BP neural network, random forest regression prediction, and the K-R model, as shown in the neural network architecture diagram in Figure 4. By utilizing PHAST to construct models and perform simulation calculations, this study conducts a comprehensive analysis of the data and organizes them based on the obtained computational results, thereby constructing a dataset. Subsequently, this dataset is used to train algorithmic models, and based on the predictive outcomes of the training process, this study explores and identifies the optimal algorithm.

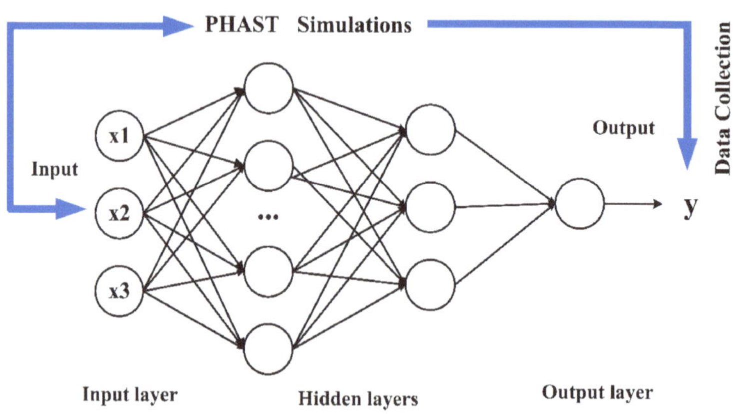

Figure 4. Neural network architecture diagram.

Regarding the BP neural network model [20], its distinct feature is that signals are propagated forward and errors are propagated backward. Through a certain number of iterations and updates, the network's predictive results reach the expected accuracy or convergence, yielding the best predictive outcome. The hyperparameters in the neural network also have a significant impact on its predictive outcome. The main hyperparameters include the learning rate, batch size, number of iterations, network structure and layers, activation function, optimizer, and so on. A larger learning rate leads to faster convergence, but there is a risk of skipping the optimal solution. The literature indicates that the batch size is set by considering the memory of the device running the code and the model's generalization ability [21]. In this model, the device configuration includes a processor of 12th Gen Intel(R) Core(TM) i5-12400F 2.50 GHz and an NVIDIA GeForce RTX 3060 GPU. Considering the dataset size, the batch size is set to 32. The number of iterations and the network structure and layers are mainly found through parameter tuning to seek the optimal results. The number of neurons in each layer of the neural network is calculated based on empirical Formulas (1)–(3) [22] and actual training parameter calculations. The settings are as shown in Table 5.

$$h = \sqrt{n+m} + c \tag{1}$$

$$h = \sqrt{nm} \tag{2}$$

$$h = 1.5 \times (n+m) \tag{3}$$

where h is the number of neurons, n is the number of input neurons, m is the number of output neurons, c is an integer, and $c \in [2, 10)$.

Table 5. BP neural network parameter settings.

	Number of Hidden Layers	Number of Neurons in Hidden Layer 1	Number of Neurons in Hidden Layer 2	Number of Iterations	Learning Rate
1	1	14	-	100	0.001
2	1	14		150	0.001
3	1	14	-	150	0.01
4	1	24	-	150	0.01
5	1	24	-	150	0.001
6	2	14	5	150	0.001
7	2	24	5	100	0.01
8	2	24	5	150	0.01
9	2	24	10	150	0.01

During the training process of the BP neural network model, a common problem encountered is the vanishing and exploding gradients. The vanishing gradient issue can lead to a slowdown or almost negligible weight updates, while the exploding gradient occurs when the gradient values between network layers exceed 1.0, resulting in an exponential growth of the gradients. This makes the gradients extremely large, causing significant updates to the network weights and thereby making the network unstable. Common activation functions include Sigmoid, Tanh, and ReLU [23], as shown in Figure 5. To address the issues of gradient descent and exploding gradients in the training process mentioned above, the ReLU function is set as the activation function to avoid the vanishing gradient. The final hyperparameter related to the BP neural network is the optimizer [24], with commonly used optimizers such as tochastic gradient descent (SGD), Momentum, Adagrad, and Adam. To avoid becoming stuck in local minima, the optimizer chosen is SGD.

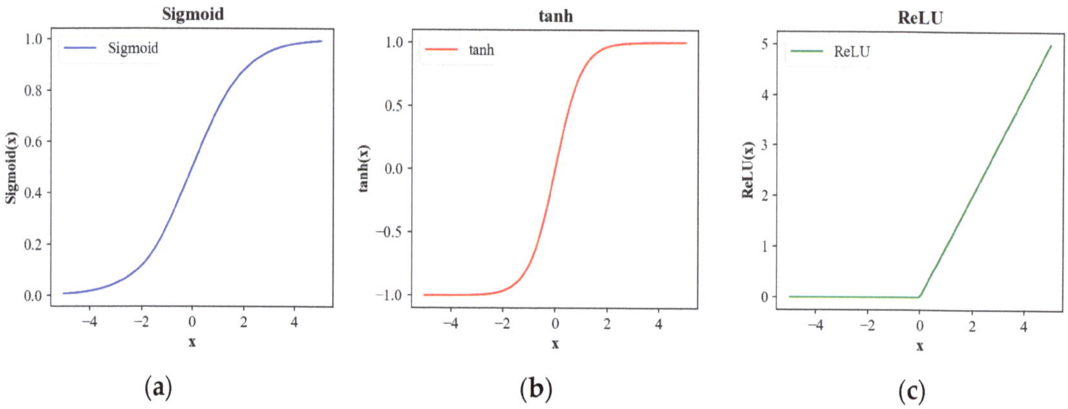

Figure 5. Illustration of activation functions. (**a**) Sigmoid activation function graph; (**b**) tanh activation function graph; (**c**) ReLU activation function graph.

2.3.2. Random Forest

Random forest [25] is an ensemble model composed of multiple decision trees. It creates multiple subsets by performing sampling with replacement on the training set and builds a decision tree based on each subset. When training each tree, random forest also introduces feature random selection, considering only a random subset of features during the splitting process at each node. During prediction, random forest aggregates the prediction results from multiple trees, commonly using methods such as taking the average (for regression problems) or voting (for classification problems) to obtain the final prediction result. In summary, decision trees split and predict data through a tree-like structure, while random forest utilizes an ensemble strategy of multiple decision trees to enhance predictive performance. Its performance is superior to that of decision tree regression prediction models.

2.3.3. Cross-Validation

Cross-validation is a commonly used model evaluation technique that involves dividing the dataset into training and testing sets. This process is repeated multiple times, allowing for the derivation of various performance metrics for the model. Common cross-validation methods include K-fold cross-validation, leave-one-out cross-validation, and stratified K-fold cross-validation [26], as well as time series cross-validation. The process of K-fold cross-validation, as depicted in Figure 6, involves randomly dividing the dataset into K mutually exclusive subsets, referred to as folds. Then, the model is trained K times, with each training iteration reserving one fold as the testing data and using the remaining K-1 folds as the training data, ensuring different folds are retained for each training iteration. After K iterations of model training, the model with the minimal estimated error on the testing set is selected, and its network architecture and parameters are retained for random forest regression prediction. This article utilizes K-fold cross-validation for model optimization and proposes a new model, the K-R model, by combining K-fold cross-validation with tree models to optimize the tree model's hyperparameters, thereby improving prediction accuracy.

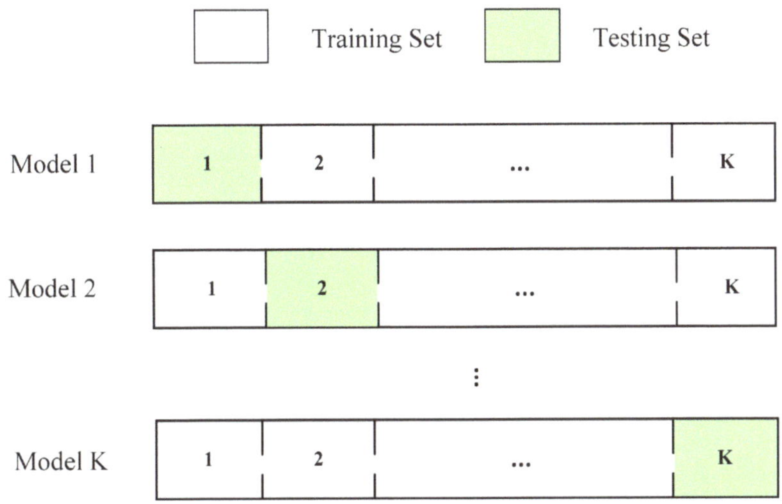

Figure 6. K-fold cross-validation.

2.4. Model Evaluation

Regarding the performance measurement of the model, the main metrics include accuracy, precision/recall, P-R curve, ROC curve/AUC, R^2, MSE, etc. This article utilizes R^2 and MSE to evaluate the model's performance, with the formulas for R^2 and MSE shown as Equations (4) and (5).

$$R^2 = \frac{\sum\limits_{i=1}^{n} (\hat{y}_i - y)^2}{\sum\limits_{i=1}^{n} (y_i - \hat{y})^2} \tag{4}$$

$$MSE = \frac{\sum_{i=1}^{n} (\hat{y}_i - y_i)^2}{n} \tag{5}$$

where $\hat{y}_i$ is the predicted output value from the neural network model, y_i is the target value obtained from PHAST, and y is the average of the target values, with n representing the number of scenarios or the quantity of data.

3. Data Preprocessing and Discussion

3.1. Data Preprocessing

In the section regarding data preprocessing and discussion, the initial step involved assessing and cleaning the data for quality. Subsequently, an in-depth exploration of the data's correlations was undertaken, employing the Pearson correlation coefficient analysis to investigate the linear relationships between variables. Additionally, attention was drawn to the presence of multicollinearity, and corresponding exploration and treatment were carried out. The following will provide a detailed introduction of the specific data analysis methods and results, along with a discussion and explanations of the findings.

3.1.1. Correlation Analysis

The correlation of data within a database is crucial to model training prediction effects; through the Person correlation coefficient [27] analysis for variables—as shown in Formula (6)—we obtain a correlation matrix to eliminate highly correlated variables. As shown in Figure 7, on the right side is a color bar for heat maps; r represents various correlation coefficients. Generally, $|r| \geq 0.8$ indicates high correlation between two variables;

$0.5 \leq |r| < 0.8$ indicates medium correlation; $0.3 \leq |r| < 0.5$ suggests low correlation; and $|r| < 0.3$ implies little to no correlation between two variables.

$$\rho_{X,Y} = \text{cov}(X,Y)/\sigma_X\sigma_Y \quad (6)$$

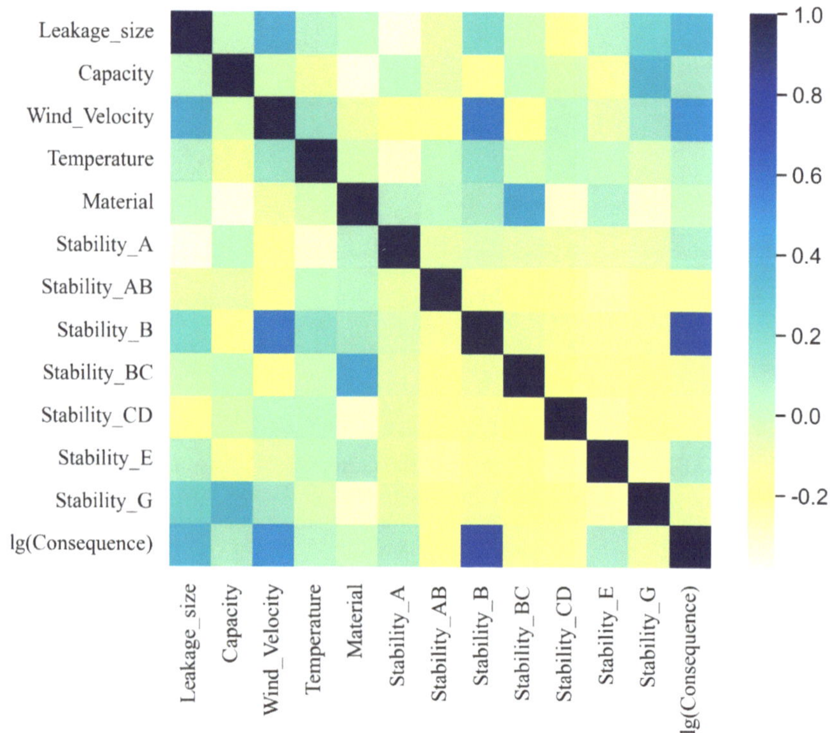

Figure 7. Thermal map of correlation coefficients in fire aftermath area prediction models.

3.1.2. Multicollinearity

Selecting the right variables is essential to the model's predictive effectiveness. We determine the input variables for the model based on expert opinions, a search of the literature, and software modeling parameters. Variance Inflation Factor (VIF) and Tolerance (Tol) [28] are widely used indicators for assessing multicollinearity between independent variables in a regression model. VIF represents the variance inflation factor, while Tol represents tolerance.

The VIF value is a measure of the correlation between independent variables. In general, a stronger correlation between independent variables indicates more severe multicollinearity. A VIF value below 10 is generally considered reasonable. Values exceeding 10 indicate strong correlation, necessitating variable screening or transformation. In Table 6, all VIF values are below 10, which is acceptable. Tol represents tolerance and is defined as 1/VIF. Smaller Tol values indicate more severe multicollinearity. Severe multicollinearity is typically indicated when Tol values are below 0.1. Specifically, for this dataset, the minimum Tol value is 0.4827, indicating no multicollinearity.

Table 6. VIF and Tol of variables.

NO	Variable	VIF	Tol
1	Leakage_pore_size	1.00	0.97
2	Capacity	1.227	0.814
3	Wind_Velocity	1.00	0.97
4	Temperature	1.00	0.972
5	Material	1.227	0.814
6	Stability	2.0716	0.4827

3.2. Discussion

The PHAST simulation was used to construct a fire database containing over thirty thousand records. This database includes 12,799 sets of jet fire data, 35,998 sets of early pool fire data, and 39,991 sets of late pool fire data. In order to train and validate the model, the training and validation sets were divided in a 7:3 ratio, and predictive analysis was performed. The BP neural network was used for training, and the hyperparameters that minimized prediction errors were selected. The structure of the BP neural network was determined to have two hidden layers, with the first hidden layer comprising 24 neurons and the second hidden layer containing 10 neurons. The learning rate was set to 0.01, the number of iterations was set to 150, and the batch size was 32.

The analysis was conducted on the random forest regression prediction model. Based on the training and debugging results of the model, it was found that the initial parameters of the random forest model with the max depth set to 2 had a large error.

Therefore, in combination with five-fold cross-validation, parameter optimization was carried out, and the K-R model was proposed. The evaluation results of the three models, including MSE and R^2 values, are shown in Table 7. It can be observed that the BP neural network model had a large prediction error for the fire dataset, but with a moderate level of accuracy. The random forest model performed well in predicting jet fires and early pool fires but had poor predictions for late pool fires. An analysis indicated that the addition of data related to catastrophic ruptured fire consequence in the late pool fire dataset led to the poor performance of the model. The optimized K-R model outperformed the first two models in predicting jet fires and early pool fires, achieving a high accuracy of 0.99 and optimal predictive performance. The predictive performance of the K-R model for late pool fires improved by approximately 12% compared to the first two algorithms, with an accuracy of 0.99, proving its improved accuracy. This demonstrates that the optimized K-R model has good predictive effects.

Table 7. MSE values and R^2 of three algorithms.

	Jet Fire		Early Pool Fire		Late Pool Fire	
Evaluation	**MSE**	**R2_score**	**MSE**	**R2_score**	**MSE**	**R2_score**
BP	0.008	0.945	0.004	0.986	0.017	0.968
RandomForest	0.010	0.945	0.010	0.954	0.084	0.888
K-R	0.0005	0.997	0.0004	0.997	0.0007	0.998

The bar chart shows the MSE and R^2 values of the predictions of the three algorithms for the three types of fires. From Figure 8a, it can be seen that the BP neural network, after model training, parameter tuning, and selecting the optimal network structure, outperforms the random forest regression prediction model. However, compared to the K-R model, the prediction results of the BP neural network are slightly inferior. In Figure 8b, the K-R model's prediction accuracy is at least 0.01 higher than that of the BP neural network, demonstrating a more superior predictive performance.

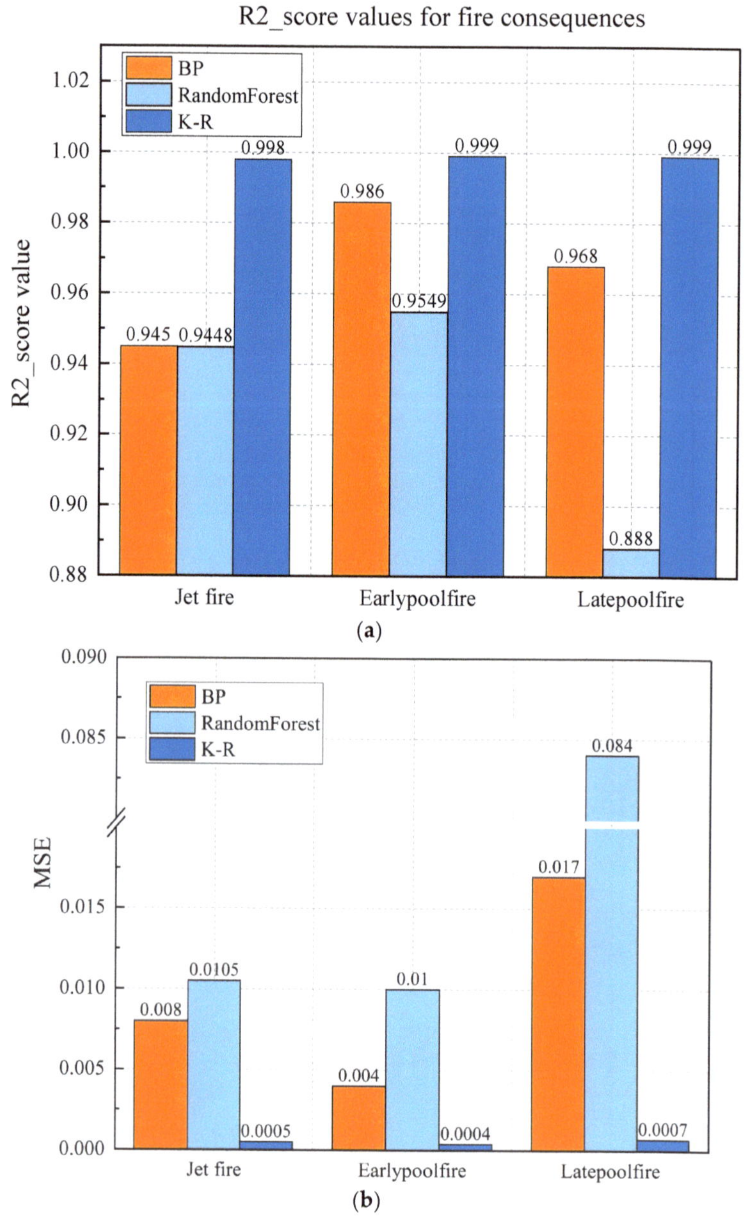

Figure 8. (**a**) Bar chart of R^2 values for fire prediction outcomes. (**b**) Bar chart of MSE values for fire prediction outcomes.

We analyzed the predicted values and actual values of three models and plotted the scatter and residual graphs. It is clear from Figure 9 that the data distribution of the K-R model is more concentrated, with a smaller range of fluctuations along the Y = X line, approximately between −0.05 and 0.05, demonstrating a better clustering effect; at the same time, the residuals of this model are relatively small.

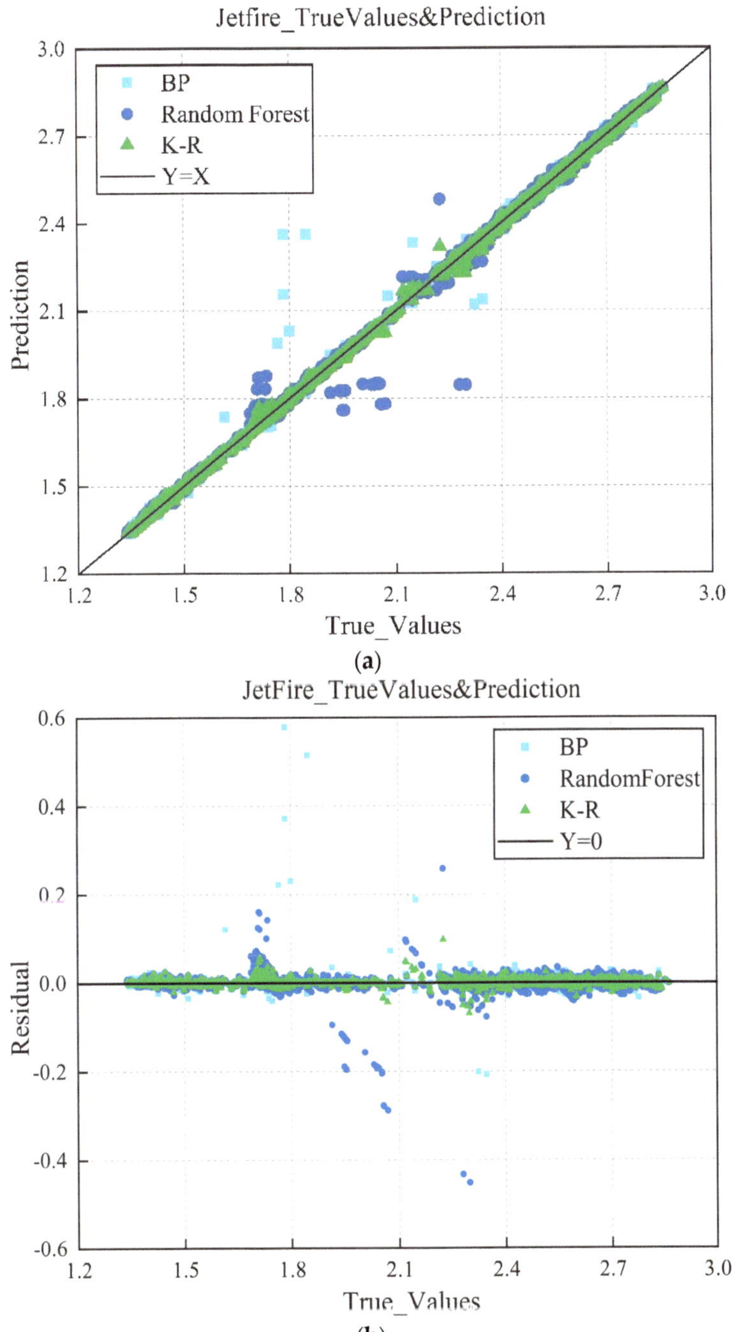

(**a**)

(**b**)

Figure 9. *Cont.*

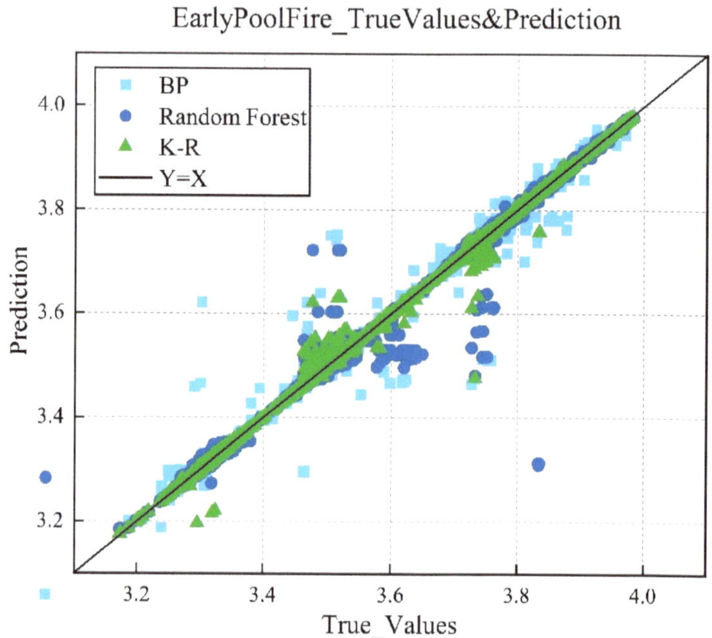

(**c**)

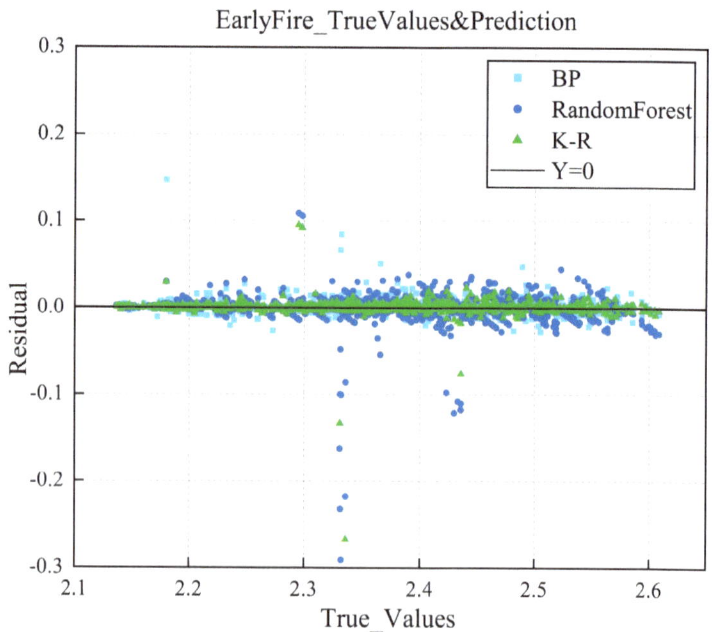

(**d**)

Figure 9. *Cont.*

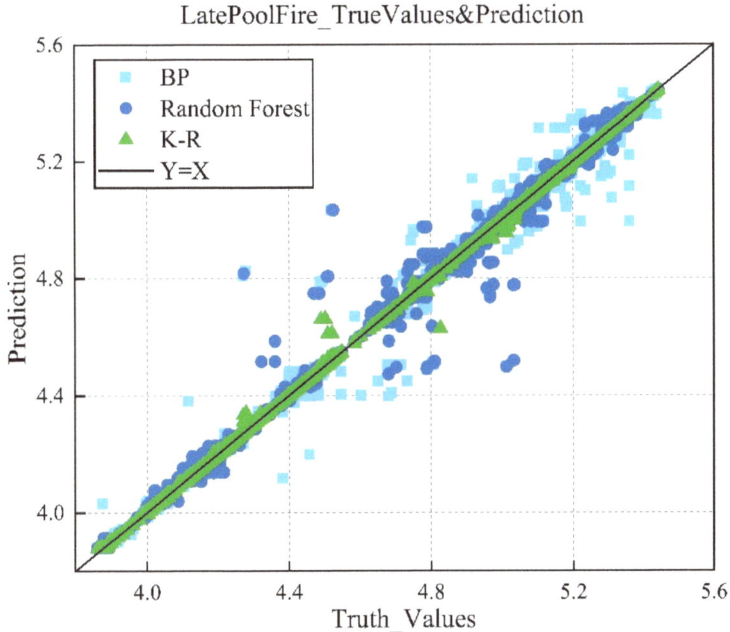

(e)

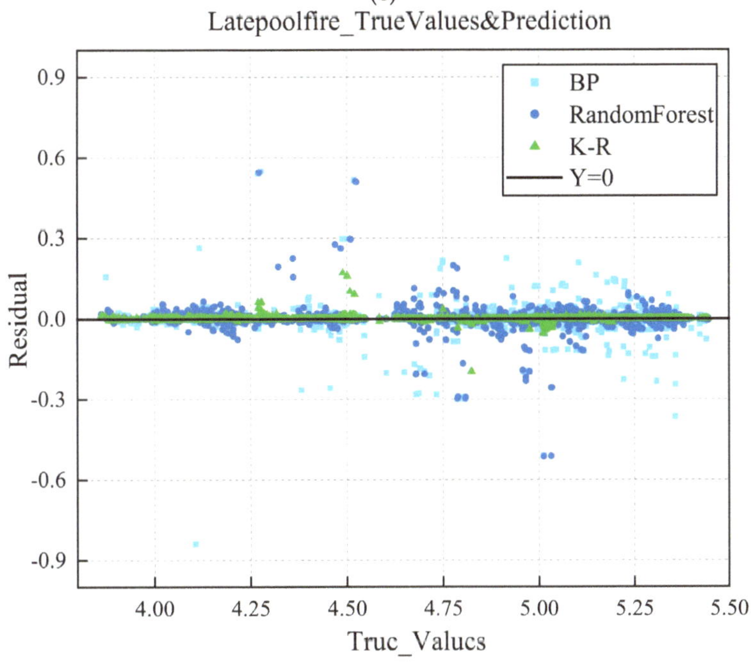

(f)

Figure 9. (**a**) Scatter plot of predicted outcomes for jet fires. (**b**) Residual plot of predicted outcomes for jet fires. (**c**) Scatter plot of predicted outcomes for early pool fires. (**d**) Residual plot of predicted outcomes for early pool fires. (**e**) Scatter plot of predicted outcomes for late pool fires. (**f**) Residual plot of predicted outcomes for late pool fires.

The predicted results of fire consequences were analyzed, and scatter plots and residual plots of the fire consequences were drawn. According to Figure 9a, the predicted values of the three models are evenly distributed along the Y = X line, but it is clear that some points of the BP and random forest models deviate from the Y = X line, indicating a significant error between the predicted and actual values. Based on the residual plot in Figure 9b, the regression effect of the K-R model is relatively better. A comparison of the scatter plots for early pool fires, shown in Figure 9c, and late pool fires, shown in Figure 9d, reveals that, compared to the scatter plot for jet fires, the three models for early pool fires, especially the BP neural network, exhibit a significant error, with many data points distributed outside the Y = X fitting line. In comparison to the three models, the K-R model demonstrates relatively good performance, although there is still room for improvement. From the scatter plots, the residual points of the BP neural network and random forest models are approximately between −0.05 and 0.05, while the absolute error of the K-R model is approximately between −0.02 and 0.02. The data points in the residual plot for late pool fires are more chaotic, with the residual points of the BP neural network and random forest models distributed between −0.03 and 0.03, and the residual points of the K-R model distributed between −0.01 and 0.01. This is attributed to the performances of the first two models.

4. Conclusions

4.1. Case Study

Based on the model database developed through the above analysis and research, a mid-hole leakage incident in a 100,000 m^3 internal floating roof gasoline storage tank in a large oil depot was analyzed. The outdoor temperature was 20 °C, the wind speed was 6.5 m/s, and the atmospheric stability was D. Through a PHAST simulation, we derived the thermal radiation intensity of an early pool fire, late pool fire, and jet fire from this oil tank, as well as the corresponding graph (Figure 10) of thermal radiation intensity and downwind distance. From Figure 10a, it can be seen that the downwind distance to 4 kW/m^2 for an early pool fire is 35.9504 m; the downwind distance to 12.5 kW/m^2 is 28.2839 m; and the downwind distance to 37.5 kW/m^2 is 17.8842 m. From Figure 10b, it can be observed that the thermal radiation effect almost disappears at a downwind distance of about 46 m. For a late pool fire, the downwind distance to reach a radiation intensity of 4 kW/m^2 is found to be 71.9381 m, and the required downwind distance to reach an intensity level of 12.5 kW/m^2 is approximately 48.9713 m, while achieving an intensity level of 37.5 kW/m^2 demands a downwind distance of around 40.7759 m, and thermal radiation intensity from late pool fires ceases around at about 100 m downwind. The thermal radiation intensity from jet fires gradually increases with increasing downwind distances—it starts declining when the downwind distance reaches around 2 m, and the rate of decline slows at about the 5 m mark, whereas such thermal radiation effects disappear once the downwind distance reaches approximately 9.2 m.

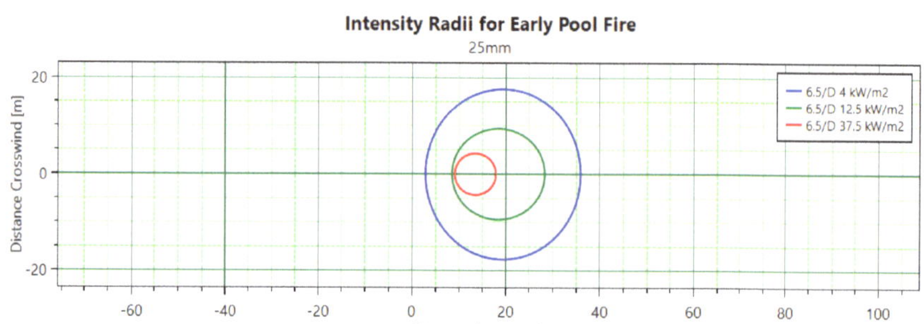

(a)

Figure 10. *Cont.*

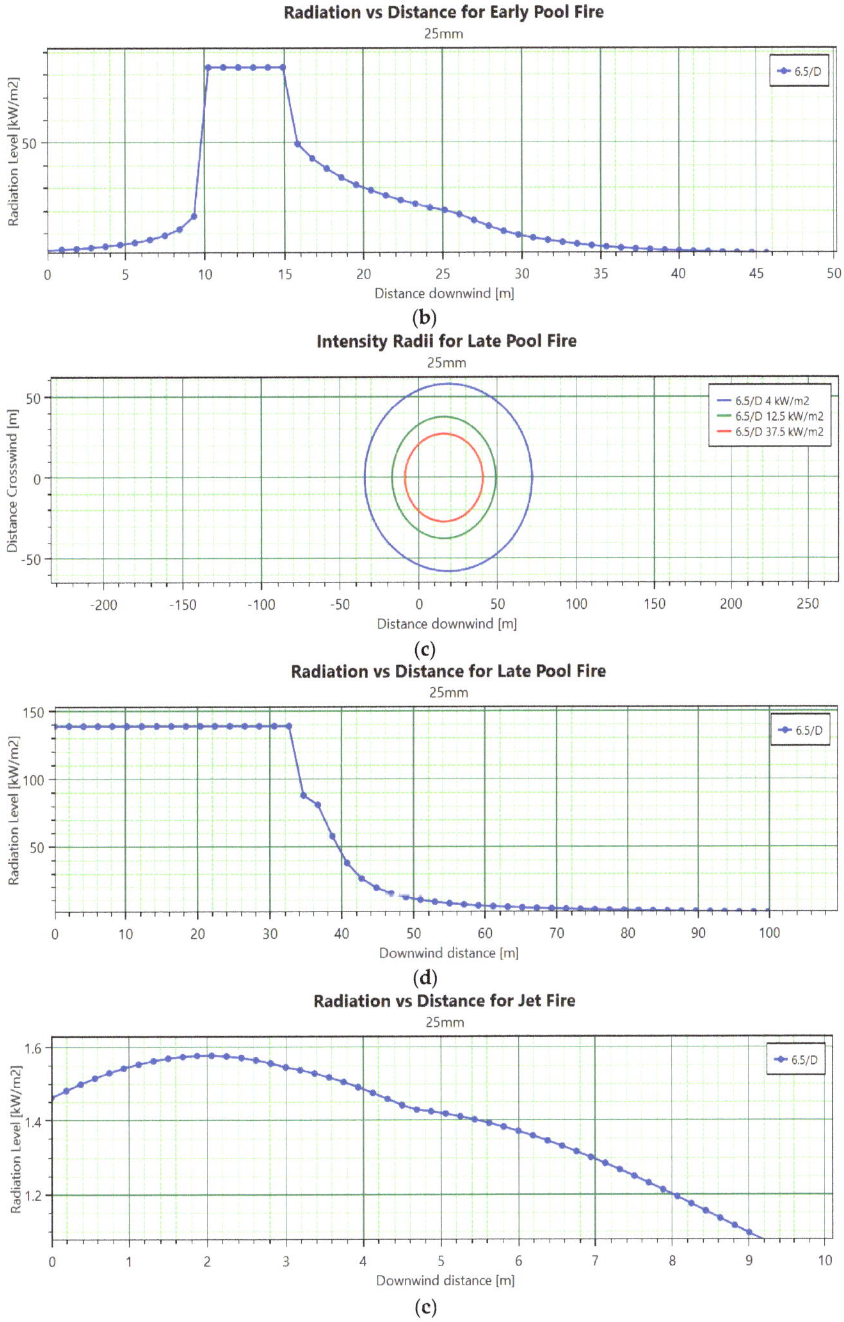

Figure 10. Related charts for pool and jet fire cases. (**a**) Early pool fire thermal radiation intensity; (**b**) Comparison chart of early pool fire thermal radiation intensity and downwind distance; (**c**) Late pool fire thermal radiation intensity; (**d**) Comparison chart of late pool fire thermal radiation intensity and downwind distance; (**e**) Comparison chart of jet fire thermal radiation distance and downwind distance.

The consequences of the tank leakage fire were predicted using the K-R model. The logarithmic value of the consequences was obtained and converted to ascertain the downwind distance of an early pool fire to a specified thermal radiation intensity, which equates to 36.15 m; for a late pool fire, this distance extends to 69.98 m. The early pool fire prediction error is 0.1996 m, and the late pool fire prediction error is 1.9581 m.

4.2. Conclusions

Incorporating case studies into the analysis offered practical insights into the model's applicability and effectiveness in real-world scenarios, reinforcing the conclusion that machine learning algorithms, notably the K-R model, are essential tools for predicting fire consequences in chemical storage tank areas. This progress signifies a new era in hazard prediction, wherein technology-based solutions have the potential to enhance emergency response planning through more informed decision making, thus potentially saving lives and preventing property damage.

Author Contributions: Data Curation, S.W.; Funding Acquisition, W.Z.; Methodology, W.Z.; Project Administration, X.G.; Software, T.W.; Supervision, T.L.; Validation, X.G.; Writing—Original Draft, S.W.; Writing—Review and Editing, X.G. All authors have read and agreed to the published version of the manuscript.

Funding: This research was funded by the National Natural Science Foundation of China for the self-extinguishing behavior of tunnel fires under the combined action of longitudinal wind and water fog curtain, grant number [52376130], Engineering Technology Research Centre for Safe and Efficient Coal Mining (Anhui University of Science and Technology), grant number [SECM2206], and the National Supercomputing Center in Zhengzhou.

Data Availability Statement: Due to privacy concerns, data cannot be provided, and this statement is hereby declared.

Conflicts of Interest: The authors declare no conflicts of interest.

References

1. Lonzaga, J.B. Time reversal for localization of sources of infrasound signals in a windy stratified atmosphere. *J. Acoust. Soc. Am.* **2016**, *139*, 3053–3062. [CrossRef] [PubMed]
2. Witlox, H.W.M.; Fernández, M.; Harper, M.; Oke, A.; Stene, J.; Xu, Y. Verification and validation of Phast consequence models for accidental releases of toxic or flammable chemicals to the atmosphere. *J. Loss Prev. Process Ind.* **2018**, *55*, 457–470. [CrossRef]
3. Wu, T.; Wei, X.; Gao, X.; Han, J.; Huang, W. Study on the risk analysis and system safety integrity of enclosed ground flare. *Therm. Sci. Eng. Prog.* **2019**, *10*, 208–216. [CrossRef]
4. Wang, K.; Zhou, M.; Zhang, S.; Ming, Y.; Shi, T.; Yang, F. Study on the Consequences of Accidents of High-Pressure Hydrogen Storage Vessel Groups in Hydrogen Refueling Stations. *J. Saf. Environ.* **2023**, *23*, 2024–2032. [CrossRef]
5. Song, X.; Su, H.; Xie, Z. Numerical Simulation Study on Leakage and Explosion of LPG Tanker [J/OL]. Engineering Blasting:1-7. Available online: http://kns.cnki.net/kcms/detail/11.3675.TD.20231030.1022.002.html (accessed on 1 November 2023).
6. Barros-Daza, M.J.; Luxbacher, K.D.; Lattimer, B.Y.; Hodges, J.L. Real time mine fire classification to support firefighter decision making. *Fire Technol.* **2022**, *58*, 1545–1578. [CrossRef]
7. Ayhan, B.U.; Tokdemir, O.B. Accident analysis for construction safety using latent class clustering and artificial neural networks. *J. Constr. Eng. Manag.* **2020**, *146*, 04019114. [CrossRef]
8. Hu, P.; Peng, X.; Tang, F. Prediction of maximum ceiling temperature of rectangular fire against wall in longitudinally ventilation tunnels: Experimental analysis and machine learning modeling. *Tunn. Undergr. Space Technol.* **2023**, *140*, 105275. [CrossRef]
9. Khan, A.A.; Zhang, T.; Huang, X.; Usmani, A. Machine learning driven smart fire safety design of false ceiling and emergency response. *Process Saf. Environ. Prot.* **2023**, *177*, 1294–1306. [CrossRef]
10. Rihan, M.; Bindajam, A.A.; Talukdar, S.; Shahfahad; Naikoo, M.W.; Mallick, J.; Rahman, A. Forest fire susceptibility mapping with sensitivity and uncertainty analysis using machine learning and deep learning algorithms. *Adv. Space Res.* **2023**, *72*, 426–443. [CrossRef]
11. Sharma, R.; Rani, S.; Memon, I. A smart approach for fire prediction under uncertain conditions using machine learning. *Multimed. Tools Appl.* **2020**, *79*, 28155–28168. [CrossRef]
12. Sun, Y.; Wang, J.; Zhu, W.; Yuan, S.; Hong, Y.; Mannan, M.S.; Wilhite, B. Development of consequent models for three categories of fire through artificial neural networks. *Ind. Eng. Chem. Res.* **2019**, *59*, 464–474. [CrossRef]
13. Wang, R.; Chen, B.; Qiu, S.; Ma, L.; Zhu, Z.; Wang, Y.; Qiu, X. Hazardous source estimation using an artificial neural network, particle swarm optimization and a simulated annealing algorithm. *Atmosphere* **2018**, *9*, 119. [CrossRef]
14. Yuan, S.; Jiao, Z.; Quddus, N.; Kwon, J.S.-I.; Mashuga, C.V. Developing quantitative structure–property relationship models to predict the upper flammability limit using machine learning. *Ind. Eng. Chem. Res.* **2019**, *58*, 3531–3537. [CrossRef]

15. Sarbayev, M.; Yang, M.; Wang, H. Risk assessment of process systems by mapping fault tree into artificial neural network. *J. Loss Prev. Process Ind.* **2019**, *60*, 203–212. [CrossRef]
16. Jiao, Z.; Ji, C.; Sun, Y.; Hong, Y.; Wang, Q. Deep learning based quantitative property-consequence relationship (QPCR) models for toxic dispersion prediction. *Process Saf. Environ. Prot.* **2021**, *152*, 352–360. [CrossRef]
17. Sathesh, T.; Shih, Y.C. Optimized deep learning-based prediction model for chiller performance prediction. *Data Knowl. Eng.* **2023**, *144*, 102120. [CrossRef]
18. Khodadadi-Mousiri, A.; Yaghoot-Nezhada, A.; Sadeghi-Yarandi, M.; Soltanzadeh, A. Consequence modeling and root cause analysis (RCA) of the real explosion of a methane pressure vessel in a gas refinery. *Heliyon* **2023**, *9*, e14628. [CrossRef] [PubMed]
19. Mahesh, B. Machine learning algorithms-a review. *Int. J. Sci. Res. (IJSR)* **2020**, *9*, 381–386.
20. Sun, B.; Liu, X.; Xu, Z.D.; Xu, D. BP neural network-based adaptive spatial-temporal data generation technology for predicting ceiling temperature in tunnel fire and full-scale experimental verification. *Fire Saf. J.* **2022**, *130*, 103577. [CrossRef]
21. Smith, S.L.; Kindermans, P.J.; Ying, C.; Le, Q.V. Don't decay the learning rate, increase the batch size. *arXiv* **2017**, arXiv:1711.00489.
22. Yan, S. Dynamic Adaptive Risk Assessment System for Individual Building Fires Based on Internet of Things and Deep Neural Networks. Master's Thesis, China University of Mining and Technology, Xuzhou, China, 2021. [CrossRef]
23. He, J.; Li, L.; Xu, J.; Zheng, C. ReLU deep neural networks and linear finite elements. *arXiv* **2018**, arXiv:1807.03973.
24. Friedman, J.H. Stochastic gradient boosting. *Comput. Stat. Data Anal.* **2002**, *38*, 367–378. [CrossRef]
25. Biau, G.; Scornet, E. A random forest guided tour. *Test* **2016**, *25*, 197–227. [CrossRef]
26. Rodriguez, J.D.; Perez, A.; Lozano, J.A. Sensitivity analysis of k-fold cross validation in prediction error estimation. *IEEE Trans. Pattern Anal. Mach. Intell.* **2009**, *32*, 569–575. [CrossRef] [PubMed]
27. Cohen, I.; Huang, Y.; Chen, J.; Benesty, J. Pearson correlation coefficient. *Noise Reduct. Speech Process.* **2009**, *2*, 1–4. [CrossRef]
28. O'brien, R.M. A caution regarding rules of thumb for variance inflation factors. *Qual. Quant.* **2007**, *41*, 673–690. [CrossRef]

fire

Article

Indoor Fire Simulation in Low-Rise Teaching Buildings Based on BIM–FDS

Mengkai Liu * and Guoquan Wang

Department of Engineering Management, Wuhan University of Science and Technology, Wuhan 430074, China
* Correspondence: mengkailiu@wust.edu.cn

Abstract: School buildings gather a large number of underage students, and the disastrous consequences of fire in such buildings are very serious, which is one of the key concerns of society in fire prevention and control. This study takes a "[" type kindergarten teaching building as the background and constructs a BIM–FDS building fire simulation model to reveal the fire smoke dispersion law under the coupling of the typical building structure and fire protection systems. The results show that the stairwells on both sides of the "[" type building are the main channels for the diffusion of fire smoke, and the asymmetry of the stairwell structure will cause apparent differences in the diffusion of smoke. Using the natural smoke exhaust in the stairwells of low-rise buildings does not aggravate the spread of smoke in the building and is conducive to smoke emissions. The high-pressure water mist system is superior to the water spray system in fire extinguishing and controlling room temperature. While it reduces smoke exhaust performance, it does not adversely affect personnel evacuation. This study systematically reveals the law of diffusion of fire smoke from "["-type teaching buildings, which can support the design of similar building structures, ventilation, fire protection, and the formulation of fire escape plans.

Keywords: fire; building information modeling; fire dynamics simulator; kindergarten; low-rise building; high pressure water mist; natural smoke exhaust

Citation: Liu, M.; Wang, G. Indoor Fire Simulation in Low-Rise Teaching Buildings Based on BIM–FDS. *Fire* **2023**, *6*, 203. https://doi.org/10.3390/fire6050203

Academic Editors: Song Lu, Changcheng Liu, Guohui Li and Paweł Wolny

Received: 12 April 2023
Revised: 5 May 2023
Accepted: 12 May 2023
Published: 14 May 2023

1. Introduction

Building fire is one of the most common sudden social security incidents. According to the U.S. Fire Administration report, fires caused 3515 deaths and 16,600 injuries in the United States in 2019 [1]. In general, fires have the characteristics of suddenness, rapid process development, and severe catastrophic consequences. The probability of successfully escaping trapped people is affected by the response speed and evacuation efficiency of themselves and rescuers. Children and students are one of the more vulnerable groups in society. Their ability to make emergency judgments and active ability in a fire emergency is severely lacking, significantly reducing their probability of successful escape. For example, in 2001, a fire broke out at night in a kindergarten in Jiangxi, China, killing 13 children and injuring one [2]. Therefore, for kindergartens and other places where children concentrate, practical fire safety emergency plans should be made to improve the protection ability and rescue efficiency for young children in disaster scenarios. A complete fire safety emergency plan should follow the unique fire laws of each type of building to be practical and targeted. Therefore, studying the fire law of specific buildings is of great significance.

The evolution process of building fires is affected by many factors, such as building structure characteristics, building materials, and fire protection systems. When formulating building fire emergency plans, copying other engineering data is impossible. It is necessary to carry out particular relevant research to determine the potential fire evolution characteristics of specific buildings to effectively improve disaster protection and mitigation. At present, the research on the fire development process and smoke diffusion law mainly adopts physical and numerical model experiments [3]. Physical model experiments consume a lot of labor and material costs in operation, and environmental parameters such as temperature, wind speed, and air humidity cannot be accurately set. Therefore, numerical simulation has become the primary way to study building fire characteristics. The numerical model mainly includes the zone model, network model, and field model. The field model describes the fire development process by calculating the change of these state parameters with time [4]. As a classical field model simulation software, FDS is widely used to study the smoke and heat transfer process in fire and the effect of the water spray system. In recent years, the advantages of BIM technology in life cycle management and data sharing have become increasingly prominent, and the combination of BIM and professional emergency management software has become particularly important. Fire simulation can be realized by relying on the simulation method in the "BIM+" framework, setting the initial simulation information, using computer program algorithms, and displaying the simulation process and results visually. For example, Schatz designed a serious game based on BIM to explore human behavior in fire evacuation [5]. Wang proposed a dynamic fire escape path planning method based on BIM [6]. In addition, BIM can correct deviations between 3D and 2D drawings, visualize building surroundings and facility locations in 3D, and improve differences when using traditional 2D fire management tools [7]. BIM not only supports 3D visualization, but its models also contain architectural information such as building materials and quantities. Fire trends are closely related to building materials, and these parameters are critical for simulation [8].

Building fire is a fire in an indoor space with unique evolution characteristics due to its structural characteristics. The horizontal and vertical structures inside the building have various forms, which form a unique indoor space, thus affecting the evolution of fire and the law of smoke diffusion. For example, the rectangular configuration contributes better to smoke ventilation design than the square and triangular configurations [9]. Kerber's physical experiments concluded that smoke takes less time to fill buildings with smaller aspect ratios (R/H) [10]. The smoke will accumulate at the closed end of the L-shaped and annular corridors and the corner of the corridor, forming an escape danger area, while the T-shaped corridor will not form a similar area [11]. The stack effect in shafts such as elevator shafts and stairwells can accelerate smoke propagation in tall buildings [12]. The concave building structure can increase the longitudinal propagation of fire and smoke, increasing the level of danger on the upper floors of the building [13].

The design of building fire protection systems is an essential part of fire emergency management. Water-based fire extinguishing systems are widely used as a relatively mature fire extinguishing technology in several locations, such as urban tunnels [14,15], warehouses [16], subways [17,18], and nuclear power plants [19]. According to the different particle sizes of sprayed water droplets, water-based fire extinguishing systems can be divided into water sprinkler systems (WSS) and fine water mist fire extinguishing systems (WMS) [20]. The traditional WSS mainly achieves the fire extinguishing effect by directly spraying and cooling the fire source. Water mist also has the same cooling effect, but its sprayed droplets are smaller in diameter, resulting in faster heat absorption and evaporation and a better cooling effect [21]. At the same time, the water mist will be suspended in the form of dense droplet particles around the ignition point, blocking the heat transfer from the fire source to the surrounding material [22]. The water vapor formed by evaporating droplets will also create a barrier around the ignition point, isolating oxygen from it [20,23].

WMS is an emerging fire extinguishing technology using water as the medium. It has the advantages of fast extinguishing speed and small water consumption and is widely used in several scenarios [24,25]. Jiang's study on the application of HPWMS in subway fires concluded that HPWMS could effectively suppress fire development and has better suppression of ambient temperature and CO concentration[18]. Liu studied the application of WMS and WSS in an indoor ventilation environment and concluded that the WMS has a shorter action time and better cooling effect, effectively reducing the risk of backfire [26]. Liu demonstrated the ability of fine water mist to bypass obstacles in complex indoor environments through half-size experiments [27]. Wang concluded that water mist curtains applied in narrow passages effectively hindered the early spread of smoke, and the CO concentration and smoke particles in the protected area were reduced to a large extent [28]. Ku studied the application of HPWMS in transformer fires to provide theoretical and technical references for the safe and stable operation of transformers [29]. The WMS nozzle parameters are critical for effective fire suppression in different application scenarios. Lee studied the application of WMS in a nuclear power plant electrical room and derived a power function relationship between the nozzle and fire source horizontal distance on fire suppression time [30]. Gui discussed the effect of nozzle characteristics parameters such as atomization cone angle, spray velocity, droplet size, and spray flow rate on fire suppression effectiveness in a naturally ventilated room [31]. Ku discussed the effect of different droplet velocities and nozzle flow rates on fire extension [29]. In addition to the aforementioned building fires, WMS still provides excellent fire suppression in restricted spaces with little water volume, such as vehicle and aircraft fires [32]. There are even a large number of applications in particular scenarios, such as oil and gas explosions [33], lithium battery fires [34], suppression of natural gas leaks [35], and jet fires caused by gas leaks [36]. However, WMS are currently less applied in densely populated public buildings such as school buildings, shopping malls, and office buildings, and targeted research on such buildings is necessary.

Although the cooling and extinguishing performance of the WMS are good, it cannot reduce the amount of smoke diffused in the room. The building smoke exhaust system can effectively reduce the smoke concentration. Common smoke extraction systems include mechanical and natural smoke extraction systems. Natural smoke exhaust is an unorganized natural smoke exhaust through building exterior windows or smoke exhaust windows without consuming mechanical power. It is the preferred smoke exhaust method for multi-story civil buildings. In multi-story buildings, vertical shaft structures such as stairwells and elevator shafts are the main stairwells for smoke diffusion. Using the natural smoke exhaust in stairwells can quickly discharge smoke and is conducive to smoke control. For example, Chen concluded in the smoke exhaust experiment of an office building that the maximum total smoke exhaust of stairwell windows is 7852 m^3, and the smoke exhaust capacity of the first-floor window is the largest at 2800 m^3 [37]. Through small-scale experiments, Ahn found that when all the stairwell windows were open, the time of smoke rising was prolonged [38]. However, using natural smoke exhaust in the stairwell will form a stack effect, increasing the risk of the upper building. For example, Su conducted a fire experiment on a residential building in Taiwan and concluded that the chimney effect could cause smoke to spread to the top floor of the building within 180 s [39]. In a 10-story residential building simulation, Philip showed that opening the fire doors reduced residents' available safe egress time (ASET) by 36% [40]. In addition, the effect of natural smoke exhaust often depends on the temperature difference between indoors and outdoors, the height difference between exhaust and air inlet, outdoor wind, and other factors. Therefore, whether the natural smoke exhaust method is used in the stairwell must be determined by the smoke diffusion simulation of a specific building.

Fire extinguishing and smoke exhaust are essential in building fire emergency management. It is challenging to simultaneously achieve fire extinguishing and smoke exhaust by relying solely on a single fire protection system. Multiple fire protection systems often need to be installed in buildings to work together. However, some studies have shown that the water mist system will reduce the smoke diffusion rate, destroy the stability of the smoke layer, and reduce the efficiency of the smoke exhaust system. Sun also found that the water mist system can effectively prevent the spread of smoke in the tunnel but does not work under mechanical ventilation [41]. The open doors and windows will increase the oxygen content of the fire and increase the fire. However, Zhou concluded that a lower mechanical exhaust rate (0.381 m^3/s)-assisted water mist system could shorten the fire extinguishing time [42]. Lee proposed that setting the activation temperature of the spray system below 85 °C can eliminate the activation time delay caused by the smoke exhaust system [43]. However, these studies mainly focus on the coupling effect of two fire protection systems in the same room. There are few studies on the influence of the spray system on the smoke exhaust system in the main evacuation path.

From the above, it can be seen that the building structure greatly influences the smoke diffusion law, and the research on the application of WMS is mostly focused on special-purpose buildings such as warehouses, subways, nuclear power plants, urban tunnels, etc. However, the research on applying WMS in low-rise buildings with complex internal structures, such as shopping malls and school buildings, is less. In addition, whether the use of natural smoke exhausts in the stairwell of low-rise buildings will aggravate the spread of smoke in the building or is beneficial to smoke emission requires the study of smoke diffusion in the building. Therefore, this study takes the actual single building of a kindergarten as the research object supported by BIM technology and uses the numerical simulation method to study the internal temperature and smoke diffusion law of the building under the joint action of the building structure, smoke exhaust system and spray system, and explores the smoke law in the stairwell of low-rise buildings, the fire suppression and smoke extraction performance of the fire protection systems, and the influence of the spray system on the smoke exhaust system in the main evacuation path. It strongly supports the fire protection system design, fire emergency plan, and emergency rescue of the kindergarten and similar building structures.

2. Research Methodology

2.1. Methods of FDS

The FDS (Fire Dynamics Simulator) is a fluid dynamics software developed by the National Institute of Standards and Technology for analyzing and simulating fire flames and smoke. The FDS uses large eddy simulation (LES) to show the flame propagation process and smoke diffusion path in fire scenarios. Since the FDS is open-access, it is widely used in fire research.

Pyrosim was developed based on the FDS, which provides a graphical user interface (GUI) for dynamic fire simulation. Based on computational fluid dynamics, the software can predict the movement, concentration, and temperature changes of toxic gases such as smoke and CO in the fire. Among many fire simulation software, Pyrosim is widely used in fire simulation because of its fast model establishment, convenient parameter setting, and location determination. Therefore, this paper chooses the Pyrosim software to simulate the fire of the model.

2.2. BIM-Based Simulation Framework

The accuracy of parameters such as building geometry, building material information, and fire source information is crucial to the numerical study of building fires. BIM integrates various information about the whole life cycle of the building and allows the numerical model to approximate the physical model as closely as possible. This study uses BIM as an information integration and visualization platform to establish a BIM–FDS-based fire simulation framework for building fire risk management. The overall framework is shown in Figure 1. The first step of the framework is to build the building model through the Autodesk Revit software and then import the building material information into the model to ensure the accuracy of the graphical and non-graphical information of the building model. In addition to the building model, fire origin information and detection equipment are equally important. Therefore, the second step is to import the Revit model into Pyrosim in DXF format and set the calculated fire parameters, grid parameters, and detection equipment in Pyrosim. Finally, the corresponding building's fire simulation results and detecting points data are obtained.

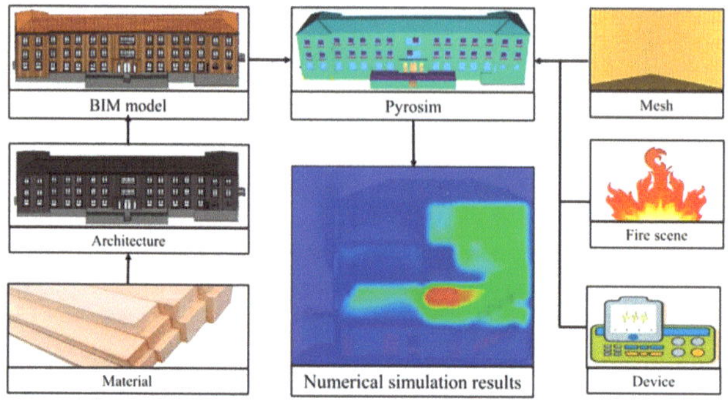

Figure 1. BIM–FDS integration framework.

3. Case Study

3.1. Fire Model Setting

This article takes a large kindergarten in Zhejiang Province, as an example. The kindergarten is "[" shaped, covers an area of 1466 m^2, with a building area of 3363 m^2, including three floors above ground and one underground floor, and the floor height is 4 m. The whole building can accommodate 240 students and 70 staff members. The first and second floors are mainly for teaching and activity areas, and the third is for the staff office area. The layout of the ground floor plan is shown in Figure 2.

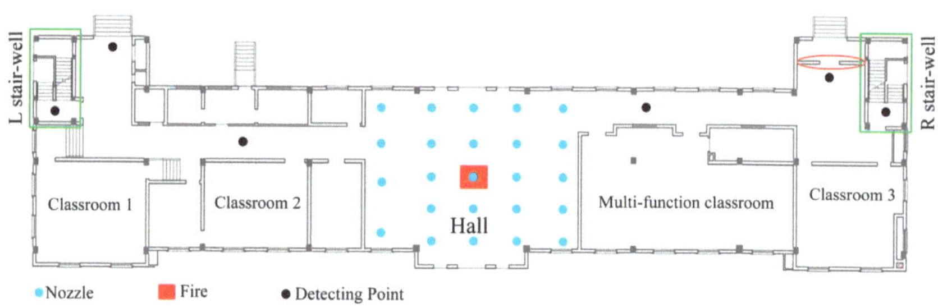

Figure 2. First floor layout plan.

3.1.1. Design of the Fire Scene

The lobby is an important public space in kindergartens. It is often used as an indoor playground or festival site, where many decorative materials are often laid out, posing a greater fire hazard. Therefore, this article assumes that the fire event is a Christmas tree made of PVC in the hall that catches fire due to a short circuit in the decorative lamp wire. It was during teaching hours at the time of the fire, and the door to the hall was closed. The development of early fires can be described using the t^2 model, the formula for which is as follows [44].

$$Q_f = \alpha t^2 \tag{1}$$

where Q_f is the heat release rate of fire, α is the fire growth coefficient, and t is the burning time.

In this paper, the fire source is set to fast fire with the power of 4 MW, and the growth coefficient is $\alpha = 0.047$. Table 1 describes the main parameters of the model and sets the other parameters to the system defaults.

Table 1. Details of environment parameters.

Settings	Parameters
HRRPUA (kW/m^2)	4000
Burner size	1 m × 1 m
Soot yield (kg/kg)	0.07 [45]
CO yield (kg/kg)	0.04 [45]
EPUMO2 (kJ/kg)	13,100
Initial temperature (°C)	20
Initial pressure (Pa)	101,300
Humidity (%)	40
Simulation time (s)	500
Initial visibility (m)	30

3.1.2. Grid Resolution Analysis

The mesh size has a significant impact on the accuracy of numerical simulation results. The more precise the meshing in the fire model, the more accurate are the numerical simulation results, but it will lead to longer calculation time. The mesh size needs to be determined according to the characteristic diameter of the fire source (D^*) as follows [46]:

$$D^* = \left(\frac{Q}{\rho_\infty c_P T_\infty \sqrt{g}} \right)^{2/5} \quad (2)$$

where D^* is a characteristic fire diameter, m; Q is the heat release rate of the fire, KW; ρ_∞ is the ambient air density, generally 1.204 kg/m^3; c_P is the specific heat capacity of ambient air, generally 1.005 kJ/(kg·K); T_∞ is the ambient air temperature, generally 293 K, g is the acceleration of gravity, generally 9.8 m/s^2.

This paper's fire source power is 4 MW, and the calculated D^* is 1.67 m. The 0.06D^*–0.25D^* recommended by the National Institute of Standards and Technology is used to obtain a reliable calculation grid size range of 0.100–0.418 m. In this interval, the grid sizes of 0.1 m, 0.2 m, 0.3 m, and 0.4 m were selected for fire temperature simulation and comparative analysis. To save time, the simulation time was set to 300 s, and the results are shown in Figure 3. It can conclude that the temperature variation curve of the grid size of 0.4 m is different from the others, while the temperature deviation caused by the grid size of 0.3 m and 0.1 m is only 3%. Therefore, in the case of similar accuracy, to minimize the simulation time, this case selected the mesh size of 0.3 m × 0.3 m × 0.3 m.

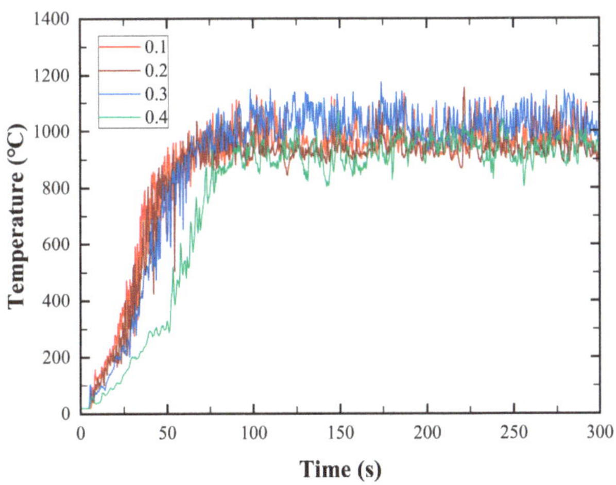

Figure 3. Comparison of temperature above the fire source among the four grid systems.

3.1.3. Configuring the Detecting Point

Most of the casualties during the fire are caused by smoke. The CO concentration, temperature, and visibility in the air are the key factors affecting the evacuation of the affected people. In order to obtain the diffusion path information of these three factors inside the building, 14 detecting points are set in this paper, and CO concentration, temperature, and visibility sensors are set at each detecting point. The National Health Commission of China has pointed out that the average height of Chinese adult men and women is 169.7 cm and 158.0 cm, respectively. Therefore, detecting points are set at the height of 1.7 m to better reflect the impact of smoke on the affected personnel. The elevation position of each detecting point is shown in Figure 4, and the plane position is shown in Figure 2.

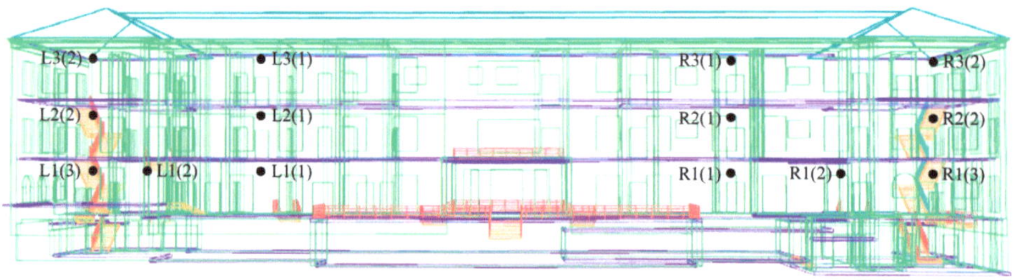

Figure 4. Distribution of detecting points.

3.2. *Scenarios of Simulation*

In this paper, HPWMS and WSS are used as the water-based fire extinguishing systems of the building, and assuming that only the sprinklers in the hall are triggered, other building locations are not reached. According to the relevant provisions of the architectural design of fire protection norms, the hall of the spray system layout is shown in Figure 2. The WSS parameters refer to the fire design drawings of the case model, and the HPWMS parameters are assumed based on the existing research in the review. The two parameters are shown in Table 2. Moreover, the activation time of the system is assumed to be the 50 s when the fire occurs.

Table 2. Spray system setting parameters.

Parameter	HPWMS	WSS
Operating pressure (MPa)	10	0.13
K-Factor	1.2	80
Jet velocity (m/s)	15	5
Median volumetric diameter (μm)	100	1000
Distribution	Constant	Constant
Spray cone angle	60°	60°

There are two different forms of window design in stairwells, fixed windows for lighting only and sliding windows for lighting and exhausting. In this paper, the above two kinds of windows are set in the stairwells on both sides of the case to study the influence of natural smoke exhaust on the smoke diffusion of such typical buildings.

HPWMS, WSS, sliding window, and fixed window are combined into four working conditions, as shown in Table 3. The influence of different water-based fire extinguishing systems and natural smoke exhaust system on fire and smoke will be obtained by comparing the detecting data of the four scenarios.

Table 3. Operating conditions.

Working Conditions	Sprinkler System	Window Type
Test 1	WSS	Fixed window
Test 2	WSS	Sliding window
Test 3	HPWMS	Fixed window
Test 4	HPWMS	Sliding window

4. Results and Analysis

4.1. Smoke Diffusion Law

The building structure will affect fire smoke during the diffusion process to form a unique smoke diffusion law. Figure 5 shows the smoke diffusion law of Test 1 within 500 s. In a short period after the fire, the fire layer and the top floor are the first to be affected by the smoke, followed by the middle floor. The two side stairwells are the main path of smoke spread, and the amount of smoke in the R stairwell is more than in the L stairwell. The same diffusion law was observed in the simulation results of the other three working conditions.

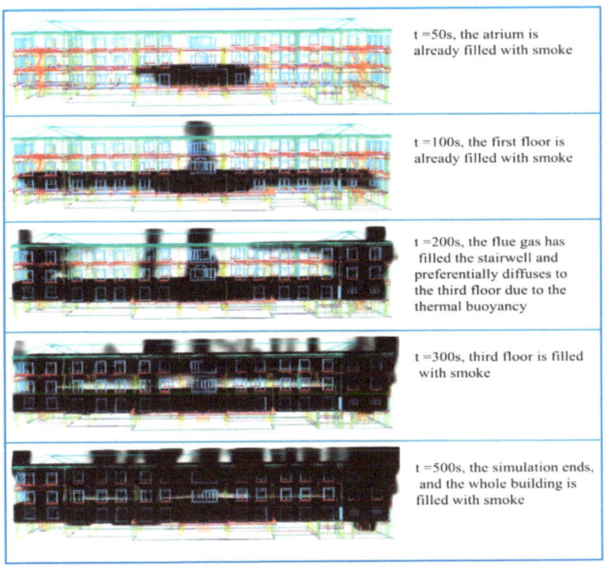

Figure 5. Smoke diffusion process (Test 1).

During the fire evacuation process, the maximum ambient temperature and CO concentration that the human body can withstand are 60 °C and 500 ppm, respectively, and the visibility should not be less than 10 m [47,48], exceeding which will pose a threat to the escape of affected persons. The time for one of the indicators to reach the critical value is the available safety egress time (ASET) under that indicator. The ASET corresponding to the air environment indicators at each detecting point was obtained statistically, as shown in Table 4. It can be seen that the difference in the ASET among the three indexes increases with the increase in smoke diffusion distance. Among them, the change in visibility is the fastest and most widely affected by smoke, which is the first factor threatening the escape of disaster victims. The CO concentration is second, and the temperature change is the slowest.

Table 4. ASET Statistics Details (Test 1).

Site	Detecting Point	Available Safety Egress Time (s)		
		CO Concentration	Temperature	Smoke Visibility
First floor	R1(1)	101	101	72
	L1(1)	93	97	76
	R1(2)	104	110	88
	L1(2)	NO *	NO	124
Stairwell	R1(3)	134	168	100
	L1(3)	NO	NO	124
	R2(2)	180	334	122
	L2(2)	NO	NO	126
	R3(2)	169	330	124
	L3(2)	NO	NO	132
Second floor	R2(1)	NO	NO	299
	L2(1)	NO	NO	392
Third floor	R3(1)	NO	NO	188
	L3(1)	398	NO	250

* NO—indicates that the critical value is not reached during the simulation period.

According to the data of the detecting points in the stairwells, the three data of each detecting point of the L stairwell are smaller than those of the R stairwell. The reason can be attributed to two points. First, two staircases in the building are located at the ends of each side of the stairwell. In case of a fire, smoke accumulates at the end of the stairwell, forming a hazardous area [11]. Second, the exits on both sides of the first-floor stairwell are set differently, as shown in Figure 2, using a red wireframe marker area. A wall is set at the right end of the stairwell exit to separate the interior from the exterior, which reduces the smoke extraction area and prevents the smoke from spreading outside in time, resulting in a large influx into the R stairwell. Therefore, when formulating fire emergency plans, teachers and students should try to avoid evacuation from the R stairwell.

4.2. *Effect of HPWMS on Smoke Diffusion*

This section selects the CO concentration and ambient temperature of Test 1 and Test 3 for quantitative comparative analysis.

4.2.1. Temperature

Figure 6 depicts the temperature change with time at 1 m above the fire source for Test 1 and Test 3 simulation conditions. It can be seen that the fire source has gone through three stages: 1. rapid growth stage; 2. attenuation stage; and 3. stabilization phase.

In the first stage, the temperature above the fire source continues to rise due to the intense combustion of combustibles. The sprinkler systems were activated after 50 s, but as the fire source was burning violently at this time, neither sprinkler system could immediately reduce the temperature around the fire source; hence, the temperature above the fire source continued to rise. The temperature curve of the two conditions at this stage tend to be consistent. In the second stage, the temperature above the fire source decreases due to the influence of the spray system. Under the action of HPWMS, the temperature in Test 3 decreased sharply at 82 s, and decreased to 68 °C at 132 s, and then entered the next stage. However, it is worth noting that the temperature curve shows obvious oscillation in the stage of 114 s–132 s. This is due to the decrease in fire temperature, resulting the decrease of evaporated droplets, which leads to a worse cooling and oxygen insulation effect. The temperature rises briefly and then decreases again, but the overall trend is downward. Under WSS, the temperature decreases slowly and enters the next stage after decreasing to 160 °C at 304 s. In comparison, HPWMS reduces the fire extinguishing time by 76.6%. In the third stage, the temperature above the fire source in both conditions fluctuates in a small range. The difference is that the temperature in Test 1 is stable at about 170 °C, while Test 3 is stable at about 75 °C.

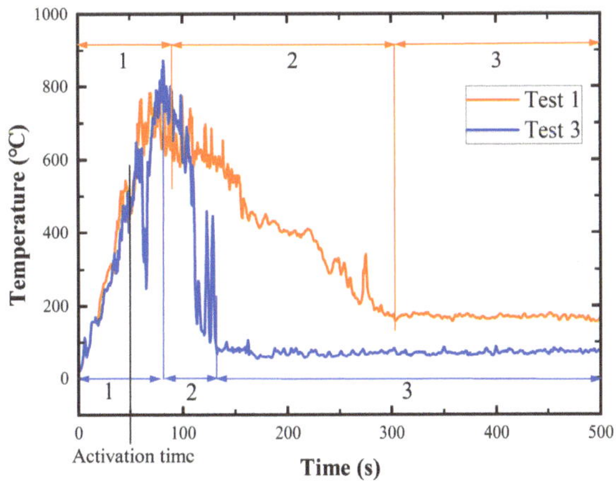

Figure 6. Temperature change at 1 m above the fire source during Test 1 and Test 3.

HPWMS shows a better effect than WSS in fire extinguishing; the cooling effect of HPWMS is also better for the ambient temperature of the first floor. Figure 7 shows the temperature distribution for the first floor at 400 s for the two working conditions. It can be seen that the temperature difference between the two conditions in the hall is enormous. Most of the Test 1 hall is about 100 °C higher than the Test 3 hall, and the temperature in the stairwell is also higher.

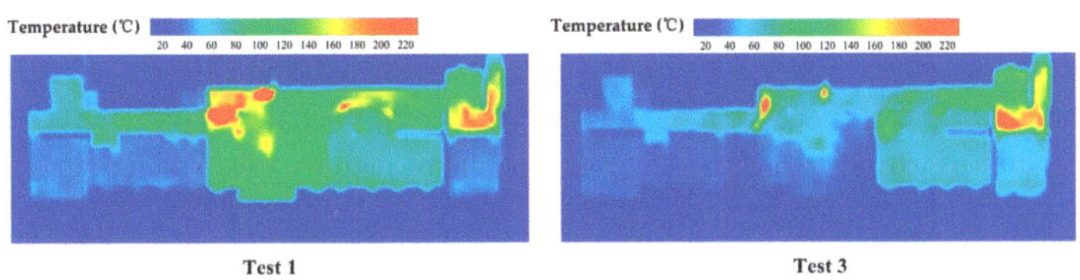

Figure 7. First floor temperature distribution during Test 1 and Test 3 at 400 s.

In summary, HPWMS has more advantages than WSS in extinguishing fire sources and controlling ambient temperature.

4.2.2. CO Concentration

As shown in Table 4, the CO concentrations at R1(2), R1(3), R2(2), and R3(2) exceeded the critical value during the simulation period. The CO concentration curves corresponding to these four detecting points are shown in Figure 8. It can be seen that the CO concentration is lower under the action of HPWMS, and the difference between the two conditions is the largest at R1(2). However, with the increase in smoke diffusion distance, the difference gradually decreases, but this inhibitory effect gradually decreases with the increase in smoke diffusion distance. This shows that HPWMS has a significant inhibitory effect on CO concentration, but this inhibitory effect gradually decreases with the increase in smoke diffusion distance, which indicates the limitation of the action range.

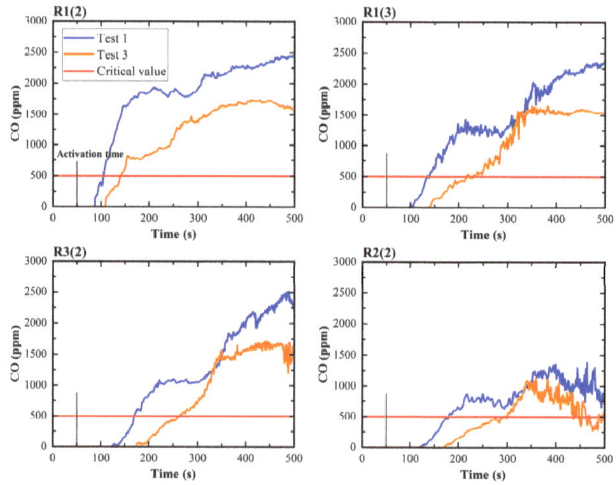

Figure 8. Variation in CO concentration during Test 1 and Test 3 at the following detecting points R1(2), R1(3), R3(2) and R2(2).

It can also be seen from the change curve of CO concentration in Figure 8 that the CO concentration curve of Test 3 has a slower growth trend, which is due to the better cooling effect of HPWMS on the smoke, weakening the thermodynamic force of the smoke and slowing down the smoke diffusion rate. As shown in the CO concentration observed in Figure 9 (Test 1), the smoke has obvious stratification. Therefore, in the process of smoke diffusion into the stairwell, a small amount of smoke enters first, and when the smoke layer falls below the height of the door frame, a large amount of smoke enters the stairwell. Therefore, the CO concentration change curve shows a stepwise growth. However, the stability of the smoke layer in Test 3 is destroyed by HPWMS, as shown in Figure 9 (Test 3), which is consistent with the conclusions of Blanchard [49] and Morlon [50]. Therefore, the CO concentration curve will continue to rise and then tends to be stable.

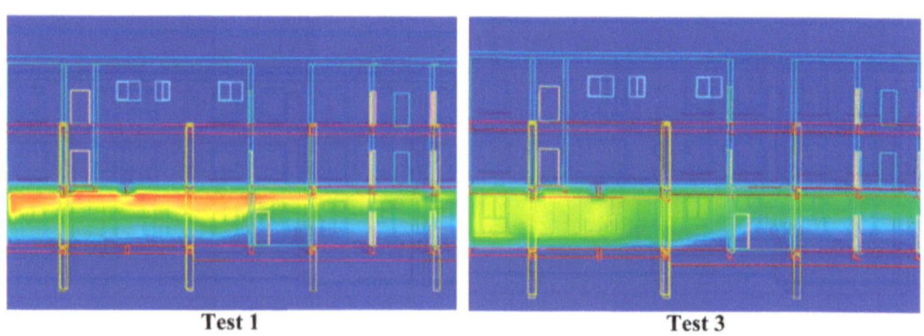

Figure 9. CO concentration distribution in the corridor during Test 1 and Test 3 at 200 s.

4.3. *Influence of Natural Smoke Exhaust Systems on Smoke Diffusion*

When sliding windows are used in stairwells, the stack effect has an important impact on the smoke diffusion. Therefore, the determination of the pressure neutral surface in the stack effect is crucial for studying the smoke diffusion law. The formula is as follows [51].

$$\frac{H_n}{H} = \frac{1}{1 + (T_S/T_O)(A_b/A_a)^2} \tag{3}$$

where H_n is the vertical distance from the neutral surface to the lower edge of the shaft, H is the height of the lateral opening of the shaft, T_S is the absolute temperature inside the shaft, T_O is the absolute temperature outside the shaft, and A_a, A_b are the areas of the upper and lower windows, respectively, and since this model uses uniform windows, the ratio is taken as 1.

The calculated height of the pressure-neutral surface of the model is 9.24 m, located at the bottom of the third floor. Therefore, in theory, the natural smoke exhaust system of the stairwell can reduce the impact of smoke on the second floor of the building. In the numerical simulation of this paper, the smoke distribution in the building under Test 1 and Test 2 also confirms this conclusion. As shown in Figure 10, under the condition of natural smoke exhaust in the stairwell, the smoke volume on the second floor of the building is significantly reduced, and most of the area is in a blank state without smoke. Based on the visibility curve shown in Figure 11, it can be concluded that the visibility during Test 2 in the second-floor corridor fluctuates within the safe range, in contrast to that observed in Test 1, and the visibility during Test 2 on the third floor decreased to the critical value later. From the above analysis, it can be concluded that using the natural smoke exhaust in the stairwell can significantly reduce the hazard of smoke to the middle floor and does not increase the amount of smoke on the top floor due to the stack effect.

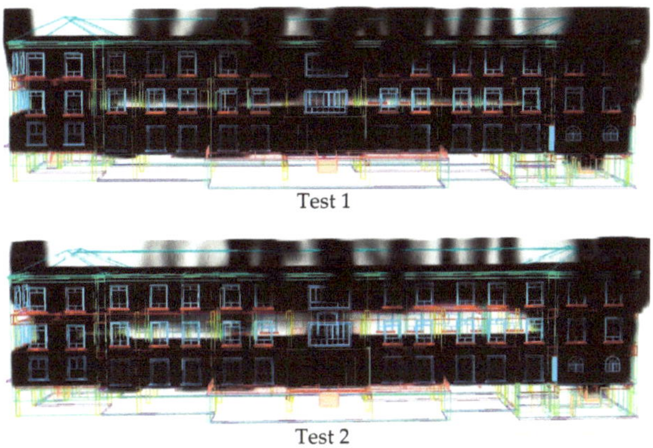

Figure 10. Comparison of smoke diffusion during Test 1 and Test 2 at 500 s.

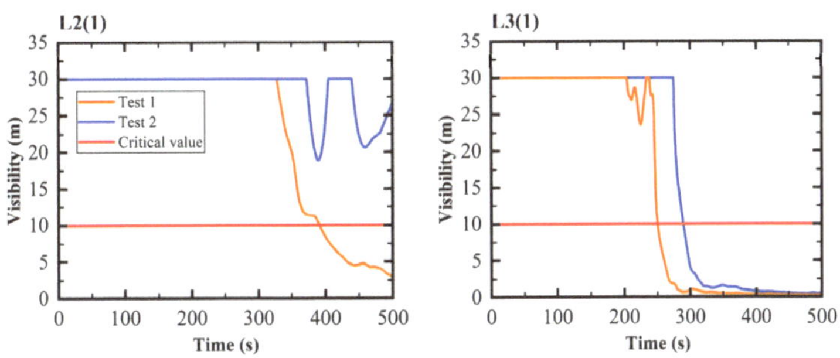

Figure 11. Visibility changes during Test 1 and Test 2 at the following detecting points: L2(1) and L3(1).

4.3.1. CO Concentration

The CO concentration data in the R stairwell are shown in Table 5. It can be seen that the CO concentration at each detecting point is reduced by more than 35% when the normally open sliding window is used in the stairwell. When only considering the influence of CO on evacuation, the ASET in the R stairwell increased significantly under the condition of the windows opening. The difference in ASET for the R3(2) is small due to the rapid diffusion of the smoke in the stairwell. In the fire, priority should be given to evacuating the affected people on the top floor. Therefore, from the above analysis, it can be concluded that the natural smoke exhaust in the stairwell has a good effect on reducing the CO concentration in the air and ensuring the evacuation safety of the affected people.

Table 5. Comparison of ASET under the effect of CO (Test 1, Test 2).

Point	Test	Average CO Concentration (ppm)	Degree of Reduction	ASET (s)	Difference Value (s)
R2(2)	1 2	613.03 358.02	41.60%	180 301	121
R3(2)	1 2	984.52 636.98	35.30%	169 203	34
R3(1)	1 2	148.92 80.68	45.82%	NO * NO	NO

* NO—indicates that the critical condition is not reached during the simulation period.

4.3.2. Temperature

The average temperature data calculated in detecting points in the corridors on the second and third floors during Test 1 and Test 2 have been collected in Table 6. Combined with the data of the corresponding detecting points in Tables 4 and 6, it can be seen that the ambient temperature of the top and middle floors of the building were less affected during the fire. In addition, the temperature difference between the two working conditions at each detecting point is small, which indicates that the natural smoke exhaust in the stairwell has a negligible effect on the temperature of the floor above the fire. Ahn explained this phenomenon in his research. He proved that opening a single window has little effect on the temperature of the stairwell. When the number of open windows exceeds five, the temperature of the stairwell will be significantly reduced [38].

Table 6. Average temperature of detecting points (Test 1, Test 2).

Detecting Point	Test 1	Test 2	Difference
R2(1)	30.56 °C	29.99 °C	0.57 °C
L2(1)	30.19 °C	30.05 °C	0.14 °C
R3(1)	35.22 °C	32.27 °C	2.96 °C
L3(1)	33.14 °C	31.22 °C	1.92 °C

4.4. *Linkage Effect of the Fire Control System*

HPWMS and natural smoke exhaust systems are active and passive fire-fighting facilities, respectively, which play different roles in different areas of the building and in different stages of the fire. The interaction between the two systems cannot be ignored. In the analysis of Section 4.2.1, HPWMS has a good cooling effect on the ambient temperature, as well as the ambient smoke. The temperature difference between smoke and air is the main driving force for smoke, while HPWMS will reduce the temperature difference between smoke and air, destabilizing the smoke layer and reducing the diffusion speed. Therefore, the efficiency of natural smoke exhaust will be reduced. As shown in the change in CO concentration on the third floor in Figure 12, the CO concentration of Test 3 and 4 (using HPWMS) was less than that of Test 2 before 300 s, but the CO concentration of Test 3 and 4 was higher than that of Test 2 after 300 s. At the end of the simulation, the concentration between Tests 3 and 4 with two different windows is similar. This shows that HPWMS has an inhibitory effect on the natural smoke exhaust. However, in the latter simulation stage, the CO concentration value in Test 4 is not significantly larger than that in Test 2, which indicates that the effect of HPWMS on natural smoke exhaust does not significantly increase the amount of smoke on the top floor in a low-rise building, such as the one evaluated in this study paper.

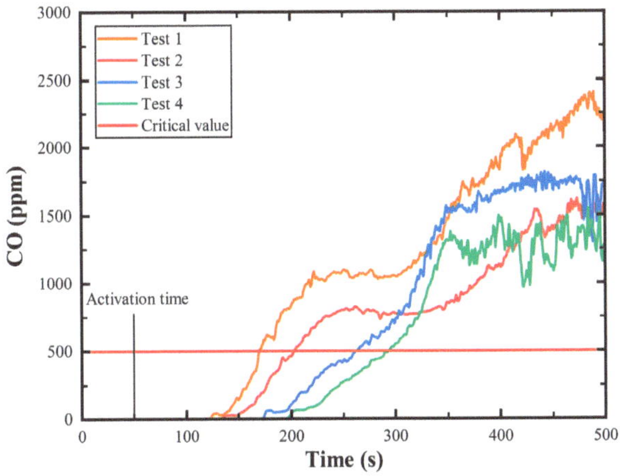

Figure 12. Variation of R3(2) CO concentration during Test 1, Test 2, Test 3, and Test 4.

Stairwells are the only way for the victims to escape, so the ASET in the stairwell is an important parameter to measure the disaster relief capacity of the fire system. Under the comprehensive consideration of CO concentration, temperature, and visibility conditions, the ASET in the R and L stairwells is shown in Figure 13. It can be seen that when the two systems are applied simultaneously, the ASET at each detecting point is increased by 50–77 s. Under this condition, the ASET at each point in Test 4 is higher than that in other conditions. HPWMS reduces the smoke temperature and slows down the diffusion rate, and has the best effect in increasing the ASET of the victims. The inhibiting effect of HPWMS on the natural smoke exhaust does not endanger the evacuation. In addition, the ASET of the L stairwell on the same floor is higher than that of the R stairwell, so the L stairwell should be the first choice when formulating the emergency plan.

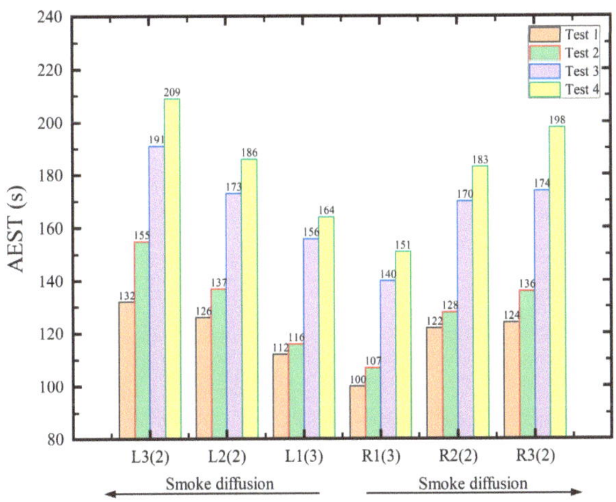

Figure 13. ASET comparison of each detecting point in the stairwell.

5. Conclusions

This study established a BIM–FDS fire simulation model based on the "["-type kindergarten building. Based on the results and analysis, the following conclusions are formulated:

1. This study verified the practical value of the BIM technology in emergency evacuation management for public buildings. A public building emergency evacuation manage-

ment model based on the BIM–FDS is established by combining accurate BIM building information data with a professional fire simulation program. The accurate smoke information and smoke diffusion law at each key node in the building evacuation path are obtained, which provides a basis for emergency management and better guides the emergency management that is difficult to predict and control.
2. This study obtained the unique smoke diffusion law in "]" buildings. In the "]" building, the fire smoke diffuses through the penetrating corridor to the stairwells on both sides and through the stairwells to the whole building. In the diffusion process the fire source and the top layer are the most vulnerable to smoke, followed by the middle floor. In the case studied in this paper, the asymmetric building structure at both ends of the corridor causes different smoke exhaust areas on both sides, resulting in a large difference in the amount of smoke diffused upward, reflecting the important influence of building structure on the smoke diffusion law and the accurate guarantee of fine BIM models for fire simulation.
3. This study demonstrates the excellent performance of HPWMS in extinguishing fire and inhibiting smoke diffusion and stairwells' natural smoke exhaust ability Comparing Tests 1 and 3 shows that HPWMS shortens the fire extinguishing time by 76.6%, reduces the smoke temperature and diffusion velocity, and reduces the average concentration of CO in the R stairwell by more than 60%. In low-rise buildings, using the natural smoke exhaust in the stairwell can timely discharge the smoke, reduce the threat of smoke to other floors, and will not increase the hazard of smoke to other floors.
4. This study demonstrates the negative effect of HPWMS on natural smoke exhaust and obtains the influence of this effect on smoke diffusion in buildings. HPWMS reduces the diffusion velocity and destabilizes the smoke layer, weakening the performance of natural smoke exhaust. However, in this case, the building floor is lower, and the difference between the natural smoke exhaust efficiency of Tests 2 and 4 is not obvious Moreover, the ASET value at each staircase of Test 4 is the largest, which indicates that the combined effect of HPWMS and natural smoke exhaust can be more beneficial for fire evacuation.

In this study, the interior's fire load and furniture decoration aspects were simplified, and the influence of fire spread on the fire outcome was not considered. In addition, this paper does not carry out escape simulation for special groups such as young children and needs to be discussed in depth in the later stage.

Author Contributions: Conceptualization, M.L.; Data curation, G.W.; Methodology, M.L.; Software, G.W.; Formal analysis, G.W.; Investigation, G.W.; Visualization, G.W.; Supervision, M.L. Writing—original draft preparation, G.W.; Writing—review and editing, M.L. All authors have read and agreed to the published version of the manuscript.

Funding: This research received no external funding.

Institutional Review Board Statement: Not applicable.

Informed Consent Statement: Not applicable.

Data Availability Statement: The data used in this study are available on request to corresponding author.

Conflicts of Interest: The authors declare no conflict of interest.

References

1. USFA, Fire Risk in 2019. Available online: https://www.usfa.fema.gov/statistics/deaths-injuries/ (accessed on 14 December 2022).
2. SAWS. *Jiangxi Radio and Television Development Center Art Kindergarten Fire Accident*; China's Work Safety Yearbook: Beijing, China, 2002; pp. 456–457.
3. Lulea, M.D.; Iordache, V.; Nastase, I. Numerical Model Development of the Air Temperature Variation in a Room Set on Fire for Different Ventilation Scenarios. *Appl. Sci.* **2021**, *11*, 11698. [CrossRef]
4. Ji, J. *Fire Dynamic in High-Rise Building*; Science Press: Beijing, China, 2020; pp. 50–51.

5. Rüppel, U.; Schatz, K. Designing a BIM-based serious game for fire safety evacuation simulations. *Adv. Eng. Inf.* **2011**, *25*, 600–611. [CrossRef]
6. Wang, J.; Wei, G.; Dong, X. A dynamic fire escape path planning method with BIM. *J. Ambient Intell. Hum. Comput.* **2021**, *12*, 10253–10265. [CrossRef]
7. Chen, X.-S.; Liu, C.-C.; Wu, I.C. A BIM-based visualization and warning system for fire rescue. *Adv. Eng. Inf.* **2018**, *37*, 42–53. [CrossRef]
8. Shi, L.; Chew, M.Y.L.; Liu, X.Y.; Cheng, X.D.; Wang, B.; Zhang, G.M. An experimental and numerical study on fire behaviors of charring materials frequently used in buildings. *Energy Build.* **2017**, *138*, 140–153. [CrossRef]
9. Doheim, R.M.; Yohanis, Y.G.; Nadjai, A.; Elkadi, H. The impact of atrium shape on natural smoke ventilation. *Fire Saf. J.* **2014**, *63*, 9–16. [CrossRef]
10. Kaye, N.B.; Hunt, G.R. Smoke filling time for a room due to a small fire: The effect of ceiling height to floor width aspect ratio. *Fire Saf. J.* **2007**, *42*, 329–339. [CrossRef]
11. Tang, J.; Shi, B.m.; Chen, K. Numerical simulation of fire smoke flow in typical structure of building corridor. *J. Saf. Sci. Technol.* **2015**, *11*, 33–37.
12. Zhang, J.; Weng, J.W.; Zhou, T.N.; Ouyang, D.X.; Chen, Q.P.; Wei, R.C.; Wang, J. Investigation on Smoke Flow in Stairwells induced by an Adjacent Compartment Fire in High Rise Buildings. *Appl. Sci.* **2019**, *9*, 1431. [CrossRef]
13. Lu, X.X.; Shi, G.Y.; Song, S.Y. Investigation on the fire behaviour and propagation rule of fire overflow in the high-rise concave building structure. *Indoor Built Environ.* **2022**, *32*, 214–229. [CrossRef]
14. Li, J.; Li, Y.F.; Bi, Q.; Li, Y.; Chow, W.K.; Cheng, C.H.; To, C.W.; Chow, C.L. Performance evaluation on fixed water-based firefighting system in suppressing large fire in urban tunnels. *Tunn. Undergr. Space Technol.* **2019**, *84*, 56–69. [CrossRef]
15. Cheong, M.K.; Cheong, W.O.; Leong, K.W.; Lemaire, A.D.; Noordijk, L.M. Heat Release Rate of Heavy Goods Vehicle Fire in Tunnels with Fixed Water Based Fire-Fighting System. *Fire Technol.* **2014**, *50*, 249–266. [CrossRef]
16. Dasgotra, A.; Rangarajan, G.; Tauseef, S.M. CFD-based study and analysis on the effectiveness of water mist in interacting pool fire suppression. *Process Saf. Environ. Prot.* **2021**, *152*, 614–629. [CrossRef]
17. Li-wei, P.A.N.; Lo, S.M.; Guang-xuan, L.; Bei-hua, C. Experimental Study of Smoke Control in Subway Station for Tunnel Area Fire by Water Mist System. *Procedia Eng.* **2011**, *11*, 335–342. [CrossRef]
18. Jiang, Q.; Gao, S.; Kong, X.X. Application analysis of high pressure water mist fire extinguishing system in subway train. *Water Wastewater Eng.* **2020**, *56*, 76–80, 85. [CrossRef]
19. Lee, J. Numerical analysis on the rapid fire suppression using a water mist nozzle in a fire compartment with a door opening. *Nucl. Eng. Technol.* **2019**, *51*, 410–423. [CrossRef]
20. Yang, P.Z.; Liu, T.; Qin, X.A. Experimental and numerical study on water mist suppression system on room fire. *Build. Environ.* **2010**, *45*, 2309–2316. [CrossRef]
21. Santangelo, P.E.; Tarozzi, L.; Tartarini, P. Full-Scale Experiments of Fire Control and Suppression in Enclosed Car Parks: A Comparison Between Sprinkler and Water-Mist Systems. *Fire Technol.* **2016**, *52*, 1369–1407. [CrossRef]
22. Li, B.Q.; Liu, B.; Tang, L.G.; Zhao, S.; Chen, Y. Application Analysis of High Pressure Fine Water Mist Extinguishing Technology in Metro Vehicles. *Electr. Driv. Loco.* **2020**, *273*, 124–128. [CrossRef]
23. Yao, B.; Chow, W.K. Extinguishment of a PMMA fire by water spray with high droplet speeds. *Int. J. Therm. Sci.* **2005**, *44*, 410–419. [CrossRef]
24. Yu, H.Z. The Ceiling Height Limits for Water Mist Protection of Two Solid Combustible Fire Hazards in Open Environment. *Fire Technol.* **2021**, *57*, 163–187. [CrossRef]
25. Zhang, P.H.; Tang, X.; Tian, X.L.; Liu, C.; Zhong, M.H. Experimental study on the interaction between fire and water mist in long and narrow spaces. *Appl. Therm. Eng.* **2016**, *94*, 706–714. [CrossRef]
26. Liu, H.; Wang, C.; De Cachinho Cordeiro, I.M.; Yuen, A.C.Y.; Chen, Q.; Chan, Q.N.; Kook, S.; Yeoh, G.H. Critical assessment on operating water droplet sizes for fire sprinkler and water mist systems. *J. Build. Eng.* **2020**, *28*, 100999. [CrossRef]
27. Liu, Y.; Wang, X.; Liu, T.; Ma, J.; Li, G.; Zhao, Z. Preliminary study on extinguishing shielded fire with water mist. *Process Saf. Environ. Prot.* **2020**, *141*, 344–354. [CrossRef]
28. Wang, Z.; Wang, X.; Huang, Y.; Tao, C.; Zhang, H. Experimental study on fire smoke control using water mist curtain in channel. *J. Hazard. Mater.* **2018**, *342*, 231–241. [CrossRef]
29. Ku, Y.; Zhang, J.; Wang, Z.; Li, Y.; Yao, H. A Numerical Study on the Extinguishing Performances of High-Pressure Water Mist on Power-Transformer Fires for Different Flow Rates and Particle Velocities. *Fld. Dyn. Mater. Process.* **2021**, *17*, 1077–1090. [CrossRef]
30. Lee, J.; Moon, J. Numerical analysis of the effect of horizontal distance between a water mist nozzle and ignition source on reduction in heat release rate. *Ann. Nucl. Energy* **2020**, *144*, 107560. [CrossRef]
31. Gui, X.; Xue, H.; Hu, Z.; Cui, Z. Influence of Water Mist Nozzle Characteristic Parameters on Oil Pool Fire Extinguishing in Confined Space. *Arab. J. Sci. Eng.* **2023**, *48*, 3441–3454. [CrossRef]
32. Papa, R.; Andrade, C.R.; Zaparoli, E.L.; Santos, L.C.D. CFD analysis of smoke transport inside an aircraft cargo compartment. *J. Braz. Soc. Mech. Sci. Eng.* **2016**, *38*, 327–334. [CrossRef]
33. Li, G.C.; Pan, C.Y.; Liu, Y.P.; Zhu, X.L.; Ni, X.M.; Zhao, X.D.; Chen, G.X.; Wang, X.S. Evaluation of the effect of water mist on propane/air mixture deflagration: Large-scale test. *Process Saf. Environ. Prot.* **2021**, *147*, 1101–1109. [CrossRef]

34. Liu, T.; Liu, Y.; Wang, X.; Kong, X.; Li, G. Cooling control of thermally-induced thermal runaway in 18,650 lithium ion battery with water mist. *Energy Convers. Manag.* **2019**, *199*, 111969. [CrossRef]
35. Liu, Y.P.; Wang, X.S.; Zhu, P.; Li, G.C.; Ni, X.M.; Zhang, J. Experimental study on gas jet suppressed by water mist: A clean control technique in natural gas leakage incidents. *J. Cleaner Prod.* **2019**, *223*, 163–175. [CrossRef]
36. Liu, Y.P.; Shen, J.X.; Ma, J.; Li, G.C.; Zhao, Z.H.; Ni, X.M.; Wang, X.S. Laser-based measurement and numerical simulation of methane-air jet flame suppression with water mist. *Process Saf. Environ. Prot.* **2021**, *148*, 1033–1047. [CrossRef]
37. Chen, Y.B. Influence of layout of windows in the staircase on natural smoke exhausting. *Fire Sci. Technol.* **2015**, *34*, 1436–1439.
38. Ahn, C.-S.; Kim, D.-Y.; Park, C.; Kim, M.-W.; Kim, T.; Bang, B.-H.; An, S.; Yarin, A.L.; Yoon, S.S. Experimental and numerical study of smoke behavior in high-rise stairwells with open and closed windows. *Int. J. Therm. Sci.* **2020**, *157*, 106500. [CrossRef]
39. Su, C.H.; Lin, Y.C.; Shu, C.M.; Hsu, M.C. Stack effect of smoke for an old-style apartment in Taiwan. *Build. Environ.* **2011**, *46*, 2425–2433. [CrossRef]
40. McKeen, P.; Liao, Z. Numerical analysis on the hazards of open stairwell doors in high-rise residential buildings. *J. Build. Eng.* **2022**, *54*, 104561. [CrossRef]
41. Sun, J.Y.; Fang, Z.; Tang, Z.; Beji, T.; Merci, B. Experimental study of the effectiveness of a water system in blocking fire-induced smoke and heat in reduced-scale tunnel tests. *Tunn. Undergr. Space Technol.* **2016**, *56*, 34–44. [CrossRef]
42. Zhou, Y.; Bu, R.W.; Zhang, X.N.; Fan, C.G.; Gong, J.H. Performance evaluation of water mist fire suppression: A clean and sustainable fire-fighting technique in mechanically-ventilated place. *J. Clean. Prod.* **2019**, *209*, 1319–1331. [CrossRef]
43. Lee, J. Numerical analysis of how ventilation conditions impact compartment fire suppression by water mist. *Ann. Nucl. Energy* **2020**, *136*, 107021. [CrossRef]
44. Zhang, T.; Cao, D.; Lv, Z. Simulation Study on the Influence of Fire Partition on Curtain Wall Temperature in Super High-Rise Buildings in China. *Complexity* **2021**, *2021*, 4124049. [CrossRef]
45. Zhang, T. Simulation Study on Fire Visibility of Typical Floor Planes of Modern Super High-Rise Office Buildings in China. *Complexity* **2020**, *2020*, 8868522. [CrossRef]
46. Meng, N.; Hu, L.; Wu, L.; Yang, L.; Zhu, S.; Chen, L.; Tang, W. Numerical study on the optimization of smoke ventilation mode at the conjunction area between tunnel track and platform in emergency of a train fire at subway station. *Tunn. Undergr. Space Technol.* **2014**, *40*, 151–159. [CrossRef]
47. Xiao, M.F.; Zhou, X.H.; Pan, X.X.; Wang, Y.A.; Wang, J.; Li, X.L.; Sun, Y.S.N.; Wang, Y.M. Simulation of emergency evacuation from construction site of prefabricated buildings. *Sci. Rep.* **2022**, *12*, 2732. [CrossRef]
48. Wang, J.G.; Zhao, S.R.; Su, J.K. Influence of different smoke exhaust modes on personnel evacuation in subway station fire. *Fire Sci. Technol.* **2021**, *40*, 1141–1145.
49. Blanchard, E.; Morlon, R.; Parent, G.; Fromy, P.; Boulet, P.; Borgiallo, D. Experimental Study of the Interaction Between Water Sprays and Smoke Layer. *Fire Technol.* **2018**, *54*, 479–501. [CrossRef]
50. Morlon, R.; Boulet, P.; Parent, G.; Morlon, R.; Lechene, S.; Blanchard, E.; Rebuffat, C.; Fromy, P.; Vantelon, J.P.; Borgiallo, D. Study of De-stratification and Optical Effects Observed During Smoke/Mist Interactions. *Fire Technol.* **2015**, *51*, 1231–1248. [CrossRef]
51. Xu, Z.S.; Jiang, X.P. *Smoke Control Engineering*; China Machine Press: Beijing, China, 2011; pp. 2–3.

fire

MDPI

Article

Dropping Fire Retardants by Helicopter and Its Application to Wildfire Prevention near Electrical Transmission Lines

Jiazheng Lu [1], Tejun Zhou [1,2,*], Chuanping Wu [1,3] and Yangyi Ou [1]

1 State Key Laboratory of Disaster Prevention and Mitigation of Power Transmission and Transformation Equipment, Changsha 410007, China
2 College of Electrical and Information Engineering, Changsha University of Science and Technology, Changsha 410007, China
3 Hunan Disaster Prevention Technology Co., Ltd., Changsha 410007, China
* Correspondence: zhoutejun1988@126.com

Abstract: Dropping fire retardants by helicopter can effectively reduce the intensity of wildfires. This study proposes a test plan for spraying different fire retardants from a helicopter bucket fire extinguisher. In this study, pure water, 10% Class AB flame retardant, 0.3% gel flame retardant, 10% Class A flame retardant, and 10% Class A flame retardant + 0.6% guar gum were each added to the bucket fire extinguishing device and sprayed on 4-layer, 6-layer, and 12-layer wood cribs. The radiation intensity, mass loss, and temperature were used as indicators to compare the burning intensity of the fire field and the difference in fire field combustion intensity after the wood cribs were ignited 1 h after natural air drying. The results showed that flame retardancy could be ranked from high to low as follows: 10% Class A flame retardant + 0.6% guar gum > gel flame retardant > 10% Class A flame retardant > Class AB flame retardant > pure water. During the long term high temperature and drought period in Hunan Province, China, from August to September 2022, a field application showed that dropping fire retardants by helicopter effectively reduced the intensity of wildfires and avoided transmission line trips due to the wildfire, which reduced the number of ground personnel required when fighting large-scale forest fires.

Keywords: wildfire near transmission line; helicopter; flame retardant

Citation: Lu, J.; Zhou, T.; Wu, C.; Ou, Y. Dropping Fire Retardants by Helicopter and Its Application to Wildfire Prevention near Electrical Transmission Lines. *Fire* **2023**, *6*, 176. https://doi.org/10.3390/fire6050176

Academic Editors: Song Lu, Changcheng Liu, Guohui Li and Paweł Wolny

Received: 11 December 2022
Revised: 4 February 2023
Accepted: 9 February 2023
Published: 25 April 2023

1. Introduction

The frequency and intensity of global extreme disasters have increased [1,2]. Furthermore, the high incidence of forest fires poses a major threat to the natural resources provided by forests and grasslands and to human health. Wildfire disasters have also caused major damage to public facilities, such as power grids and communications [3–6]. For example, in California and other regions where large-scale wildfires occurred every year from 2017 to 2020, the high temperature and smoke from wildfire disasters near transmission lines led to a decline in the insulation performance of air gaps, which caused a large number of transmission lines to trip, resulting in large-scale power outages. On average, there are more than 70,000 wildfires in China every year and they have been one of the major causes of power transmission failure over the past 10 years. In 2013, the 1000 kV Changnan Line tripped several times due to wildfires and the power outage lasted 60 h in total. In August 2013, extreme dry weather caused 10 wildfire trips in the Hunan Power Grid area over 13 days from 5 August to 17 August [7]. In August 2022, countries across the globe experienced extreme drought with surging electricity loads. The wildfire in Banan District, Chongqing, China, caused the 500 kV Luonan Line 1 and Line 2 of the outgoing line from Banan Power Plant to trip and shut down for 40 h, which affected power grid operations. The United States, China, and other countries have carried out research on prevention and control technologies, such as prediction, monitoring, and extinguishing wildfires near power transmission lines [8–12]. Kal'avský, Peter, et al. [8] investigated the effect of water mist on the development of a long-gap discharge and provided guidance for the safe application of water mists to extinguish wildfires near high-voltage transmission lines, and Trakas proposed stochastic mixed integer programming with quadratic constraints to increase the resiliency of a distribution system threatened by wildfire [11]. Firefighting is the most effective way to prevent transmission lines from tripping due to wildfires. However, fires around transmission lines can be difficult to control due to high mountains, precipitous paths, and extremely high emergency time limits. It takes a long time to extinguish the fire using ground fire trucks and backpack firefighting equipment, which means that the fire spread can become larger. Helicopters are important items of equipment that can be used to extinguish forest fires because they can overcome terrain restrictions. In addition, firefighters have no direct contact with the fire scene and are less exposed to danger. There are three advantages to using helicopters and fire retardants to extinguish wildfires near transmission lines: (1) During large-scale wildfires, the fire safety risk to ground personnel is relatively high. On 30 March 2019 and 2020, wildfires occurred in Liangshan and Sichuan, China, resulting in the deaths of 30 firefighters and 19 firefighters, respectively. Dropping fire retardant water agents by helicopter can reduce the intensity of wildfires, which is conducive to ground fire safety and improves the safety of ground firefighters. (2) The number of helicopters available for firefighting is limited when dozens of wildfire disasters occur near the transmission line, which means that it is impossible to quickly put out on-site wildfires in a short period of time. Spraying flame retardants onto the combustibles between the fire site and the line can form an isolation zone to prevent the wildfire from spreading to the transmission line and reduce the risk of large-scale power outages caused by the simultaneous or sequential shutdown of multiple lines due to wildfires. (3) Existing studies have shown that the main factors causing line tripping by wildfires are the high temperature of the flame, flame ionization, and distortion to the electric field caused by the smoke and dust, which are likely to cause streamer discharge [13,14]. Spraying flame retardants on the surface of combustibles in corridors along both sides of the transmission lines can reduce the intensity of wildfires, reduce ionization, and further reduce tripping by wildfires. Therefore, it is important to study helicopter spraying of flame retardants to reduce the intensity of wildfires.

Helicopters and retardants could be used to pretreat areas at high risk of fire ignitions to make them fire resistant [15,16]. There have been many international studies on the formulation of flame-retardant water agents, their flame-retardant mechanism [17–20], and their impact on the environment [21–23]. However, few studies have quantitatively analyzed and evaluated the flame retardant effect on forest fires. Some previous studies on the subject include the following: Ref. [24] measured the solid density and the gas temperature to determine the effect of the chemical agent Phos-Chek, a commercial flame retardant widely used by firefighters in North America, on pyrolysis and flame-spread on untreated wood samples and wood samples treated with Phos-Chek; Ref. [25] describes the application scenarios for long-term retardants, fire suppressant foam, wet water, and water enhancer (gel); and Ref. [15] describes the effect of a sprayable, environmentally benign, viscoelastic fluid comprising biopolymers and colloidal silica that was sprayed using a backpack onto a layer of grass taped to a wood slab. The nozzle was placed ~30 cm away from the grass and sprayed in bursts. Temperature and normalized area burned over time were used to demonstrate the amount needed to prevent the spread of fire. However, the existing literature does not directly compare the fire intensity changes after spraying with flame retardants and the tests were conducted by spraying flame retardants onto the ground. The distribution of flame retardants sprayed on the ground differs greatly from those sprayed by helicopters. Therefore, it is not possible to apply the existing conclusions to the protection of transmission lines on site using a helicopter.

Thus, the main purpose of this study was to undertake in-flame retarding tests using helicopter bucket spraying of an extinguishing agent. The aims were to evaluate how well different fire retardants at different potential fuel concentrations decreased fire intensity and slowed the advance of a fire by measuring the temperature, heat flux density, and mass loss of forest fuel and by analyzing the fire retarding mechanism in combination with the chemical components of the extinguishing agent.

This paper consists of the following sections: Part I provides an overview; Part II describes the in-flame retarding test program using helicopters; Part III discusses the in-flame retarding test data; Part IV provides an interpretation of the data and presents the mechanism; Part V describes an application based on the test results; and Part VI is a summary, as shown in Figure 1.

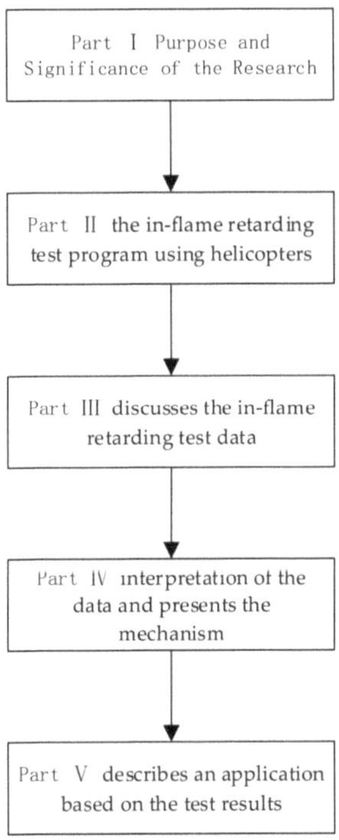

Figure 1. Flowchart for linking research parts.

2. Test Program

2.1. Test Combustibles

Wood cribs are often used for various tests requiring repeatable combustion temperatures, such as fire-extinguisher performance (ANSI/UL 711) [26,27]. Each crib was made of 1A wooden strips that were 40 mm × 40 mm × 500 mm in size, as stipulated in GB4351.1-2005 Portable Fire Extinguishers. The pine species *Pinus massoniana* Lamb was selected as the raw material for the wood strips. The size and layout of wood crib were all consistent with the literature [28] and are not described in detail in this paper and the 12-layer wood crib was shown in Figure 2.

The wood crib combustibles were placed on an iron mesh support that was 1200 mm × 1200 mm × 250 mm in size, and the ground under the wood crib combustibles was laid out with square fuel-oil basins that were 700 mm × 700 mm × 100 mm in size. Four liters of blue high-42 octane Avgas 100# aviation gasoline was used as the ignition fuel.

Figure 2. Photograph showing the construction of the 12-layer wood cribs.

2.2. Fire Scene Layout and Measurement Scheme

After the helicopter fire extinguishing device had sprayed the flame-retardant water agent at the same height, speed, and spraying conditions in the same trip, Crib#1, Crib#2, and Crib#3 were ignited after 1 h of natural drying to observe their burning characteristics. The main aim was to measure the flame retardant effect of the flame-retardant water agent sprayed by the helicopter fire extinguishing device. Among them, the structure size parameters, flight parameters, and distribution of the ground flame-retardant water agent released by the helicopter were all consistent with the literature [28] and are not described in detail in this paper. The radiation intensity [29], mass loss [30], and temperature [31] are regarded as the burning intensity features of wood crib fires. Therefore, this study mainly focused on the burning rate, radiation intensity, and temperature of the wood crib fire after the flame-retardant water agent was sprayed on the wood cribs. These features were used to characterize the intensity of combustion and to reflect the differences in the flame retardant effect of each agent on the wood cribs.

The tests used K-type thermocouples, the diameter of each thermocouple was 3 mm, the probe was bare, and the response time was one second. The uncertainty in the temperature of the Type-K thermocouple wire is given by the manufacturer as ±2.2 °C with a 95% confidence interval. The expanded uncertainty for the thermocouple when the temperature change is from 0 °C to 1250 °C is 1.5% and the change from −200 °C to 0 °C is 4.0% with a coverage factor of 2, which corresponds to a confidence interval of 95%.

The positions of the thermocouple arrays are marked as Array#1, Array#2, and Array#3 in Figure 3. The positions and heights of thermocouples for measuring the temperature of 4-layer wood cribs, 12-layer wood cribs and 6-layer wood cribs are shown in Figures 4–6, respectively. Inside the wood crib, the thermocouples were relatively densely arranged with one thermocouple set on every two layers with a height difference of 0.08 m (the height of two layers of wooden strips). Two thermocouples were set on the upper part of the wood crib at a height of 1.0 m and 1.7 m.

An electronic scale was placed under the fire field of Array#1, Array#2, and Array#3. The electronic scale under the wood crib was used to measure the mass changes to the wood crib fire site during the burning process to obtain the mass loss rate of the wood cribs. The size of the electronic scale was 1200 mm × 1200 mm with a measuring range of 300 kg, division values of 100 g, and an accuracy of ±0.02%. The data were directly recorded by U disk, and the recording frequency was 1 time/second.

The tests used a Model C-3500 flame intensity calorimeter (America ITI company, USA) to obtain the heat fluxes. It was designed to be inserted directly into a flame front for the instantaneous determination of impinging heat energy. The length of the probe was 3 m and the probe diameter was 25.4 mm. The response time was less than 0.1 s, the max flux density was 31.9 kW/m^2, maximum operating temperature was 1920 °C nominal sensitivity was 314 w/m^2.uV, and the accuracy was 5%. The Portable Handheld Data Logger was a DaqPRO 5300 with a recording frequency of 1 time/second. There was only one ITI Model C-3500 flame intensity calorimeter in the laboratory, so it was installed at a height of 10 cm above the top surface of wood Crib#2, which was where the fire field intensity was greatest. The arrangement positions and heights of the thermocouple arrays, radiation heat flow meters, and the electronic scales are shown in Table 1.

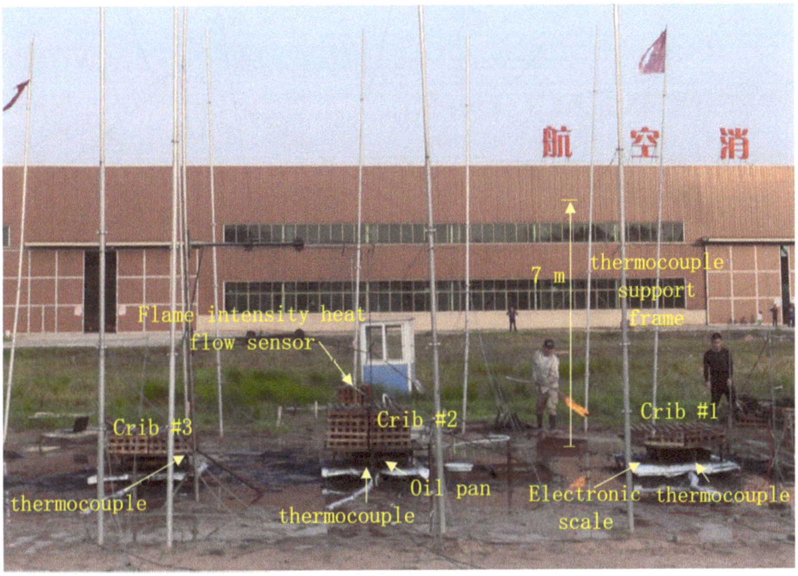

Figure 3. Actual layout of the flame-retardant test and measurements of temperature, radiant heat flux, and weight.

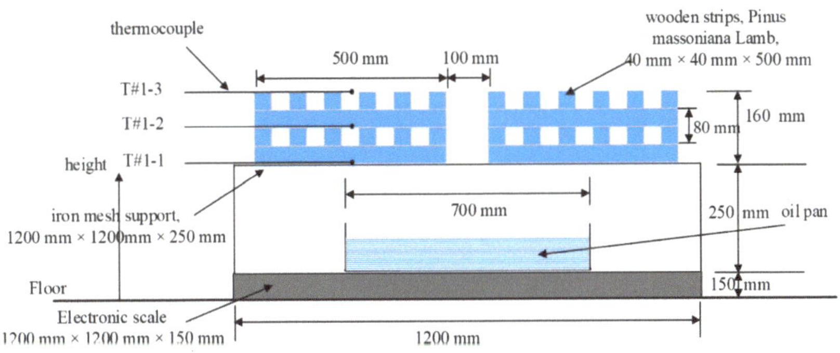

Figure 4. Diagram of the 4-layer wood crib apparatus.

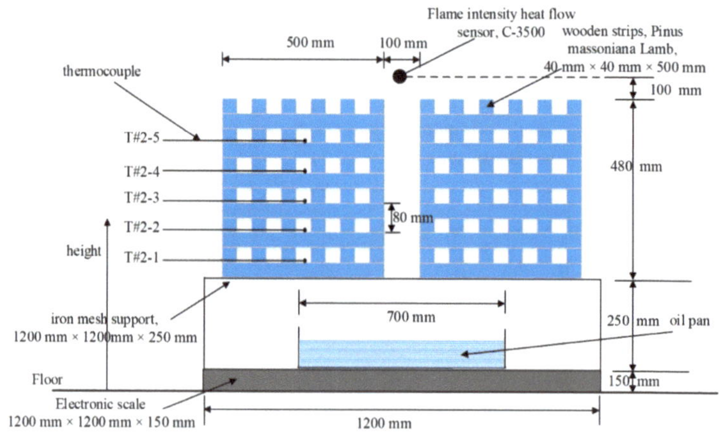

Figure 5. Diagram of the 12-layer wood crib apparatus.

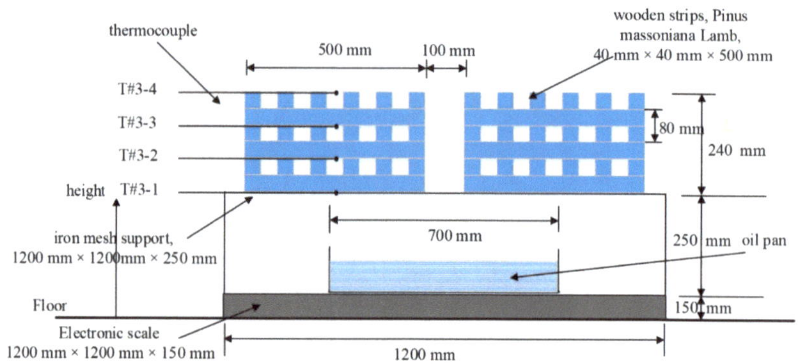

Figure 6. Diagram of the 6-layer wood crib apparatus.

Table 1. Layout of the thermocouples/scale and the flame intensity calorimeter in the fire retardants experiment.

Serial No.	Thermocouple No.	Measuring Height/m	Inside or Above the Wood Crib	Location No.
1	Scale#1	0.0	Underneath the wood crib	
2	T#1-1	0.25	Inside the wood crib	
3	T#1-2	0.33	Inside the wood crib	
4	T#1-3	0.41	Inside the wood crib	Array#1
5	T#1-4	1.0	Above the wood crib	
6	T#1-5	1.7	Above the wood crib	
7	Scale#2	0.0	Underneath the wood crib	
8	T#2-1	0.29	Inside the wood crib	
9	T#2-2	0.37	Inside the wood crib	
10	T#2-3	0.45	Inside the wood crib	
11	T#2-4	0.53	Inside the wood crib	
12	T#2-5	0.61	Inside the wood crib	Array#2
13	T#2-6	1.0	Above the wood crib	
14	T#2-7	1.7	Above the wood crib	
15	T#2-8	2.7	Above the wood crib	
16	RadiationHeatFlowMeters	0.98	Above the wood crib	
17	Scale#3	0.0	Underneath the wood crib	
18	T#3-1	0.25	Inside the wood crib	
19	T#3-2	0.33	Inside the wood crib	
20	T#3-3	0.41	Inside the wood crib	Array#3
21	T#3-4	0.48	Inside the wood crib	
22	T#3-5	1.0	Above the wood crib	
23	T#3-6	1.7	Above the wood crib	

Three cameras were used during the test. Two were vertical to the fire field. Their short focal length lens was used to record whether the area covered by the flame-retardant water agent completely covered the three wood cribs in the flight direction. The long focal length camera and was used to record the combustion process after the flame-retardant water agent was sprayed on them. The third camera was parallel to the fire field to record whether the agent sprayed on the area completely covered the wood cribs in the perpendicular direction to the flight.

2.3. Fire Extinguishing Schemes

Pure water (Agent#1), 10% Class AB flame retardant (Agent#2), 0.3% gel flame retardant (Agent#3), 10% Class A flame retardant (Agent#4), and 10% Class A flame retardant + 0.6% guar gum (Agent#5) were used as the flame-retardant media to compare their flame-retardant properties when applied with a helicopter. The main components and descriptions of the various flame-retardant media are shown in Table 2.

Table 2. Main components of the flame-retardant water agents.

No.	Flame Retardant Type	Weight Ratio of Flame Retardant to Water	Flame Retardant No.	Flame Retardant Components
1	Control group			The control group is not sprayed with any water agent
2	Pure water	/	Agent#1	Pure water
3	Class AB flame retardant	10%	Agent#2	Main components: 0.1~2% fs-1157 fluorocarbon surfactant; 1~3% surface active betaine (α-Dodecyldimethyl betaine); 0.5~2% corrosion inhibitor (1H-Benzotriazole); The rest is water
4	Gel flame retardant	0.3%	Agent#3	Main components: Benzoin-SA Complex 50%; Polyacrylamide 50%;
5	Class A flame retardant	10%	Agent#4	Main components: 23 wt.% potassium chloride; 52~65 wt.% ammonium carbonate; 12~25 wt.% disodium hydrogen phosphate; The rest are flame retardants adjusted to a specific gravity of 1.1 by water
6	Class A flame retardant + 0.6% guar gum	10%	Agent#5	Main components: 23 wt.% potassium chloride; 52~65 wt.% ammonium carbonate; 12~25 wt.% disodium hydrogen phosphate; 0.6 wt.% Guar gum; The rest are flame retardants adjusted to a specific gravity of 1.1 by water

Adding a thickener to the fire extinguishing agent can improve the wind resistance of the fire extinguishing agent and make it easier to adhere to the surface of the wooden pile, thereby improving the utilization rate of the fire extinguishing agent. Guar gum is a simple and readily available thickener. Therefore, this study applied guar gum to the flame-retardant water agents. The viscosity of the fire extinguishing agent at 0.6 wt.% was 190 mPa.s, which satisfies the above two conditions for improving the utilization rate.

The viscosities of Agent#3, Agent#4, and Agent#5 were tested and curves of the viscosity values for the fire extinguishing agents as a function of temperature were obtained, as shown in Figure 7. The viscosity of Agent#5 which mixed with guar gum decreased with the increase in temperature. The viscosities of Agent#3 and Agent#4 fire extinguishing agents without guar gum hardly changed with the increase in temperature. When the temperature of the Agent#5 fire extinguishing agent was lower than 20 °C, the viscosity of the fire extinguishing agent reached 190 mPa.s, but the viscosity dropped sharply when the temperature of the fire extinguishing agent was between 40 °C and 60 °C. When the temperature of the fire extinguishing agent reached 60 °C, its viscosity dropped to 55 mPa.s. It can be seen that the viscosity of Agent#5, when mixed with guar gum, was very sensitive to temperature.

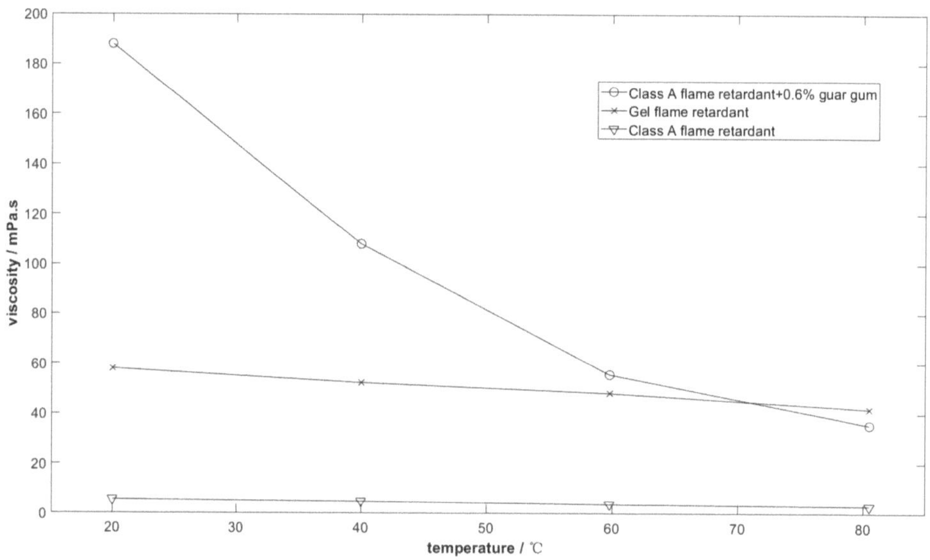

Figure 7. Viscosity of the fire extinguishing agents.

The test for flame retardancy by helicopter was divided into five steps:

(1) The helicopter carried the bucket fire extinguishing device to spray the flame-retardant water agent on the wood cribs three times at a speed of 20 km/h and at a height of 32 m, as shown in Figure 8;
(2) The wood cribs were allowed to naturally air-dry for 1 h;
(3) A certain amount of water and oil was added to the pilot oil pan to ignite the pilot oil under the wood cribs;
(4) The temperature, radiant heat flux density, and mass loss in the fire field were measured;
(5) A ground water pipe was used to extinguish the remaining fire in the wood crib when the Crib#2 wood crib fire field collapsed.

Figure 8. Helicopter high-altitude spraying of flame retardants. (**Left**): gel fire-extinguishing agent; (**Right**): class AB fire-extinguishing agent.

3. Flame Retardant Test

3.1. Comparative Testing of Flame Retardancy When Different Wood Crib Sizes Were Used

Mass Loss, Radiation Intensity and Temperature of the Wood Crib in the Control Group

Direct burning was performed on the surface of the wood crib where no spraying flame-retardant water agent was applied, and the mass loss, temperature, and radiant heat flux were measured after the wood crib was ignited. This treatment was the control group and was used to compare the differences in the parameters after helicopter spraying the flame-retardant water agents listed in Table 2.

Figure 9 shows the fire field burning conditions of the wood crib fire model when no flame-retardant water agent was applied. Screenshots at three time points were selected to show the burning conditions of the wood crib fire. These time points were 300 s (after burning of the wood stack pilot oil pan), 600 s (during the stable combustion stage), and 870 s (the minimum collapse time of the wood stack under the six flame retardant conditions).

Figure 9. Combustion conditions of the fire scenes in the model without spraying flame retardants onto the wood crib. First subgraph: 180 s; second subgraph: 600 s; third subgraph: 870 s.

The mass loss, radiant heat flux density, and temperature change curves for the crib without any flame-retardant water agent are shown in Figure 10. The change curves for temperature, radiant heat flux density, and the mass of the wood cribs are represented by three subgraphs. The three subgraphs from top to bottom represent the changes process for Crib#1, Crib#2, and Crib#3, respectively. The maximum mass loss refers to the average value of the mass loss over 5 consecutive seconds, and the average value of the radiant heat flux value refers to the average value of the radiant heat flux density of Crib#2 from the burnt of the oil to the collapse of the wood crib. This was to avoid the influence of any mutation values caused by interference factors during the data acquisition process. It can be seen from the mass loss, radiant heat flux density, and temperature curves that after the pilot oil pan was ignited, the 100# aviation gasoline in the oil pan and the wood crib started to burn violently, releasing large amounts of heat. The temperature and radiant heat flux density curves increased rapidly and the quality gradually decreased, reaching a peak at about 90 s. After about 180 s, as the pilot fuel in the oil pan was consumed, the temperature measured by the thermocouples at all heights dropped and the wood crib entered the free combustion stage.

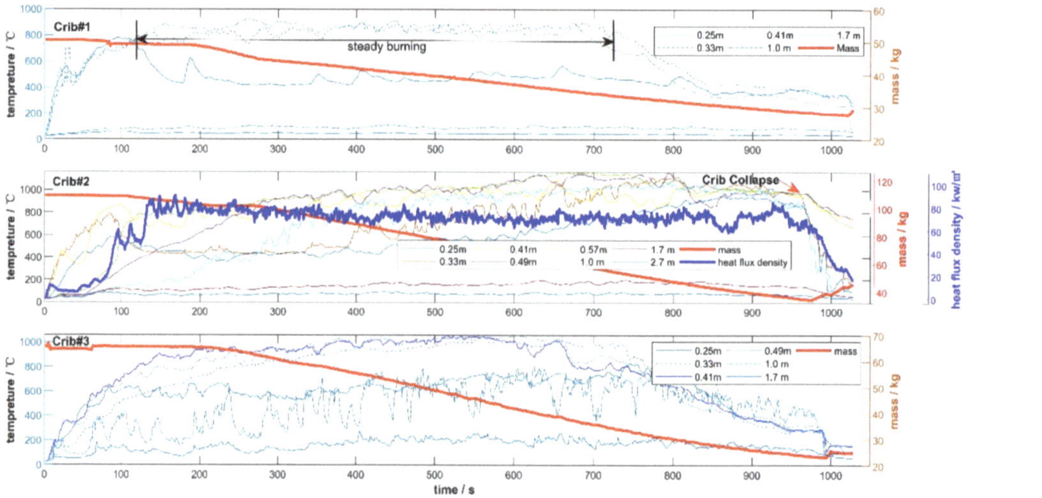

Figure 10. Temperature, radiant heat flux density, and mass change curves for the wood cribs not sprayed with fire extinguishing agent.

The temperature curve for the Crib#1 wood crib fire was in the stable combustion stage from 180 to 720 s, with a maximum temperature of 928 °C, and was when the wood crib mass loss was most rapid. The average wood crib mass loss during the stable combustion stage was 0.026 kg/s. After 720 s of the test, the combustion entered the decay stage, the temperature measured by the thermocouples gradually decreased, and the rate of mass decline eased.

The temperature curve for the Crib#2 wood crib fire was in the stable combustion stage from 200 to 800 s, with a maximum temperature of 1156 °C. The radiant heat flux value reached 110 kW/m^2, which was when the wood crib mass loss was most rapid. The average wood crib mass loss in the stable combustion stage was 0.115 kg/s. After 1360 s, the combustion entered the decay stage, the temperature measured by the thermocouple gradually decreased, and the rate of mass decline eased. After 2106 s, a small number of the wood strips fell off due to the loss of supporting force (carbonization) and the crib lost its original shape.

The temperature curve for the Crib#3 fire was in the stable combustion stage from 180 to 740 s and reached a maximum temperature of 1056 °C, which was when the wood crib mass loss was most rapid. The average wood crib mass loss in the stable combustion stage was 0.053 kg/s. After 740 s, the combustion entered the decay stage, the temperature measured by the thermocouple gradually decreased, and the rate of mass decline eased.

The data measured by the thermocouples and radiant heat flux density meter after the collapse of the wooden crib was no longer the measurement value of the assumed position in the design plan so ground water pipes were used to extinguish the residual fire associated with the wooden cribs after their collapse. At this time, the temperatures of the

Crib#1, Crib#2, and Crib#3 fires decreased rapidly, their quality increased rapidly under the effect of water spraying from the ground, and their data was of no practical significance. However, due to the limited number of test personnel, when ground water pipes were used to extinguish the residual wood crib fire, water was not sprayed on Crib#1, Crib#2, and Crib#3 fires at the same time. This resulted in a rapid drop in temperature and a rapid mass increase at the three fire sites that did not occur at the same time. This is reflected in the subsequent analysis charts.

3.2. Tests on Flame-Retardant Water Agents When Sprayed from a Helicopter

3.2.1. Coverage by the Flame-Retardant Water Agent When Sprayed from a Helicopter

In order to intuitively describe the coverage of the flame-retardant water agent on wood crib fires when a helicopter sprayed the flame-retardant water agent on three separate occasions, the flame-retardant agent has been shown overlapping the water agent applied to Crib#1, Crib#2, and Crib#3, as shown in Figure 11. To better record the track of the wood crib and flame-retardant water agent, the coverage area of flame-retardant water agent was assumed to be approximately oval, and the size of wood crib has been enlarged. The sub-graphs showing the coverage by each agent are in the following order from top to bottom: Crib#1, Crib#2, and Crib#3 in each subgraph.

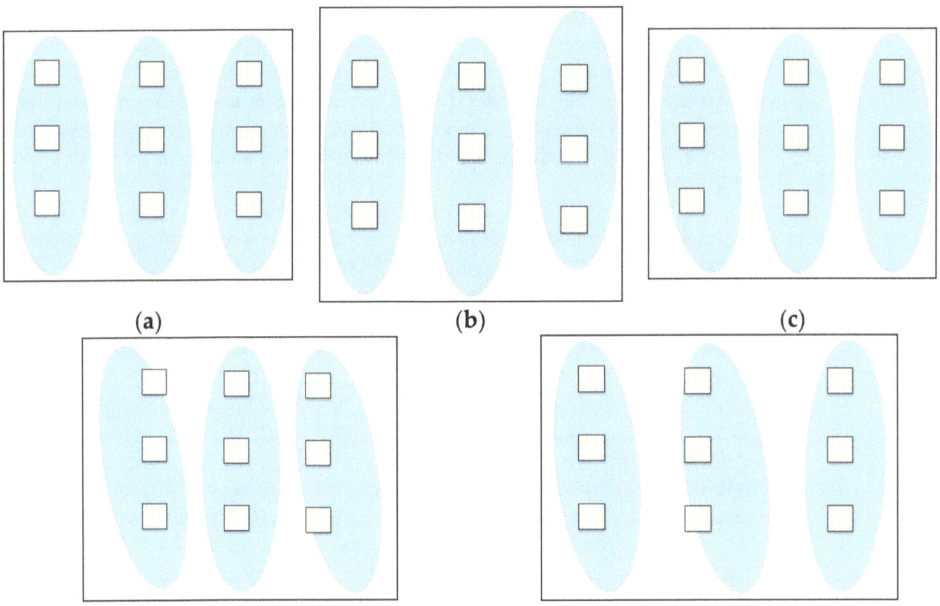

(a) (b) (c) (d) (e)

Figure 11. Coverage areas for each retardant agent dropped from the helicopter: (**a**) Agent#1, (**b**) Agent#2, (**c**) Agent#3, (**d**) Agent#4, and (**e**) Agent#5. Crib#1, Crib#2, Crib#3 from top to bottom.

It can be seen from Table 3 that the weight increases due to the Class AB flame retardant and the gel flame retardant are larger than those for the other flame-retardant water agents and more of the flame-retardant water agents are absorbed. This is mainly related to the active ingredients in Class AB flame retardants. The flame-retardant water agent can quickly penetrate the surface of the wood crib and enter the interior. The gel flame retardant has poor fluidity, good adhesion performance, and can adhere to the surface of the wood crib. Therefore, the loss of the flame-retardant water agent with gel is less than that of other flame-retardant water agents.

Table 3. Weight changes after flame-retardant water agents were sprayed by a helicopter.

No.	Flame Retardant	Crib#1 Initial Weight/kg	Crib#1 Weight after Spraying Fire-Extinguishing Agent/kg	Crib#2 Initial Weight/kg	Crib#2 Weight after Spraying Fire-Extinguishing Agent/kg	Crib#3 Initial Weight/kg	Crib#3 Weight after Spraying Fire-Extinguishing Agent/kg
1	Control Group	42.3	/	115.5	/	57.2	/
2	Pure Water	40.5	46.2	121.6	129.75	58.3	67.9
3	Class AB Flame Retardant	40.2	47.35	117.4	122.4	59.6	66.65
4	Gel Flame Retardant	39.9	55.25	126.3	148.1	58.5	68.4
5	Class A Flame Retardant	38.9	42.1	120.1	126	55.6	56.9
6	0.6% Thickened Class A Flame Retardant	41.3	47	125.8	138.2	58.0	66.1

3.2.2. Pure Water (Agent#1)

The pure water (Agent#1) flame retardant test was carried out at a temperature of 29 °C. Before spraying the pure water (wind speed generated by the helicopter rotor was not considered), the wind speed was 2.1 m/s. The natural wind speed and humidity when the wood cribs were ignited were 2.4 m/s and 52%, respectively. Figures 12 and 13 show the fire field burning conditions of the helicopter bucket using Agent#1 to protect the wood crib fire model.

Starting from 150 s, the fire field maintained a relatively stable combustion level for about 180 s. At this stage, wood crib combustion was relatively stable, there was uniform heat release, and the wood crib mass loss increased.

The temperature curve for the Crib#1 fire was in the stable combustion stage from 250 to 1100 s with a maximum temperature of 623 °C, which was when the wood crib mass loss was most rapid. The average wood crib mass loss in the stable combustion stage was 0.0160 kg/s.

The temperature curve for the Crib#2 fire was in the stable combustion stage from 250 to 1100 s, with a maximum temperature of 1110 °C. The radiant heat flux value reached 114 kW/m^2, the wood crib mass loss was most rapid at this point, and the average wood crib mass loss during the stable combustion stage was 0.1220 kg/s. After 1100 s, the combustion entered the decay stage, the temperature measured by the thermocouple gradually decreased, and the rate of mass decline eased. At 1120 s after igniting the wood crib, a small number of the wood strips fell off due to the loss of supporting force (carbonization) and the wood crib lost its original shape.

The temperature curve for the Crib#3 fire was in the stable combustion stage from 250 to 1050 s. It reached a maximum temperature of 917 °C when the wood crib mass loss reached its peak. The average wood crib mass loss in the stable combustion stage was 0.0432 kg/s. At 1050 s after igniting the wood crib, the combustion entered the decay stage, the temperature measured by the thermocouple gradually decreased, and the rate of mass decline eased.

Figure 12. Combustion conditions of the fire scenes in the model using Agent#1 taken from the helicopter after spraying flame retardants onto the wood crib. First subgraph: 180 s; second subgraph: 600 s; third subgraph: 870 s.

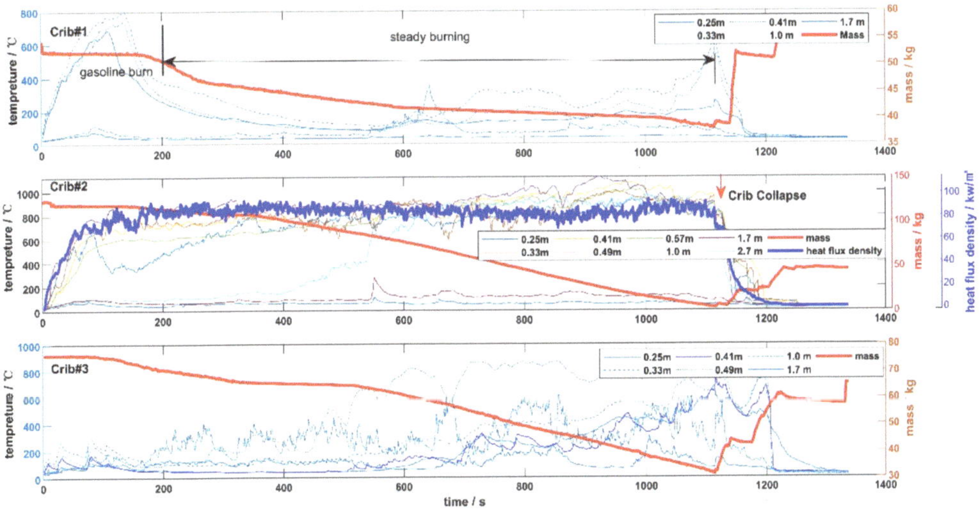

Figure 13. Temperature, radiant heat flux density, and mass change curves for the wood cribs after the helicopter sprayed Agent#1 fire extinguishing agent.

3.2.3. Class AB Flame Retardant (Agent#2)

The flame-retardant test of the Class AB flame retardant (Agent#2) was carried out at 32 °C. Before spraying the flame-retardant water agent (wind speed generated by the helicopter rotor was not considered), the wind speed was 2.4 m/s. The natural wind speed and humidity natural wind speed when the wood cribs were ignited were 1.8 m/s and 56%, respectively. Figures 14 and 15 show the fire field burning conditions after the helicopter had sprayed Agent#2 flame-retardant water agent to protect the wood crib model. The strong wind generated by the helicopter rotor when applying the flame retardant and during flight meant that the flame-retardant water agent became foam and dissipated. Therefore, the amount of the agent applied was less than that for the other flame retardants. In addition, Agent#2 covered a wider area than Agent#1, Agent#3, and Agent#4 when carrying out firefighting, which meant that the depth of the flame-retardant water agent per unit area was smaller.

Class AB flame retardant is used for oil fire extinguishing and as a flame retardant. This means that when Class AB flame retardant is sprayed by helicopter onto the combustion plate, the wood crib ignition process can be hindered. Therefore, 100# aviation gasoline was added after all the Class AB flame retardant remaining in the combustion plate had been poured out.

The temperature curve for the Crib#1 fire was in the stable combustion stage from 300 to 1186 s. It reached a peak temperature of 887 °C when the wood crib mass loss was at its most rapid. The average wood crib mass loss during the stable combustion stage was 0.023 kg/s. At 1186 s after igniting the wood crib, the combustion entered the decay stage, the temperature measured by the thermocouple gradually decreased, and the rate of mass decline eased.

The temperature curve for the Crib#2 fire was in the stable combustion stage from 300 to 1360 s. It reached a peak temperature of 1067 °C. At this point, the ignition process of the radiant heat flux value reached 100 kW/m^2 and the wood crib mass loss was most rapid. The average wood crib mass loss during the stable combustion stage was 0.1047 kg/s. After 1360 s, the combustion entered the decay stage, the temperature measured by the thermocouple gradually decreased, and the rate of mass decline eased. At 2106 s after igniting the wood crib, a small number of the wood strips fell off due to the loss of supporting force (carbonization), which meant that the wood crib lost its original shape.

The temperature curve for the Crib#3 fire was in the stable combustion stage from 300 to 1260 s. It reached a peak temperature of 990 °C, which was when the wood crib mass loss was most rapid. The average wood crib mass loss during the stable combustion stage was 0.0429 kg/s. At 1260 s after igniting the wood crib, the combustion entered the decay stage, the temperature measured by the thermocouple gradually decreased, and the rate of mass decline eased.

Figure 14. Combustion conditions of the fire scenes in the model using Agent#2 taken from the helicopter after spraying the flame retardant onto the wood crib. First subgraph: 180 s; second subgraph: 600 s; third subgraph: 870 s.

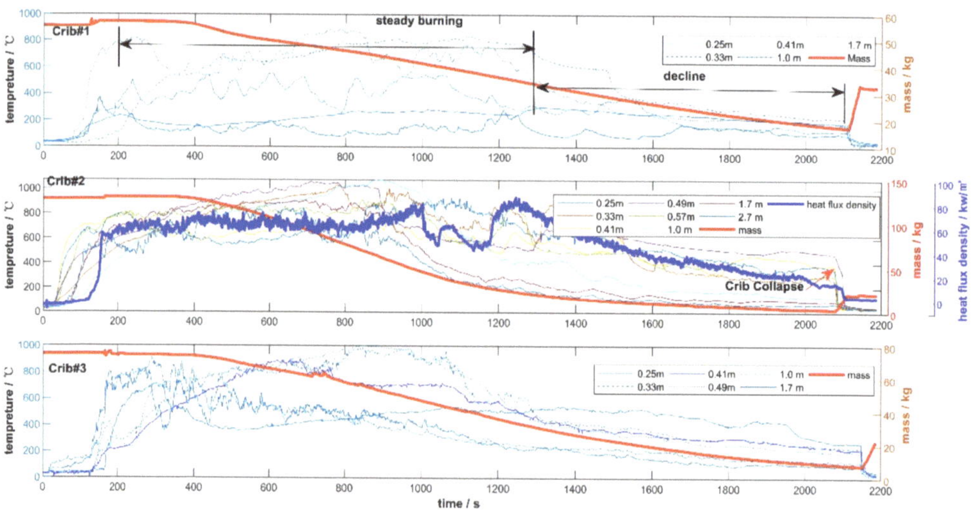

Figure 15. Temperature, radiant heat flux density, and mass change curves for the wood cribs after the helicopter had sprayed Agent#2 fire extinguishing agent.

3.2.4. Gel Flame Retardant (Agent#3)

The flame retardant test of gel flame retardant (Agent#3) was carried out at 29 °C. Before spraying the flame-retardant water agent (wind speed generated by the helicopter rotor was not considered), the wind speed was 1.9 m/s. The natural wind speed and humidity natural wind speed when the wood crib was ignited were 2.2 m/s and 48%, respectively. Figures 16 and 17 show the fire field burning conditions after the helicopter sprayed the crib using Agent#3 to extinguish the non-uniform wood crib fire model.

Figure 16. Combustion conditions of the fire scenes in the model using Agent#3 taken from the helicopter after spraying the flame retardant onto the wood crib. First subgraph: 180 s; second subgraph: 600 s; third subgraph: 870 s.

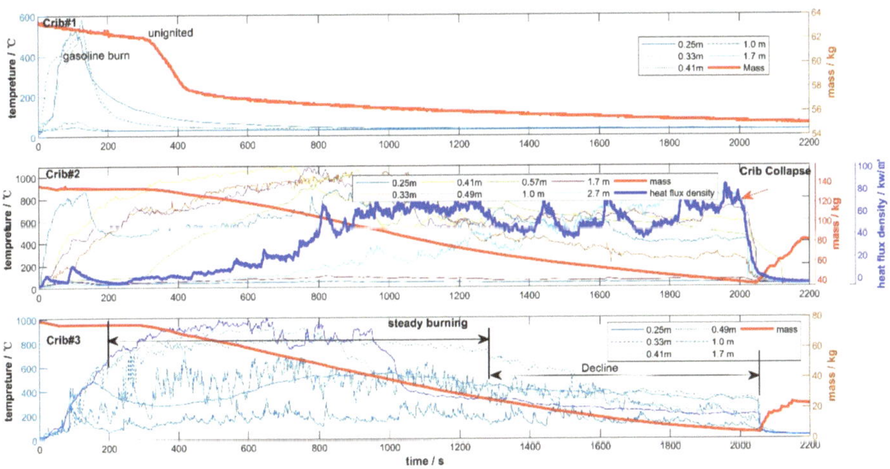

Figure 17. Temperature, radiant heat flux density, and mass change curves for the wood cribs after the helicopter had sprayed Agent#3 fire extinguishing agent.

The temperature curve for the Crib#1 fire was in the ignition stage from 0 to 127 s and the peak temperature was 583 °C. Since the wooden crib only had four layers, it could be completely covered by the gel sprayed by the helicopter. Thus, the wood crib did not ignite, the fire field was extinguished, and the pilot gasoline burned out after 127 s.

The temperature curve for the Crib#2 fire was in the stable combustion stage from 280 to 1340 s when a peak temperature of 1091 °C was reached. During this stage, the radiant heat flux value reached 102 kW/m^2 and the wood crib mass loss was most rapid. The average value of the mass loss of the wood crib in the stable combustion stage was 0.0799 kg/s. After 1500 s, the combustion entered the decay stage, the temperature measured by the thermocouple gradually decreased, and the rate of mass decline eased. At 2010 s after igniting the wood crib, a small number of the wood strips fell off due to the loss of supporting force (carbonization), which meant that the crib lost its original shape.

The temperature curve for the Crib#3 fire was in the stable combustion stage from 200 to 1250 s and reached a peak temperature of 997 °C, which was when the wood crib mass loss was most rapid. The average wood crib mass loss during the stable combustion stage was 0.0465 kg/s. At 1250 s after igniting the wood crib, the combustion entered the decay stage, the temperature measured by the thermocouple gradually decreased, and the rate of mass decline eased.

The gel flame retardant has good adhesion performance and can effectively prevent the wood cribs from burning when it adheres to the surface of the wood crib. In addition, Crib#1 had fewer layers and the gel flame retardant completely covered it, so that the wood crib did not ignite.

3.2.5. The 10% Class A Flame Retardant (Agent#4)

The flame-retardant test of 10% Class A flame retardant (Agent#4) was carried out at 31 °C. Before spraying the flame-retardant water agent (wind speed generated by the helicopter rotor was not considered), the wind speed was 1.6 m/s. The natural wind speed and humidity natural wind speed when the wood crib was ignited were 1.5 m/s and 44%, respectively. Figures 18 and 19 show the fire field burning conditions after the helicopter had sprayed the cribs using Agent#4 to extinguish the non-uniform wood crib fire model.

Figure 18. Combustion conditions of the fire scenes in the model using Agent#4 taken from the helicopter after spraying the flame retardants onto the wood crib. First subgraph: 180 s; second subgraph: 600 s; third subgraph: 870 s.

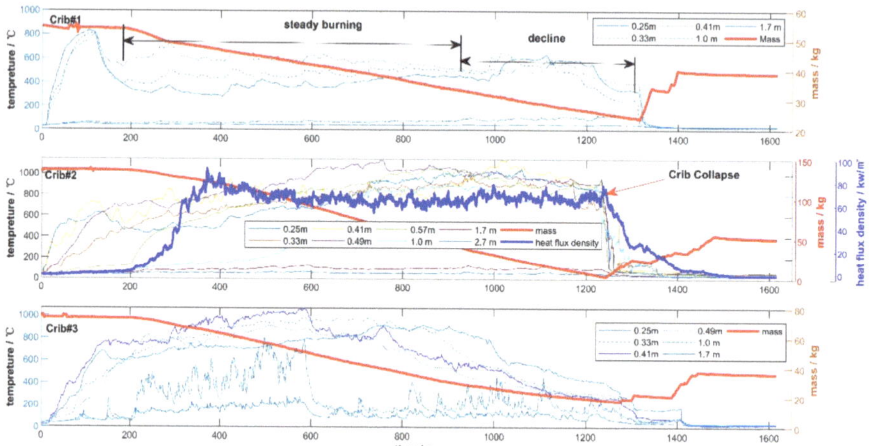

Figure 19. Temperature, radiant heat flux density, and mass change curves for the wood cribs after the helicopter had sprayed Agent#4 fire extinguishing agent.

The temperature curve of the Crib#1 fire was in the stable combustion stage from 200 to 900 s when it reached a peak temperature of 729 °C and the wood crib mass loss was most rapid. The average wood crib mass loss during the stable combustion stage was 0.0266 kg/s. At 900 s after igniting the wood crib, the combustion entered the decay stage, the temperature measured by the thermocouple gradually decreased, and the rate of mass decline eased.

The temperature curve for the Crib#2 fire was in the stable combustion stage from 300 to 1120s when it reached a peak temperature of 1128 °C. At this point, the radiant heat flux value reached 94 kW/m^2 and the wood crib mass loss was most rapid. The average value of the mass loss of the wood crib during the stable combustion stage was

0.0867 kg/s. After 1230 s, the combustion entered the decay stage, the temperature measured by the thermocouple gradually decreased, and the rate of mass decline eased. At 1230 s after igniting the wood crib, a small number of the wood strips fell off due to the loss of supporting force (carbonization), which meant that the crib lost its original shape.

The temperature curve for the Crib#3 fire was in the stable combustion stage from 200 to 920 s when it reached a peak temperature of 1058 °C and the wood crib mass loss was most rapid. The average wood crib mass loss during the stable combustion stage was 0.0522 kg/s. At 920 s after igniting the wood crib, the combustion entered the decay stage, the temperature measured by the thermocouple gradually decreased, and the rate of mass decline eased.

3.2.6. The 10% Class A flame Retardant + 0.6% Guar Gum (Agent#5)

The flame-retardant test of the 10% Class A flame retardant + 0.6% guar gum (Agent#5) was carried out at 30 °C. Before spraying the flame-retardant water agent (wind speed generated by the helicopter rotor was not considered), the wind speed was 1.8 m/s. The natural wind speed and humidity natural wind speed when the wood crib was ignited were 2.1 m/s and 51%, respectively. Figures 20 and 21 show the fire field burning conditions after the helicopter had sprayed the cribs using Agent#5 to extinguish the continuous and uniform wood crib fire model.

Figure 20. Combustion conditions of the fire scenes in the model using Agent#5 taken from the helicopter after spraying flame retardants onto the wood crib. First subgraph: 180 s; second subgraph: 600 s; third subgraph: 870 s.

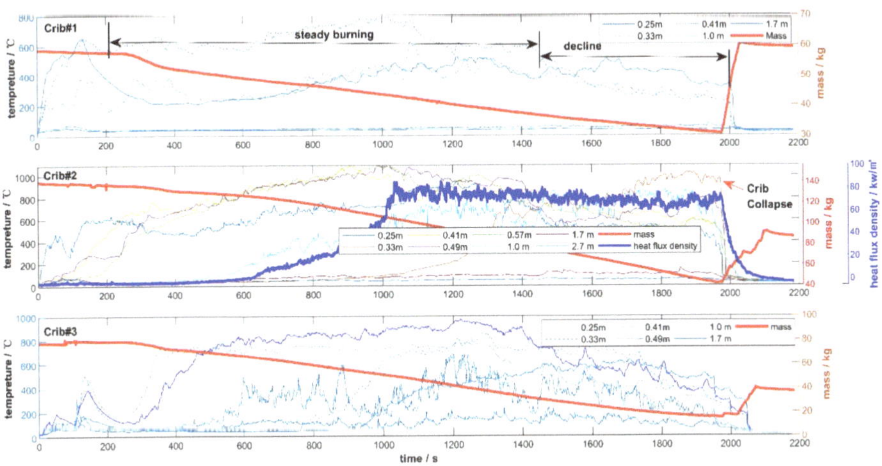

Figure 21. Temperature, radiant heat flux density, and mass change curves for the wood cribs after the helicopter had sprayed Agent#5 fire extinguishing agent.

The temperature curve for the Crib#1 fire was in the stable combustion stage from 250 to 1430 s when it reached a peak temperature of 798 °C and the wood crib mass loss was most rapid. The average wood crib mass loss during the stable combustion stage was

0.0222 kg/s. At 1430 s after igniting the wood crib, the combustion entered the decay stage the temperature measured by the thermocouple gradually decreased, and the rate of mass decline eased.

The temperature curve for the Crib#2 fire was in the stable combustion stage from 250 to 1750 s when it reached a peak temperature 1080 °C. At this point, the radiant heat flux value reached 87 kW/m^2 and the wood crib mass loss was most rapid. The average value for wood crib mass loss in the stable combustion stage was 0.0630 kg/s. After 1750 s the combustion entered the decay stage, the temperature measured by the thermocouple gradually decreased, and the rate of mass decline eased. At 1980 s after igniting the wood crib, a small number of the wood strips fell off due to the loss of supporting force (carbonization), which meant that the crib lost its original shape.

The temperature curve for the Crib#3 fire was in the stable combustion stage from 350 to 1480 s when it reached a peak temperature of 967 °C and the wood crib mass loss was most rapid. The average wood crib mass loss during the stable combustion stage was 0.0418 kg/s. At 1480 s after igniting the wood crib, the combustion entered the decay stage, the temperature measured by the thermocouple gradually decreased, and the rate of mass decline eased.

4. Discussion and Analysis

Table 4 and Figures 22–25 summarize the highest temperatures, highest radiant heat flux densities, the average radiant heat flux values, and the average mass losses during the stable combustion stage under the action of the different flame retardants (including no flame retardant sprayed).

Table 4. Summary of flame retardant performance indices of the wood crib fire sites.

Crib No.	Project	Control Group	Agent#1	Agent#2	Agent#3	Agent#4	Agent#5
Crib#1	Maximum temperature (°C)	928	623	887	583	729	798
	Average mass loss in stable combustion stage (kg/s)	0.026	0.0160	0.023	/	0.0266	0.0222
Crib#2	Maximum temperature (°C)	1156	1110	1067	1091	1128	1080
	The maximum value of radiant heat flux (kw/m^2)	110	114	100	102	94	87
	The average value of radiant heat flux (kw/m^2)	87.95	101.56	74.61	43.14	63.9	45.77
	Average mass loss in stable combustion stage (kg/s)	0.115	0.1359	0.1047	0.0799	0.0867	0.0630
Crib#3	Maximum temperature (°C)	1056	917	990	997	1058	967
	Average mass loss in stable combustion stage (kg/s)	0.053	0.0432	0.0429	0.0465	0.0522	0.0418

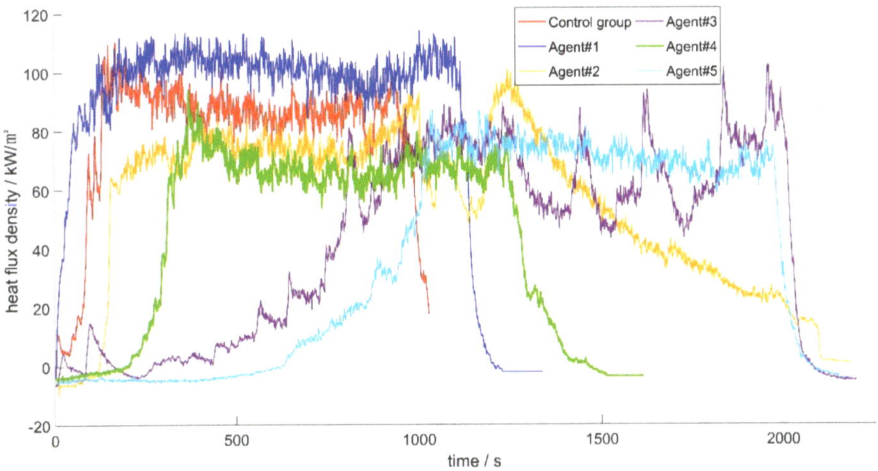

Figure 22. Differences in heat flux density after the helicopter had sprayed the flame retardants.

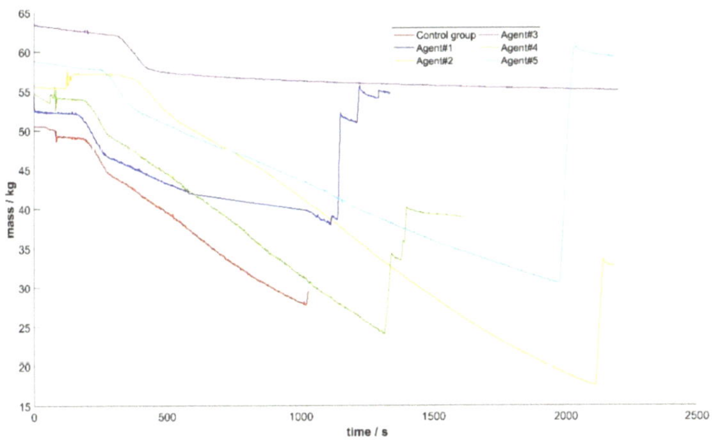

Figure 23. Differences in crib#1 mass losses after the helicopter had sprayed the flame retardants.

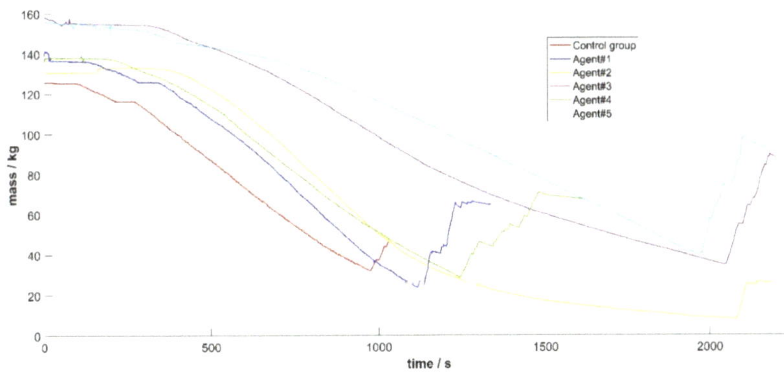

Figure 24. Differences in crib#2 mass losses after the helicopter had sprayed the flame retardants.

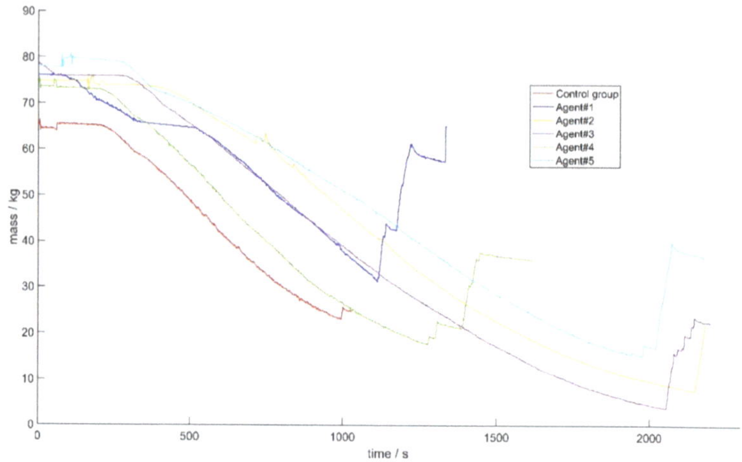

Figure 25. Differences in crib#3 mass losses after the helicopter had sprayed the flame retardants.

When the H125 helicopter flies at a height of 30 m and sprays 660 kg of flame-retardant water agent at a speed of 20 km/h, the following conclusions can be made after analyzing the combustion process and the above data:

(1) Compared with not spraying any flame-retardant water agent, spraying pure water, 10% Class AB flame retardant, 10% Class A flame retardant, or gel flame retardant can reduce the radiation intensity and mass loss due to the wood crib fire to a certain extent, thus prolonging the burning time of the wood crib fire;

(2) With regard to the mass loss index, the flame retardant properties from high to low are as follows: 10% Class A flame retardant + 0.6% guar gum > gel flame retardant > 10% Class A flame retardant > Class AB flame retardant > pure water. In terms of the radiant heat intensity index, the flame retardant properties from high to low are as follows: gel flame retardant $\approx$ 10% Class A flame retardant + 0.6% guar gum > 10% Class A flame retardant > Class AB flame retardant > pure water. Although the use of a flame-retardant water agent has little effect on retarding and reducing the temperature inside the combustibles, the temperature above the combustibles decreased significantly due to the reduction in fire intensity;

(3) When a wildfire spreads and expands, the water on the surface and inside of the surrounding combustibles will evaporate when the adjacent combustibles burn, thus reducing the effect of flame retardants, especially when the helicopter cannot completely cover the combustible surface. These uncovered surfaces are the weak points that lead to the spread of fires, and in this case, it is often impossible to stop the wildfire by spraying flame-retardant water agents;

(4) The temperature analysis of the thermocouples at 1 m, 1.7 m, and 2.7 m height shows that the temperature at the three heights can be effectively reduced, but the temperature inside the wood crib is not reduced.

In addition, the following fire prevention measure is also proposed: since the flame retardancy of Crib#1 is obviously higher than that of Crib#2, regular removal of any surface vegetation and the humus layer will reduce the thickness of the surface combustibles and enhance flame retardancy.

The mechanisms utilized by the different flame retardants are as follows:

(1) Pure water can only moisten the wood crib. The pure water in and on the wooden crib does not have a flame retardant role when it is completely evaporated by the standing and ignition processes. Water volume limitations and the short absorption time mean that the wood crib cannot be completely covered, resulting in a poor flame retardant effect;

(2) Class AB flame retardant has poor wind resistance, which means that its diffusion area is wider than those of Agent#1, Agent#3, and Agent#4. Furthermore, the amount of flame-retardant water agent per unit area is also less. When dropped from the air, the foam covers a wide area and helps limit the spread of the fire. Once dispersed on a fire, the foam absorbs heat from combustion while the bubble structure slowly releases water, which is absorbed by wood fuels. Foam improves the effectiveness of water by (1) helping water soak deeper and more quickly into forest fuels, such as wood, brush, and wood debris; and (2) slowing the evaporation of water held within the foam;

(3) The gel flame-retardant water agent has good water absorption performance and poor fluidity, which means that that the water utilization rate is high. The principle consists of two components: One is that super absorbent particles absorb water (hundreds of times their own weight) in a chemical-physical process called hydration. The stacked and water-filled "bubbles" greatly enhance the thermal protection performance of the flame-retardant water agent. The second one is to prevent the flame-retardant water agent from turning into steam in the superheated air above the wood crib fire and being taken away by the high-temperature smoke plume gas. The adhesive properties of the gel-based Agent#3 slow down the evaporation process, enabling more product to reach the fire source through hot air [32]. These two aspects need to be taken into account when gel flame retardant is sprayed from a helicopter. When the amount of flame-retardant water agent is less than that required for fire retardant (that is, the fire field intensity is large, for example, with the 12-layer wood crib when it is still fully burning after the flame retardant has been applied), the fire retardant performance is poor because the gel-based Agent#3 cannot cover all the burning points. When the amount of flame-retardant water agent is sufficient relative to the fire retardant requirements (that is, the intensity of the fire field is small; for example, a 4-layer wood cribs cannot be ignited after the flame retardant has been applied), a colloid can form on the surface of the wood strip that wraps around it to prevent the combustibles from being ignited. If there is not complete coverage (100%) on all surfaces, then gels are useless as the exposed area can catch fire and burn right through a structure. Therefore, helicopters should spray more gel flame retardants to cover the surface of combustibles as much as possible. A ground coating should be used as far as possible to protect wooden structures and improve helicopter spraying efficiency at reducing the intensity of the fire field;

(4) Class A flame retardant contains large amount of salts. When $(NH_4)_2CO_3$ decomposes, Class A flame retardant absorbs 48 kJ/mol more heat than pure water and evaporates water to rapidly cool down. It also generates inert gases, such as NH_3 and CO_2, to isolate oxygen. The phosphoric acid, metaphosphoric acid, and polymetaphosphoric acid produced in the chemical reactions can react with carbonaceous compounds and generate a dense and flame retardant coating over the surface of combustible materials, which can effectively delay the re-ignition time and reduce the fire intensity after re-ignition [33];

(5) Class A flame retardant + guar gum is a mixture of Agent#2 and Agent#3 and has both chemical flame retardant and physical flame retardant effects. It is consistent with the flame retardant principle outlined by Ref. [12] and its flame retardancy is relatively better than single flame retardants.

5. Applications

From August to September 2022, Hunan Province suffered the strongest drought since 1961 and industrial and residential electricity demand was extremely high. The load in Hunan, Jiangsu, and other provinces increased by more than 18% compared to the same period in 2021. Therefore, there was a great contradiction between electricity consumption and power supply. At the same time, the dry weather and low moisture content of combustibles meant it was extremely easy to ignite them and the risk of wildfires in dense channels was high. In Shimen County, Changde, and Hunan, China, the crossing lines include the ±800 kV Fufeng Line #1494–1495, ±800 kV Jinsu Line #2104–2105, and ±800 kV Qishao Line #3964–3965, which are typical dense transmission lines. In order to ensure the efficient transmission of electricity during the summer peak season, on 22 August, the transmission load of the Fufeng Line was 4.76 million kW, that of the Jinsu Line was 6.83 million kW, and that of the Qishao Line was 6 million kW. A fire in the dense channel could very easily have caused multiple UHV transmission lines to trip at the same time, with considerable risk to the power grid. Since 23 August 2022, one H125 fire fighting helicopter has been deployed in central Hunan after reports of frequent wildfires.

On the afternoon of 25 August 2022, a forest fire broke out in Changde, as shown in Figure 26, which seriously threatened the operation of UHV lines. The vegetation on the site, which was mainly Chinese pine with a high oil content and an extremely high heat of combustion, was dense. At that time, the temperature was as high as 40 °C and the humidity was only 20%, which meant that the vegetation had a low moisture content. The fire spread rapidly up the mountain and an extreme fire tornado formed under agitation by the wind, which was incredibly dangerous. The burned area reached 220 hectares and the open fire area reached 45 hectares. The fire site was steep and inaccessible to ground personnel and there were no natural or artificial isolation zones.

Figure 26. Monitoring map for the local location of on-site wildfires. Left image shows an infrared image of wildfire disasters and the right hand picture shows the visible light image of a wildfire disaster.

From the 26th to the 27th, the combined on-site terrain and meteorological conditions at 18:40 before sunset led to a helicopter being used to spray Class A flame retardant + guar gum when a downhill fire began in the canyon, as shown in Figures 27 and 28. A total of 85 barrels of flame-retardant water agent, 68 tons in total, were sprayed for 4.5 h to suppress the fire site at the smoke point. When the temperature decreased and the humidity increased at night, ground personnel used the No. 2 tool and a multi-stage water pump to relay a water supply to the fire site to clean up the smoke point at a fixed point. This completely extinguished the fire, avoided the burning of houses and a large amount of economic vegetation, ensured the safety of the lives and property of the local people, and effectively reduced the threat of wildfires near sub-transmission lines that were associated with important transmission lines.

During the summer peak season of 2022, the Hunan Power Grid dispatched a helicopter 64 times, sprinkled 844 buckets of water, and the cumulative firefighting time was 83.5 h. They successfully extinguished or timely blocked 41 wildfires on transmission lines, such as the 500 kV Changmin and Jinhong Lines and the 220 kV Tizhong Line, as shown in Table 5. There were no 220 kV wildfire tripping accidents on the transmission lines with voltage levels of 220 kV or above.

Figure 27. Adding a flame retardant to a firefighting helicopter.

Figure 28. An example of using a helicopter to a spray flame-retardant water agent to quickly extinguish a large-scale wildfire on site. The picture on the left shows that the flame retardant is sprayed to form an isolation belt and the picture on the right shows that the wildfire was quickly extinguished at the spraying position without further spread.

Table 5. Fire extinguishing and the application of flame retardants to forest fires in the Hunan Power Grid area during the Summer peak season in 2022.

Location	Date	Flight Time	Number of Buckets	Amount of Flame Retardant (kg)	Fire Extinguishing Line
Changde	26–27 August	13 h 51 min	93	2258	Fire in Shimen County
Loudi	4 September	59 min	5	80	500 kV Jinhong Line #003
Loudi	5 September	2 h 21 min	18	80	500 kV Changmin Line #179–#180, 220 kV Qunkang Line
Loudi	6 September	2 h 44 min	27	50	500 kV Changmin Line #179–#181
Loudi	7 September	4 h 33 min	53	100	500 kV Changmin Line #044–#046
Loudi	8 September	8 h 56 min	97	150	Forest fire in Weishan Township, Xinhua County, Loudi City
Loudi	1 September	1 h 6 min	8	30	500 kV Changmin Line #192
Loudi	11 September	3 h 34 min	50	80	220 kV Tizhong Line #27
Yongzhou	14 September	5 h 18 min	16	80	Forest fire in Huangjiangyuan Village, Yongzhou
Yongzhou	15–17 September	18 h 45 min	92	450	
Loudi	19 September	1 h 41 min	13	50	500 kV Hongmin Line 2#74–#75
Loudi	23 September	49 min	5	50	500 kV Jinhong Line I #003, 220 kv Jinti Line III #5
Loudi	24 September	3 h 32 min	30	100	220 kV Hongbao Line #59
Loudi	26 September	2 h 2 min	13	50	220 kV Tiqun Line I #78–79
Loudi	27 September	1 h 35 min	8	40	220 kV Liankang Line 1#7–#8
Loudi	30 September	4 h 4 min	37	100	220 kV Tihe Line I #13–#14

6. Conclusions

In this study, an H125 helicopter was used to carry out flame retardant tests with 660 kg of different flame-retardant agents. The flame retardant effect of the different flame-retardant water agents on the wood cribs was observed with conclusions drawn as follows:

(1) Compared to not spraying any flame-retardant water agent, pure water, Class AB flame retardants, Class A flame retardants, gel flame retardants, etc., can reduce the intensity of a wood crib fire to a certain extent;

(2) The mass loss index results showed that the flame retardancy from high to low was 10% Class A flame retardant + 0.6% guar gum > gel flame retardant > 10% Class A flame retardant > Class AB flame retardant > pure water. The radiant heat intensity index results showed that flame retardancy from high to low was gel flame retardant ≈ 10% Class A flame retardant + 0.6% guar gum > 10% Class A flame retardant > Class AB flame retardant > pure water;

(3) Based on the flame-retardant properties of the different flame retardants tested in this study, they were applied in Loudi, Changde, and other cities in Hunan Province to prevent and control wildfire disasters near transmission lines. They effectively ensured the safe operation of dense power grid channels during the high-incidence period for wildfires in the power grid area and under extreme dry weather conditions;

(4) In this study, only laboratory experiments were carried out; the impact of ladder combustible on the flame-retardant properties has not been studied. This method can be studied more thoroughly in future wildfire experiments.

Author Contributions: Conceived and designed the paper: J.L. Data collection: T.Z. Testing and data analysis: T.Z. and Y.O. Wrote the paper: T.Z. and C.W. All authors have read and agreed to the published version of the manuscript.

Funding: This work was supported by the State Grid Major S&T Project (No. 5216A0210041).

Institutional Review Board Statement: Not applicable.

Informed Consent Statement: Not applicable.

Data Availability Statement: The data presented in this study are available on request from the corresponding author.

Acknowledgments: The authors would like to thank the Editor and the reviewers whose comments and suggestions have been very helpful in improving the quality of this paper.

Conflicts of Interest: The authors declared that they have no conflict of interest to this work.

References

1. Jolly, W.M.; Cochrane, M.A.; Freeborn, P.H.; Holden, Z.A.; Brown, T.J.; Williamson, G.J.; Bowman, D.M. Climate-induced variations in global wildfire danger from 1979 to 2013. *Nat. Commun.* **2015**, *6*, 7537. [CrossRef] [PubMed]
2. Halofsky, J.E.; Peterson, D.L.; Harvey, B.J. Changing wildfire, changing forests: The effects of climate change on fire regimes and vegetation in the Pacific Northwest, USA. *Fire Ecol.* **2020**, *16*, 1–26. [CrossRef]
3. Moein, C.; Ansari, B.; Mohagheghi, S. Vulnerability assessment of the power grid against progressing wildfires. *Fire Saf. J.* **2015**, *73*, 20–28.
4. Hamed, M.; AghaKouchak, A. Increasing exposure of energy infrastructure to compound hazards: Cascading wildfires and extreme rainfall. *Environ. Res. Lett.* **2019**, *14*, 104018.
5. Wu, C.; Lu, J.; Chen, B.; Liu, Y.; Zhou, T.; Zhou, T. Control strategy for large-scale fires near power transmission lines and its application. *IET Gener. Transm. Distrib.* **2021**, *15*, 704–715. [CrossRef]
6. Wolff, M. The San Francisco Department of Public Health's Assessment of the Vulnerability of Hospitals and Care Facilities to Extreme Heat and Wildfire Smoke, and Identification of Best Practices to Prepare for These Impacts. *AGU Fall Meet. Abstr.* **2019**, *2019*, GH33A-09.
7. Lu, J.; Zhou, T.; Wu, C.; Li, B.; Tan, Y.; Zhu, Y. Fault statistics and analysis of 220 kV and above power transmission line in province-level power grid. *High Volt. Eng.* **2016**, *42*, 200–207. (In Chinese) [CrossRef]
8. Kal'avský, P.; Petríček, P.; Kelemen, M.; Rozenberg, R.; Jevčák, J.; Tomaško, R.; Mikula, B. The efficiency of aerial firefighting in varying flying conditions. In Proceedings of the 2019 International Conference on Military Technologies (ICMT), Brno, Czech Republic, 30–31 May 2019.
9. Plucinski, M.P.; Gould, J.S.; McCarthy, G.; Hollis, J. *The Effectiveness and Efficiency of Aerial Firefighting in Australia*; Bushfire CRC: Melbourne, Australia, 2007; Available online: http://hdl.handle.net/102.100.100/126629?index=1 (accessed on 10 January 2023).
10. Bae, T.-H.; Lee, S.-Y. Study on efficient operation on aerial fire fighting helicopter in forest fire. *Fire Sci. Eng.* **2015**, *29*, 45–52. [CrossRef]
11. Trakas, D.N.; Hatziargyriou, N.D. Optimal distribution system operation for enhancing resilience against wildfires. *IEEE Trans. Power Syst.* **2017**, *33*, 2260–2271. [CrossRef]
12. Zhou, T.; Lu, J.; Wu, C.; Lan, S. Numerical calculation and analysis of water dump distribution out of the belly tanks of firefighting helicopters. *Safety* **2022**, *8*, 69. [CrossRef]
13. Jazebi, S.; De Leon, F.; Nelson, A. Review of wildfire management techniques—Part I: Causes, prevention, detection, suppression, and data analytics. *IEEE Trans. Power Deliv.* **2019**, *35*, 430–439. [CrossRef]
14. Yue, Y.S.; Zou, Y.H.; Huang, F.Y.; Gong, Z.X.; Wang, C. Study on the flashover characteristics of ±500 kV DC transmission lines caused by forest fire under reduced-voltage operation. In Proceedings of the 2017 EPTC Power Transmission and Transformation Technology Conference, Chongqing, China, 15–16 March 2017.
15. Yu, A.C.; Lopez Hernandez, H.; Kim, A.H.; Stapleton, L.M.; Brand, R.J.; Mellor, E.T.; Bauer, C.P.; McCurdy, G.D.; Wolff, A.J., III; Chan, D.; et al. Wildfire prevention through prophylactic treatment of high-risk landscapes using viscoelastic retardant fluids. *Proc. Natl. Acad. Sci. USA* **2019**, *116*, 20820–20827. [CrossRef]
16. Fiedler, N.C.; Lacerda, G.R.; Ramalho, A.H.C.; Berude, L.C.; Neves, F.P.; Rodrigues, C.K. Firefighting combat with fire retardants at different concentrations. *Floresta* **2019**, *50*, 1107–1112. [CrossRef]
17. Nabipour, H.; Shi, H.; Wang, X.; Hu, X.; Song, L.; Hu, Y. Flame retardant cellulose-based hybrid hydrogels for firefighting and fire prevention. *Fire Technol.* **2022**, *58*, 2077–2091. [CrossRef]
18. Bae, T.-H.; Lee, S.-Y.; Son, J.-H. Study on ground water drop pattern and coverage level by helicopter. *Fire Sci. Eng.* **2013**, *27*, 54–61. [CrossRef]
19. Legendre, D.; Becker, R.; Alméras, E.; Chassagne, A. Air tanker drop patterns. *Int. J. Wildland Fire* **2013**, *23*, 272–280. [CrossRef]
20. Vernol, B.; Downs, M. Development of a navigation/situation display to improve aerial fire fighting safety and efficiency. In Proceedings of the Human Factors and Ergonomics Society Annual Meeting, San Diego, CA, USA, 9–13 October 1995; SAGE Publications: Los Angeles, CA, USA, 1995; Volume 39.
21. Koufopoulou, S.; Michalopoulos, C.; Tzamtzis, N.; Pappa, A. Impact of a long term fire retardant (Fire Trol 931) on the leaching of Na, Al, Fe, Mn, Cu and Si from a Mediterranean forest soil: A short-term, lab-scale study. *Bull. Environ. Contam. Toxicol.* **2014**, *92*, 708–713. [CrossRef]

22. Crouch, R.L.; Timmenga, H.J.; Barber, T.R.; Fuchsman, P.C. Post-fire surface water quality: Comparison of fire retardant versus wildfire-related effects. *Chemosphere* **2006**, *62*, 874–889. [CrossRef]
23. Paul, M.J.; LeDuc, S.D.; Lassiter, M.G.; Moorhead, L.C.; Noyes, P.D.; Leibowitz, S.G. Wildfire induces changes in receiving waters A review with considerations for water quality management. *Water Resour. Res.* **2022**, *58*, e2021WR030699. [CrossRef]
24. Boigné, E.; Bennett, N.R.; Wang, A.; Ihme, M. Examining the effect of fire retardant on the combustion of wood via X-ray computed tomography. In Proceedings of the 12th US National Combustion Meeting, College Station, TX, USA, 24–26 May 2021.
25. National Interagency Fire Center. "Interagency Standards for Fire and Aviation Operations" (PDF). Available online https://www.nifc.gov/sites/default/files/redbook-files/RedBookAll.pdf (accessed on 1 January 2023).
26. McAllister, S.; Finney, M. Burning rates of wood cribs with implications for wildland fires. *Fire Technol.* **2016**, *52*, 1755–1777 [CrossRef]
27. Smith, P.G.; Thomas, P.H. The rate of burning of wood cribs. *Fire Technol.* **1970**, *6*, 29–38. [CrossRef]
28. Zhou, T.; Lu, J.; He, L.; Wu, C.; Luo, J. Experiments in aerial firefighting with and without additives and its application to suppress wildfires near electrical transmission lines. *J. Fire Sci.* **2022**, *40*, 313–346. [CrossRef]
29. Hermouet, F.; Guillaume, É.; Rogaume, T.; Richard, F.; El Houssami, M. Experimental determination of the evolution of the incident heat flux received by a combustible during a cone calorimeter test: Influence of the flame irradiance. *J. Fire Sci.* **2021**, *39*, 119–141. [CrossRef]
30. Gupta, V.; Torero, J.L.; Hidalgo, J.P. Burning dynamics and in-depth flame spread of wood cribs in large compartment fires *Combust. Flame* **2021**, *228*, 42–56. [CrossRef]
31. Konishi, T.; Kikugawa, H.; Iwata, Y.; Koseki, H.; Sagae, K.; Ito, A.; Kato, K. Aerial firefighting against urban fire: Mock-up house experiments of fire suppression by helicopters. *Fire Saf. J.* **2008**, *43*, 363–375. [CrossRef]
32. João, B.; Bordado, C.M.; Gomes, J.F.P. New technologies for effective forest fire fighting. *Int. J. Environ. Stud.* **2007**, *64*, 243–251.
33. Alba, À.; Pastor, E.; Planas, E. Different scales for studying the effectiveness of long-term forest fire retardants. *Prog. Energy Combust. Sci.* **2008**, *34*, 782–796.

fire

Article

An Experimental Study on the Transportation Characteristics of Perfluoro(2-methyl-3-pentanone) in a Straight Pipe

Xiaomin Ni [1,*], Ye Chen [1], Qiurui Huang [1], Chenxi Zhao [1], Songyang Li [2], Jiahui Huang [2] and Jian Wang [1]

1 State Key Laboratory of Fire Science, University of Science and Technology of China, Hefei 230026, China
2 AECC Commercial Aircraft Engine Co., Ltd., Shanghai 200241, China
* Correspondence: nxmin@ustc.edu.cn

Abstract: Gaseous fire suppressants are usually stored in a vessel via pressurization, and then discharged out through pipelines. The flow behaviors of the agents in pipes greatly affect its dispersion in space, as well as the fire extinguishing results. Here, an experimental study was carried out on the transportation characteristics of perfluoro(2-methyl-3-pentanone) ($C_6F_{12}O$) in a horizontal straight pipe with the temperature and pressure recorded synchronously. At a filling pressure of 1800 kPa and a filling density of 517 kg·m^{-3}, the agent release was completed in 2.0 s with the pipeline pressure peak of 1145 kPa and the pipeline temperature nadir of −10.6 °C. In comparison to that of bromotrifluoromethane (CF_3Br) under the same conditions, the temperature and pressure curves of $C_6F_{12}O$ exhibited similar varying trajectories but a much smaller amplitude, which could be ascribed to their different thermophysical properties. When keeping the other conditions unchanged, raising the filling pressure $C_6F_{12}O$ reduces the discharge duration and the pipeline temperatures. Increasing the filling density extends the discharge duration, but shows little influence on the pipeline temperatures. The results were expected to provide useful information for the model validation and engineering design of a $C_6F_{12}O$ fire-suppressing system with a predictable performance.

Keywords: $C_6F_{12}O$; fire suppressant; transportation; pipe; pressure; temperature

Citation: Ni, X.; Chen, Y.; Huang, Q.; Zhao, C.; Li, S.; Huang, J.; Wang, J. An Experimental Study on the Transportation Characteristics of Perfluoro(2-methyl-3-pentanone) in a Straight Pipe. *Fire* **2023**, *6*, 156. https://doi.org/10.3390/fire6040156

Academic Editor: Grant Williamson

Received: 7 February 2023
Revised: 21 March 2023
Accepted: 12 April 2023
Published: 14 April 2023

1. Introduction

Great efforts have been devoted to looking for ideal halon replacers since the use of CF_3Br was banned, by the Montreal Protocol and its subsequent revisions, for its effect on the ozone layers. Perfluoro(2-methyl-3-pentanone), denoted as $C_6F_{12}O$, has become one of the most promising halon replacers with satisfactory environmental friendliness, as well as fire extinguishing efficiency and toxicity [1]. Since being proposed by 3 M as a new kind of clean fire suppressant, numerous studies have been conducted on $C_6F_{12}O$, including studies on its physicochemical properties [2,3], fire extinguishing capacity [4,5], material compatibility [6,7], thermal decomposition products [8], etc. These studies provide valuable directions on the model computation and practical applications of $C_6F_{12}O$ in the context of real fire suppression. In 2003, $C_6F_{12}O$ was recognized as an available halon alternative and substitute in the significant new alternatives policy (SNAP) program of the United States Environment Protection Agency (EPA) [9]. It was also listed in the fire extinguishing system design standards of ISO 14520 and NFPA 2001 [10,11]. Currently, $C_6F_{12}O$ is widely used in the areas of electrical and electronic cabinets, ships, libraries, etc.

Similar to many other gaseous fire suppressants of halons and hydrofluorocarbons (HFCs), $C_6F_{12}O$ is usually stored in a vessel via pressurization. Once the valve was actuated, the pressurized agent was ejected into the piping manifold and then distributed into space. The transportation characteristics of the fire suppressant in the pipes showed a great influence on its atomization and its distribution in space, as well as in its fire extinguishing results [12]. Therefore, understanding the state of $C_6F_{12}O$ in pipes is indispensable for creating an effective fire suppression system design.

Although extensive research has been carried out on $C_6F_{12}O$, as mentioned above, seldom have the transportation behaviors of $C_6F_{12}O$ in pipes been seen. Recently, Fan reported the flow characteristics of $C_6F_{12}O$ in a complex pipeline with multi-branches [13]. By comparing the experimental data with hydraulic calculations, it was deduced that $C_6F_{12}O$ only conformed to single-phase flows in the initial upstream section of the pipeline, but presented with a two-phase flow state in the following main pipe and downstream section of the pipeline. In comparison to that of bromotrifluoromethane (CF_3Br), which has been tested within detailed studies under various conditions [14–17], investigations on $C_6F_{12}O$ have been scarce up till now. More accurate experiments are still necessary for a better understanding of the flow behaviors of $C_6F_{12}O$ in pipes.

In this work, the discharge process of $C_6F_{12}O$ in a horizontal straight pipe was, in a full scale, experimentally studied. The pressures and temperatures in the vessel and pipes were recorded with a high sampling rate. Meanwhile, CF_3Br was tested as a counterpart for comparison since it has always served as the baseline for screening halon replacer candidates. The effects of source pressure and the filling density on the flow parameters of $C_6F_{12}O$ were explored by a series of contrast trials. The transportation characteristics of $C_6F_{12}O$ in the pipe were summarized based on experimental observations and its thermophysical properties.

2. Materials and Methods

Figure 1 schematically illustrates a typical experimental setup of the agent discharge system. A spherical vessel with the volume of 3.89 L was connected with a horizontal straight pipe through a solenoid valve. The pipe was 2.0 m long with an inner diameter of 14 mm. A convergent nozzle with an inner diameter of 12 mm was also equipped at the end of the pipe.

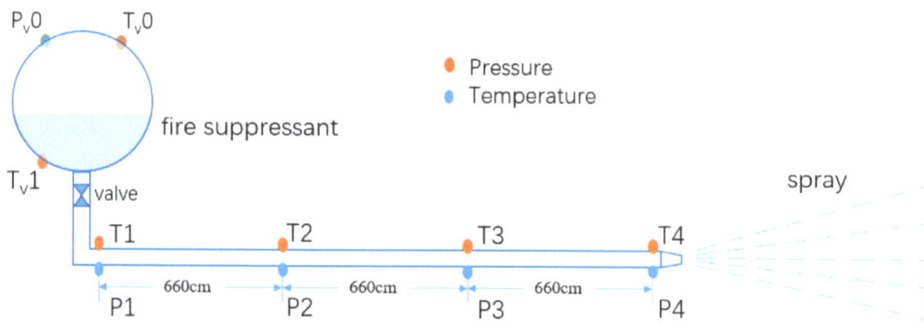

Figure 1. Schematic illustration of the experimental setup.

The pressure (P) and temperature (T) data in the vessel and the pipe were synchronously traced by thermocouples and pressure sensors with a sampling frequency of 4096 Hz. As shown in Figure 1, two thermocouples were installed on the top (T_v0) and the bottom of the vessel (T_v1), respectively. The other four thermocouples were anchored equidistantly on the pipe with a spacing of 0.66 m (T1–T4). There were five pressure sensors installed on the system. One located the upper part of the vessel (P_v1) and four other ones positioned on the pipe (P1–P4). At each measuring point of the pipe, the thermocouple and the pressure sensor were arranged face to face. The measuring errors of temperature, pressure and agent mass were estimated as 1.0%, 0.5% and 1.0%, respectively. The errors were mainly caused by the instruments and the discharge system. In addition to these, the error of agent mass also originated from the loss in the filling process. The pressure sensors, thermocouples and the electronic scale were all calibrated before measurements were taken.

$C_6F_{12}O$ and CF_3Br with a purity above 99.5% were chosen as the fire suppressants. As is the case in a typical experiment, the vessel was first pumped in order for it to become a vacuum. Then, it was filled with a certain amount of liquid $C_6F_{12}O$, then pressurized by nitrogen until reaching the desired pressure. After that, the P–T recording system was started. Once the fast-opening solenoid valve was actuated, the agent in the vessel was forced out into the straight pipe and discharged from the nozzle into an unconfined space. The test of CF_3Br was conducted in the same discharge system under similar conditions. Table 1 lists the tests with different filling pressures and filling densities. The filling pressure is the source pressure in the vessel that was achieved by nitrogen pressurization. The filling density was defined as the mass agent of the liquid divided by the vessel volume. All the tests were conducted at a room temperature of about 20 °C.

Table 1. The discharge tests of the $C_6F_{12}O$ and CF_3Br used.

Test Sequence	Agent	Agent Mass (kg)	Filling Density ($kg{\cdot}m^{-3}$)	Filling Pressure (kPa)
1	$C_6F_{12}O$	2.01	517	1822
2	CF_3Br	2.02	519	1846
3	$C_6F_{12}O$	2.12	545	1910
4	$C_6F_{12}O$	2.08	535	2540
5	$C_6F_{12}O$	2.14	550	3230
6	$C_6F_{12}O$	1.02	262	1799
7	$C_6F_{12}O$	2.05	527	1812
8	$C_6F_{12}O$	3.02	776	1813

3. Results and Discussion

3.1. Pressures and Temperatures in the Discharge System

Figure 2a shows, in test 1 of $C_6F_{12}O$, all of the pressure and temperature profiles as a function of time. The plots of the vessel pressure (P_v0) and the pipeline pressure (P1–P4) are located in the lower part of the pattern. Upon the beginning of the agent spray, P_v0 dropped continuously from 1822 kPa to atmospheric pressure. The time length of such a process was defined as the discharging duration (t_d). From the trace of P_v0 versus time, the t_d of test 1 was determined as 2.0 s. Different to the uninterrupted dropping of P_v0, the pipeline pressure plots (P1–P4) exhibited an asymmetric hump shape, which could be roughly divided into three stages of "rapid increase, gentle decrease and continuous decrease". Taking P1 as an example, in the first stage of 0–0.23 s (marked in yellow), P1 increased rapidly from zero to a peak pressure of 1145 kPa, at a high rising rate of 4978 $kPa{\cdot}s^{-1}$. Then, it declined gently with fluctuations at an average rate of 488 $kPa{\cdot}s^{-1}$ in the time range of 0.23–0.63 s (marked in blue). In the final stage of 0.63–2.0 s (marked in pink), P1 decreased continuously from 950 kPa to zero at an average decreasing rate of about 693 $kPa{\cdot}s^{-1}$. The peak pressure decreased farther down the pipe from P1 (1145 kPa) to P4 (803 kPa) with the gap between the neighboring points of 100–150 kPa. Meanwhile, the time by which the four measuring points reached the peak pressure increased upstream from P4 (0.18 s) to P1 (0.23 s), with an interval of about 15–20 ms.

The temperature curves of test 1 are located in the upper part of Figure 2a. The two T plots of the vessel that are at the top decreased very slowly, with a total reduction of about 3.3 °C for the upper gas (T_v0) and only 0.5 °C for the liquid below (T_v1). The piping temperature plots (T1–T4) were kept nearly flat in the period of 0–0.63 s. In the following stage, they declined quickly to levels below zero, showing a total reduction of about 26–30 °C. The nadir point of T1 was as low as −10.6 °C.

Figure 2b presents the P–T profiles from the trial of CF_3Br (test 2), which followed trajectories that are similar to those of $C_6F_{12}O$, albeit with distinct details. From the monotonically declining curve of P_v0, the t_d of CF_3Br was determined as 3.0 s, which is much longer than that of $C_6F_{12}O$. Likewise, the piping pressure plots were hump shaped with a quick rise and a slow drop. The peak pressures of P1–P4 were in the range of 1300~1570 kPa, which were about 400–500 kPa higher than those of $C_6F_{12}O$. However, the spacing of the peak values of P1–P4 (60 kPa) was only half of that of $C_6F_{12}O$.

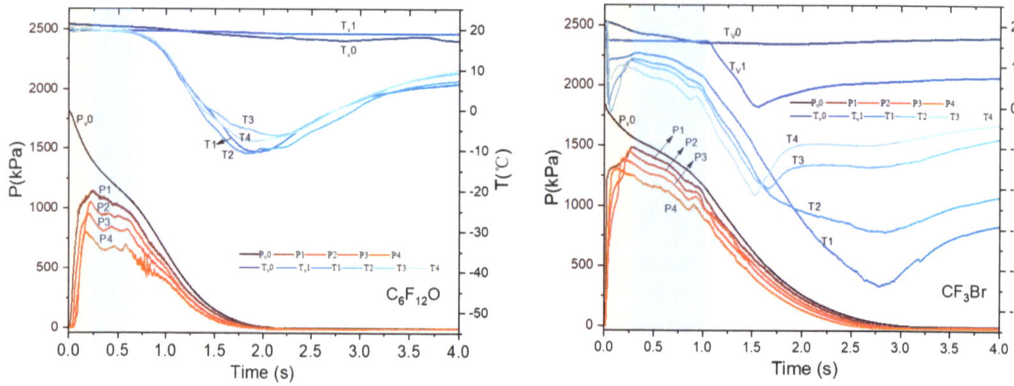

Figure 2. Pressure and temperature profiles versus time in the tests: (**a**) $C_6F_{12}O$ and (**b**) CF_3Br.

In the vessel of CF_3Br, T_v0 dropped by 5 °C in the range of 0–1.0 s. Then, it was kept nearly flat until the end of the agent's release. Meanwhile, T_v1 remained constant within the initial stage and then decreased rapidly to zero (1.0–1.5 s) at a rate of about 32 $°C·s^{-1}$. There were two big drops in the piping temperature traces (T1–T4). The first occurred at the beginning of discharge with a drop of 9~21 °C, and the second was in the latter part of discharge with a reduction of 25–40 °C. At the end of the agent's release, all the temperature points in the pipe fell below −20 °C. T1 even reached the lowest point of −44 °C, which is significantly lower than that of $C_6F_{12}O$.

3.2. *Transportation Characteristics of $C_6F_{12}O$ in the System*

The nonlinear changes in Figure 2 imply complex flow behaviors with respect to the suppressants during the rapid release process. As shown above, the P–T profiles of $C_6F_{12}O$ exhibited a varying trend that was similar to CF_3Br but with a much lower amplitude, which could be ascribed to their different thermophysical properties. Table 2 listed some of the typical parameters of the two agents, revealing their remarkable differences regarding vapor pressure and boiling points. At a room temperature and atmospheric pressure, the $C_6F_{12}O$ was a liquid, while the CF_3Br was a gas. When stored in the vessel and fed through a pipe, the agents experienced phase changes between gas and liquid with the pressure changing. Based on the P–T data acquired, the transportation properties of $C_6F_{12}O$ in the system were analyzed comparatively with CF_3Br as a reference.

Table 2. The typical parameters of $C_6F_{12}O$ and CF_3Br [2,18].

Agent	Molecular Weight	Boiling Point (°C)	Vapor Pressure (kPa)	Liquid Density ($kg·m^{-3}$, 22 °C)
$C_6F_{12}O$	316	49.2	40	1.60
CF_3Br	149	−57.8	1620	1.57

3.2.1. Variations in the Vessel

Both $C_6F_{12}O$ and CF_3Br were stored as liquids in a vessel under a nitrogen pressurization of about 1800 kPa. The liquid located at the lower part of the vessel with nitrogen dissolved, while the ullage was a mixture of nitrogen and agent vapor. As the system was depressurized, a dissolution of nitrogen and vaporization of liquefied agent occurred, which counteracted the pressure decline in the vessel and extended the discharge duration. As is confirmed by Figure 3a, both P_v0 plots were characteristics of the pressure offset in the middle stage with reduced dropping rates. For the $C_6F_{12}O$ with a lower vapor pressure and higher boiling point, the pressure compensating effect was relatively weak since there was little vaporization and what gas there was mainly came from the process of nitrogen separating from the liquid. As to the CF_3Br with a much higher vapor pressure and lower boiling point, the effect of pressure offset was stronger from the combined contributions of nitrogen dissolution and agent vaporization. Therefore, the P_v0 of CF_3Br decreased at a slower rate than that of $C_6F_{12}O$.

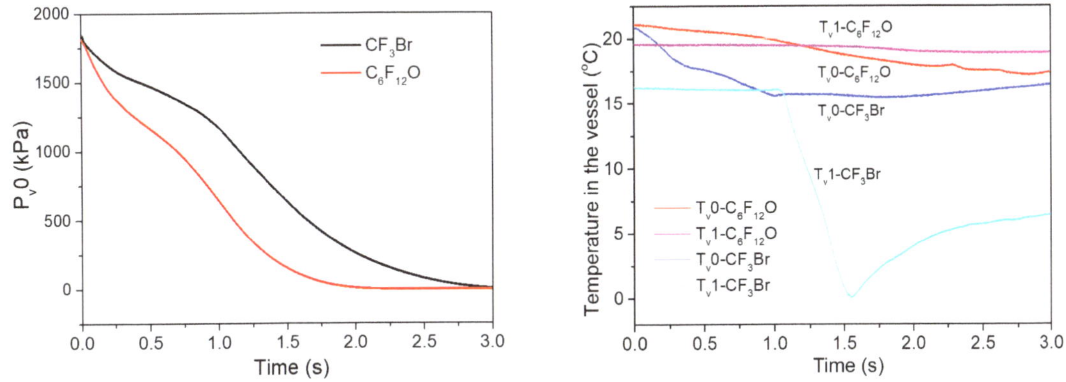

(a) (b)

Figure 3. Comparison of the P–T profiles in the vessel of the two tests: (**a**) P_v0; (**b**) T_v0 and T_v1.

Figure 3b shows the variation of temperatures in the vessel during release process of CF_3Br. The T_v0 first decreased by 5 °C in the initial 1.0 s, and this was due to gas expansion and liquid vaporization. Then, it reached a plateau, indicating a new single-phase gas status in the vessel. The vessel filled with "liquid, vapor and nitrogen" transformed into one filled with the gaseous mixture of "nitrogen and vapor". Much different to that of CF_3Br, the T_v0 of $C_6F_{12}O$ decreased steadily and slowly along with only gas expansions and with little phase change.

The T_v1 of CF_3Br initially did not change much since it was submersed in the liquid [19]. As the liquid in the vessel ran out, it experienced a vaporization of the liquid and an expansion of the remaining gas in the vessel, which showed a big temperature reduction of about 15 °C. The T_v1 of $C_6F_{12}O$ remained constant in the first stage, similar to that of CF_3Br for comparable reasons. In the latter part with gas discharge, only a small temperature drop of 0.3 °C was displayed. When considering its low vapor pressure and high boiling point, there were little phase changes taking place in the vessel of $C_6F_{12}O$. Therefore, the reduction in T_v1 was minimal during the whole process.

3.2.2. Variations in the Pipe

Initially, the pipe was full of air when at an atmospheric pressure. Once the valve was opened, the ullage pressure in the vessel drove the liquid $C_6F_{12}O$ out from the vessel into the pipe. Before the agent entered the pipe, the pressure wave propagated forward, which made the pressure sensors in the pipe first detect a pressure increase [20]. When the pressure wave arrived at the nozzle, it was restricted by the convergent nozzle and a pressure bounce back was triggered [21]. Thus the air in the pipe was compressed by the pressures from the two ends, which led to a temperature rise. As seen in Figure 4, with T2 and P2 serving as examples, in this very initial stage of only about 30 ms, both the temperature and the pressure in the pipe increased.

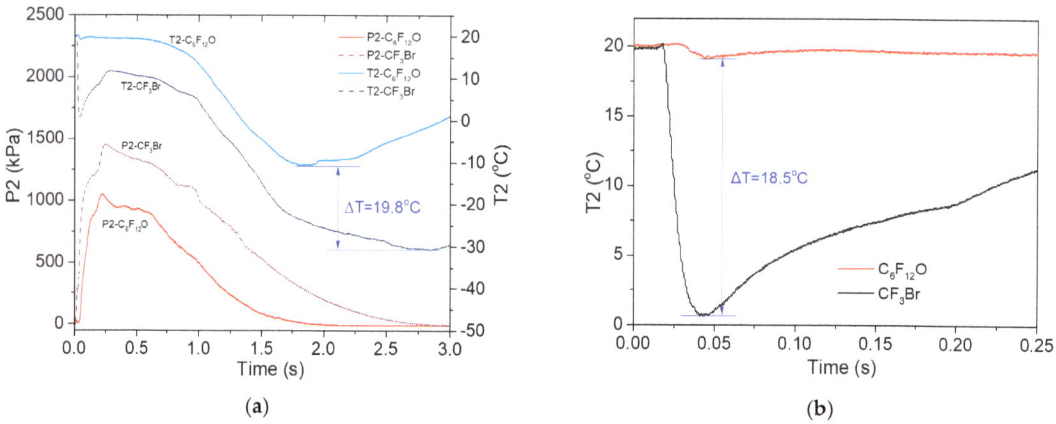

Figure 4. Comparison of the P–T profiles of the two tests: (**a**) P2 and T2; (**b**) T2.

After this short period, the pressurized liquid $C_6F_{12}O$ flew into the pipe. For the pressure difference, the liquid agent entering the pipe would evaporate immediately, together with the escape of the dissolved nitrogen. Such a process further improved the pressure in the pipe, but led to a temperature decrease due to the endothermic process of vaporization. With more fluid flowing into the pipe, both P2 and T2 went up again. When the fluid filled the pipe, P2 attained the peak value. In this very short stage with the pipeline pressure increasing, the agent in the pipe existed mainly in a gaseous state.

With the discharge ongoing, the $C_6F_{12}O$ that flowed into the pipe entered a steady state under the co-effects of the driving force of nitrogen and the friction resistance from the pipe wall. The T2 curve remained stable and the P2 decreased gently. During this middle stage, the fluid in the pipe existed mainly as a liquid, as well as a small amount of nitrogen and $C_6F_{12}O$ vapor, which is typical of a gas-liquid two-phase flow. As the pressure dropped, the mass ratio of gas to liquid in the pipe increased. With the liquid $C_6F_{12}O$ gradually being consumed, the release process enters the final period with the agent in the pipe being in the form of nitrogen and $C_6F_{12}O$ vapor. The expansion of gas brought about a big temperature decrease. At the end of the discharge, the pressure and temperature curves simultaneously went downward to the nadirs.

The P2 and T2 profiles of CF_3Br are also presented in Figure 4a for the purpose of comparison. The comparable trajectory implies similar phase changes as were described above. However, there are also some remarkable differences. During the whole release process, the P2 of CF_3Br remained above that of $C_6F_{12}O$, and the T2 of CF_3Br stayed below that of $C_6F_{12}O$. Moreover, the T2 of CF_3Br exhibited two much bigger temperature drops. Under the experimental conditions, the CF_3Br vapor was liquefied during the nitrogen pressurization, which contained superheat [20]. When the CF_3Br liquid entered the pipe, intense flashing vaporization occurred, resulting in the first temperature reduction. As is shown in Figure 4b, the nadir temperature of CF_3Br was 18.5 °C lower than that of $C_6F_{12}O$. In the following stage with an evolving gas-liquid two-phase flow, the CF_3Br in the pipe displayed a higher pressure and lower temperature than that of $C_6F_{12}O$. When it came to the final stage with a gas expansion, the second big temperature drop appeared, with the gap between the two T2 nadirs of 19.8 °C.

In the test of CF_3Br, it was also observed that P1 had the highest peak pressure, and that P4 hit the peak pressure first out of all the four piping pressures that were similar to $C_6F_{12}O$. Considering fluid behaviors in the pipe, such phenomena may be attributed to the restriction of the convergent nozzle. It was known that the strength of pressure wave waned along the pipe. As the nearest point to the valve, P1 sensed the highest peak pressure for the least pressure loss. On the other hand, as the nearest point to the nozzle,

P4 first showed the peak value because the bounce back of the pressure wave reached P4 prior to the other three measuring points.

3.3. Effect of the Filling Pressure on the Flow Behaviors of $C_6F_{12}O$

Figure 5a displays the P_v0 plots from the three tests at the filling pressures of 1910, 2540 and 3230 kPa, whilst keeping the mass of the $C_6F_{12}O$ similar, respectively. With the filling pressure increasing, the P_v0 declined at a faster rate. As is seen in the pattern inset in the upper part of Figure 5a, the t_d approximately decreased linearly with the source pressure rising. This was different to the result of CF_3Br that was reported by Jia (which was inset in the lower part of Figure 5a), where the injection duration first decreased sharply and then decreased steadily when the release pressure reached a certain value [17]. In their tests of CF_3Br, flashing vaporization contributed the most to the pressure compensation in the vessel, which led to the elongation of t_d. The proportion of such contribution was relatively high at low filling pressures, but dropped at higher filling pressures. However, in the current experiments of $C_6F_{12}O$, the pressure compensation was much smaller, and the source pressure dominated the discharge duration. Further, more tests at varied filling pressures were needed to validate the result.

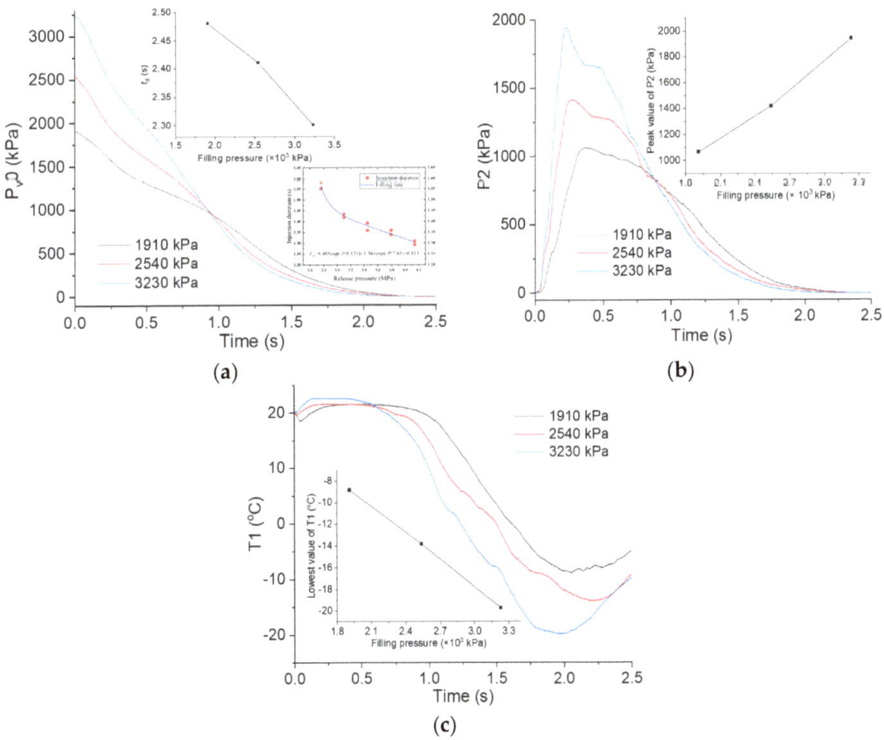

Figure 5. P–T profiles of the $C_6F_{12}O$ tests at different filling pressures: (**a**) P_v0, the pattern inset in the upper part showed t_d at different filling pressures, pattern inset in the lower part was due to the CF_3Br from ref. [17]; (**b**) P2, the pattern inset showed the peak values of P2 at different filling densities; (**c**) T1, the pattern inset showed the T1 nadir at different filling pressures.

Figure 5b,c presents the P–T variations in the pipe of $C_6F_{12}O$ at the three filling pressures. In Figure 5b, with P2 as an example, the peak values of P2 increased with the filling pressure rising, while the piping temperature decreased with the source pressure increasing. As shown in Figure 5c, at the P_v0 of 1910, 2540 and 3230 kPa, T1 showed the lowest values of −8.8, −13.8 and −19.7 °C, respectively. Similarly, a nearly linear relationship was exhibited in the patterns inset in Figure 5b,c. For the current system, the rapid release could be regarded as an adiabatic expansion process. The source pressure in the vessel was the driving force for the agent transportation. With the agent flowing, the pressure energy would transform into dynamic energy and heat energy. Higher filling pressures not only accelerated the fluid flow and shortened the discharge time, but also led to a lower temperature in the pipe.

3.4. Effect of the Filling Density on the Flow Behaviors of $C_6F_{12}O$

Figure 6a shows the changes that were tracked in P_v0 in the $C_6F_{12}O$ tests at different filling densities. With the agent mass increasing, the t_d correspondingly increased (as is seen in the pattern inset in Figure 6a). Meanwhile, the peak pressure of the pipe went down. As shown in Figure 6b, with P2 as an example, the peak value decreased from 1170 kPa to 956 kPa when the filling density increased from 262 to 776 kg·m^{-3}, which also displayed a nearly linear varying trend (as is seen in the pattern inset in Figure 6b). Figure 6c gave the temperature traces of T2 in the three tests. With the filling density rising, the time that the plot began decreasing was postponed, but the decreasing rate and the nadir value remained unchanged. The lowest point was around −10 °C for all the three T2 plots.

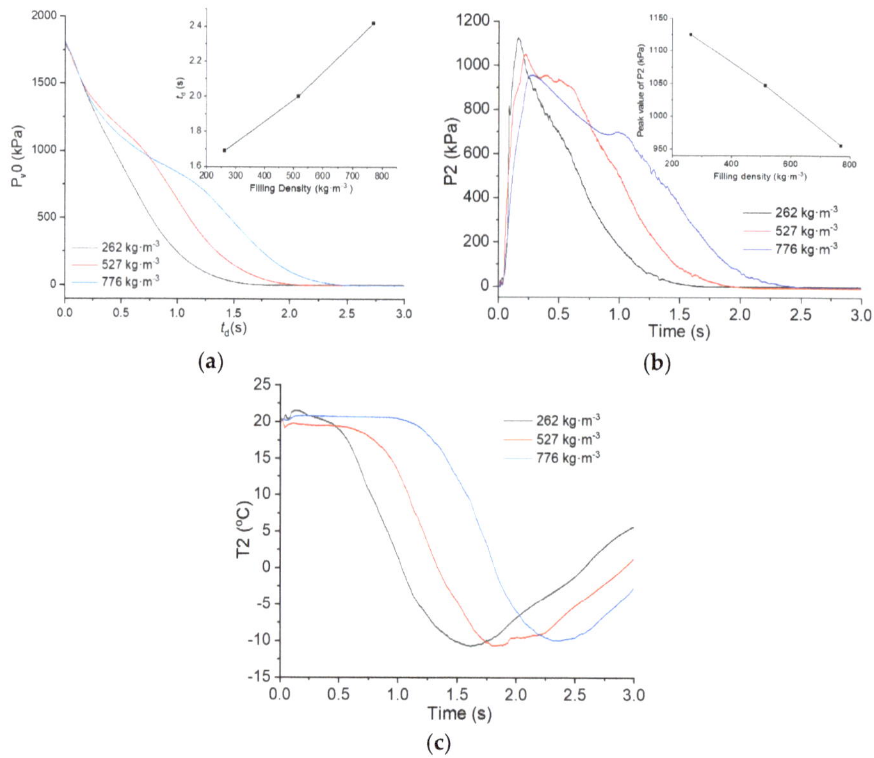

Figure 6. P–T profiles of the $C_6F_{12}O$ tests at different filling pressures: (**a**) P_v0, the pattern inset showed the t_d of the tests at different filling densities; (**b**) P2, the pattern inset showed the peak values of P2 at different filling densities; (**c**) T2.

4. Conclusions

In this work, the transportation characteristics of $C_6F_{12}O$ in a straight pipe during a rapid release process were studied by tracking the temperature and pressure variations simultaneously. The effects of the filling pressure and the filling density on the flow behaviors of $C_6F_{12}O$ were examined through a series of tests. Based on the experimental data, the main conclusions were drawn as follows:

- Under the driving of pressurized nitrogen, the $C_6F_{12}O$ was sprayed out of the pipe very rapidly. In the process of just a few seconds, the agent in the pipe experienced complex changes within the three stages of "gas, gas-liquid mixture and gas". The intermediate stage exhibited characteristics of gas-liquid two-phase flow, which is where the fluid was dominated by liquid $C_6F_{12}O$ together with a small amount of vapor and nitrogen;
- Upon the release of $C_6F_{12}O$, the pressure in the vessel went steadily downward until it reached zero, while the vessel temperature just showed a minor drop of several degrees Celsius. The pipeline pressure plots exhibited an asymmetric hump shape, which contained three stages of "rapid increase, gentle decrease and continuous decrease". In the first two stages, the piping temperature remained stable. However, a big drop of about 30 °C in the piping temperature occurred in the third stage of gas discharge;
- In comparison to that of the CF_3Br released under similar conditions, the temperature and pressure curves of $C_6F_{12}O$ exhibited similar trajectories, but with much lower varying amplitudes. Moreover, such differences can be mainly ascribed to their different vapor pressures and boiling points;
- With other conditions being the same, raising the filling pressure in the vessel of the $C_6F_{12}O$ reduced the discharge duration and the pipeline temperatures. Increasing the filling density extended the discharge duration, but showed little influence on the pipeline temperatures.

The experimental data were expected to validate the computer code predictions of $C_6F_{12}O$ in pipeline transportations, which will be useful for the design of a $C_6F_{12}O$ fire suppressing system in the areas of aircraft, shops, libraries, etc.

Author Contributions: Conceptualization, S.L. and J.W.; methodology, Q.H. and C.Z.; data curation, Q.H., Y.C. and X.N.; validation, C.Z. and J.H.; formal analysis, Y.C. and X.N.; writing—review and editing, X.N.; supervision, J.W. and S.L. All authors have read and agreed to the published version of the manuscript.

Funding: This research was funded by the National Natural Science Foundation of China (U1933126), the National Science and Technology Major Project (J2019-VIII-0010-0171) and the Fundamental Research Funds for the Central Universities of China (WK2320000053).

Institutional Review Board Statement: Not applicable.

Informed Consent Statement: Not applicable.

Data Availability Statement: Not applicable.

Conflicts of Interest: The authors declare no conflict of interest.

References

1. Xing, H.; Lu, S.; Yang, H.; Zhang, H. Review on Research Progress of $C_6F_{12}O$ as a Fire Extinguishing Agent. *Fire* **2022**, *5*, 50. [CrossRef]
2. 3M Company. *3MTM Novec1230TM Fire Protection Fluid*; 3M Company: Saint Paul, MN, USA, 2020.
3. Mclinden, M.O.; RPerkins, A.; Lemmon, E.W.; Fortin, T.J. Thermodynamic Properties of 1,1,1,2,2,4,5,5,5-Nonafluoro-4-(trifluoromethyl)-3-pentanone: Vapor Pressure, (p, ρ, T) Behavior, and Speed of Sound Measurements, and an Equation of State. *J. Chem. Eng. Data* **2015**, *60*, 3646–3659. [CrossRef]
4. Kim, A.; Crampton, G.; Kanabus-Kaminska, M. *Performance of Novec1230 in Electronic Facility Fire Protection*; NRCC-53526; National Research Council Canada: Montreal, QC, Canada, 2010.
5. Pagliaro, J.L.; Linteris, G.T. Hydrocarbon Flame Inhibition by $C_6F_{12}O$ (Novec 1230): Unstretched Burning Velocity Measurements and Predictions. *Fire Saf. J.* **2017**, *87*, 10–17. [CrossRef]
6. Li, Y.; Zhang, X.; Tian, S. Insight into the Compatibility Between $C_6F_{12}O$ and Metal Materials: Experiment and Theory. *IEEE Access.* **2018**, *6*, 58154–58160. [CrossRef]
7. Zhang, X.; Lan, J.; Tian, S.; Rao, X.; Li, X.; Yuan, Z.; Jin, X.; Gao, S.; Zhang, X. Study of Compatibility between Eco-friendly Insulating Medium $C_6F_{12}O$ and Sealing Material EPDM. *J. Mol. Struct.* **2021**, *1244*, 130949. [CrossRef]

8. Ditch, B.D.; Rivers, P.E.; Thomas, S.D. Thermal decomposition product testing with C6F-ketone. In Proceedings of the Halon Options Technical Working Conference (HOTWC), Albuquerque, NM, USA, 24–26 April 2001; pp. 349–354.
9. Environmental Protection Agency. *Protection of Stratospheric Ozone: Listing of substitutes for Ozone-depleting Substances, Final Rule and Proposed Rule*; United State Environmental Protection Agency: Washington, DC, USA, 2003.
10. *ISO 14520-1*; Gaseous Fire Extinguishing Systems-Physical Properties and System Design, Part 5: FK-5-1-12 Extinguishant. ISO: Geneva, Switzerland, 2023.
11. National Fire Protection Association (NFPA). *Standard on Clean Agent Fire Extinguishing Systems*; National Fire Protection Association (NFPA): Quincy, MA, USA, 2001.
12. Womeldorf, B.; Mitchell, M.; Grosshandler, M. Selection of a Simulant of CF_3Br for Use in Engine Nacelle Certification Tests. In Proceedings of the Halon Options Technical Working Conference (HOTWC), Albuquerque, NM, USA, 9–11 May 1995; pp. 197–210.
13. Fan, W.; Gao, Z.; Wang, D.; Yang, Y.; Gao, Y. Study on the Flow Characteristics of Fire Extinguishing Agent FK-5-1-12 in the Pipeline Release Process. *Fire Mater.* **2022**, *46*, 376–387. [CrossRef]
14. Williamson, H.V. Halon 1301 flow in pipelines. *Fire Technol.* **1976**, *12*, 18–32. [CrossRef]
15. Elliott, D.G.; Garrison, P.W.; Klein, G.A.; Moran, K.M. *Flow of Nitrogen Pressurized Halon 1301 in Fire Extinguishing Systems*; JPL Publication: Pasadena, CA, USA, 1984.
16. Kim, J.; Baek, B.; Lee, J. Numerical analysis of flow characteristics of fire extinguishing agents in aircraft fire extinguishing systems. *J. Mech. Sci. Technol.* **2009**, *23*, 1877–1884. [CrossRef]
17. Jin, J.; Pan, R.; Li, Q.; Chen, R. Effects of Release Pressure on the Transportation Characteristics in Pipeline and the Diffusion Behaviors in Enclosure Space of Typical Gas Extinguishing Agent. *J. Loss Prevent. Proc.* **2021**, *70*, 104441. [CrossRef]
18. Bromotrifluoromethane. Available online: https://webbook.nist.gov/cgi/cbook.cgi?ID=C75638&Mask=4 (accessed on 6 December 2022).
19. Tuzla, K.; Palmer, T.; Chen, J.C.; Sundaram, R.K.; Yeung, W.S. *Development of Computer Program for Fire Suppression Fluid Flow, Report Number ITF 2000-1, Vols. I and II*; Institute of Thermo-Fluid Engineering & Science, Lehigh University: Bethlehem, PA, USA, 2000.
20. Jin, J.; Hua, M.; Yan, P.; Zhang, H.; Li, Q.; Pan, X. The Transport and Diffusion Characteristics of Superheated Fire Extinguish Agent Released via Different Nozzles in a Confined Space. *Saf. Sci.* **2020**, *129*, 104787. [CrossRef]
21. Singh, J.; Zerpa, L.E.; Partington, B.; Gamboa, J. Effect of Nozzle Geometry on Critical-subcritical Flow Transitions. *Heliyon* **2019**, *5*, e01273. [CrossRef] [PubMed]

fire

MDPI

Article

Simulation Study on Suppressing Shielded Fires by Water Mist Systems

Azad Hamzehpour *, Vittorio Verda and Romano Borchiellini

Department of Energy (DENERG), Politecnico di Torino, Corso Duca degli Abruzzi 24, 10129 Torino, Italy; vittorio.verda@polito.it (V.V.); romano.borchiellini@polito.it (R.B.)

* Correspondence: azad.hamzehpour@polito.it

Abstract: This article presents a numerical analysis of the performance of three water mist fire suppression systems, with different characteristics, on shielded fires as representing more realistic fire scenarios in an enclosure. A diesel pool fire with a peak heat release rate (HRR) value of 75 kW is covered by an obstacle above it with different shielding conditions to investigate the influence of the obstacle size and the vertical distance between the obstacle and the nozzle on the efficiency of the water mist systems. The obstacle prevents a large number of droplets from directly reaching the fuel surface and flames. The modeling and numerical analysis of this study were carried out by the fire dynamics simulator (FDS) programming tool, and the designed model was validated against the experimental data for both dry and wet tests. The results show that two of the analyzed mist sprays could successfully extinguish the shielded fires in a short time with an obstacle size of 25 cm × 25 cm and 50 cm × 50 cm, placed at two heights. However, the third mist system had a longer extinguishing time compared to the first two nozzles. All three nozzles failed to suppress the fire covered by the largest obstacle (1 m × 1 m). The shielding conditions and nozzle characteristics can affect the performance of water mist systems to some extent.

Keywords: water mist system; shielded fire; fire suppression systems; CFD simulation; FDS

Citation: Hamzehpour, A.; Verda, V.; Borchiellini, R. Simulation Study on Suppressing Shielded Fires by Water Mist Systems. *Fire* **2023**, *6*, 129. https://doi.org/10.3390/fire6040129

Academic Editors: Song Lu, Changcheng Liu, Guohui Li and Paweł Wolny

Received: 8 February 2023
Revised: 17 March 2023
Accepted: 20 March 2023
Published: 23 March 2023

1. Introduction

Due to the increasing number of fire incidents and hazards in different spaces, the development of active fire suppression systems has been at the center of attention recently. Fire suppression systems are effective tools to control and extinguish fires preventing death and injury, and severe damage to structures. Water-based fire suppression systems, including water mist and water sprinkler systems, have been developed to control fires effectively. According to NFPA 750 [1], water mist systems are classified into various types based on different criteria, including system application, nozzle type, system operation method, and system media type. Water mist systems can also be categorized as low-pressure (pressure $\leq$ 12.1 bar), intermediate-pressure (12.1 bar < pressure < 34.5 bar), and high-pressure (34.5 bar $\leq$ pressure) systems. The working pressure of the nozzles can change the mist droplet size and dominant fire-extinguishing mechanisms.

The fire extinguishing performance of water mist systems has been discussed in recent years. Although experimental fire tests (large-scale or reduced-scale experiments) are generally very costly, their obtained results are valuable and significant in order to design more effective fire suppression systems. There have been several experimental tests analyzing the performance of water mist systems in various spaces like tunnels and enclosures [2,3]. Laser-based experimental tests have been carried out to study water mist and nozzle characteristics. These techniques, such as that facilitated by a phase doppler particle analyzer (PDPA), have been developed to measure droplet size distribution and to find out the optimal droplet size [4,5]. Laser experiments have been employed to measure different characteristics, such as Sauter mean diameter (SMD, D32), velocity, cumulative volume diameter (CVM), and volumetric median diameter (VMD, Dv,50). For instance, the characteristics of a low-pressure twin-fluid water mist system were measured using a particle/droplet image analyzer (PDIA) technique to find out the optimum conditions for extinguishing small pool fires [4]. Many studies have performed other laser-based techniques to investigate the spray behavior and droplet size measurements [6–8].

In addition to the characterization studies, many efforts have been made to assess and optimize the performance of water mist systems for different fire scenarios [9–12] Yinshui et al. [13] suggested two ways to improve the suppression performance of water mist systems. The first method is to decrease the flow rate and the spray angle and increase the water discharge duration. The second way is to increase the flow rate and the spray angle and reduce the water discharge duration. Other parameters, like the ventilation condition in tunnels, can also impact the performance of water mist systems. Fan et al. [14] investigated the effect of longitudinal ventilation in a tunnel fire, and they proposed the optimal condition of a water mist system.

As mentioned before, due to the cost of experimental tests and engineering constraints, numerical tools have been developed to simulate different fire scenarios. The fire dynamics simulator (FDS) developed by the National Institute of Standards and Technology (NIST-USA) is a popular open-source computational fluid dynamics (CFD) tool in the fire science literature, and numerous research works have employed FDS to simulate different fire scenarios and water mist systems [15–17]. The FDS prediction on the effectiveness of mist application for different fire scenarios has been discussed and challenged by these researchers. The details of the FDS modeling of the current work will be demonstrated in the next section. Several papers have focused on the performance of water mist systems in extinguishing fire scenarios in different environments like tunnels [18–20]. The immediate temperature drop after using water mist systems was seen in an FDS modeling of an enclosed-space fire suppression [21]. Although there are several successfully validated models in the literature proving the capability of FDS for extinguishing simulations, Kim and Ryou [22] reported their failure to simulate the suppression stage.

In real fire scenarios, the fire can be blocked by obstacles preventing mist droplets from directly reaching flames and fuel surfaces. These obstacles above the fire can represent the ceiling of cars and trains in road or railway tunnel fires or any other kind of shields in enclosure fires. Thus, investigating the interaction between mist droplets and the shielded fire is important. Fire extinguishing mechanisms, including endothermic cooling, oxygen displacement, thermal radiation attenuation, and kinetics disturbance, are involved in suppressing class A and B fires to some extent [23]. Ferng and Liu [15] demonstrated that the dominant extinguishing mechanism for larger droplets is fuel surface cooling, and for finer droplets, oxygen displacement and evaporation are used. However, the dominant extinguishing mechanism in shielded fire suppression scenarios should be further discussed. The general fire-extinguishing mechanisms are depicted in Figure 1. There is a limited number of research studies focusing on shielded fires and suppression systems [24–26]. They have tried to assess the plume–spray thrust ratio and the interaction between the upward fire plume thrust and the downward water mist thrust defined by Alpert [27].

Our current knowledge about shielded fires and proper extinguishing systems is very limited. The shielded fire is a more realistic fire scenario, and the complex physics happening in the area of shielded fire suppression is still unknown. In the current work, three different water mist systems, including low-pressure and high-pressure single-orifice nozzles, are employed to numerically investigate their capabilities to extinguish or control the shielded fire in an enclosure using FDS. Three different obstacle sizes, two different vertical distances between the obstacle and the nozzle, and different nozzle characteristics are considered the variables of this paper. The validation of the FDS model against the experimental data and the grid sensitivity analysis are reported in detail. A total number of 23 cases are defined, and the results are evaluated in terms of the extinguishing time, the HRR evolution, and the temperature fields and the capability of water mist systems to control fires with various shielding conditions is discussed. This research work contains valuable outputs that are useful for advancing mist technology and performance-based designs.

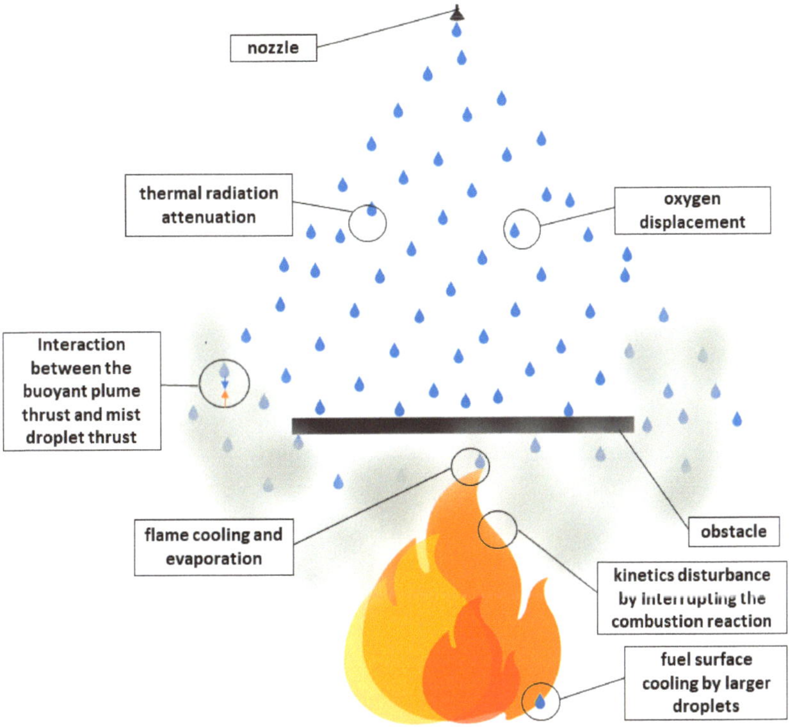

Figure 1. General fire-extinguishing mechanisms in the presence of water mist spray.

2. Methodology

2.1. Numerical Procedure in FDS

One of the most reliable tools in fire-related topics is FDS (version 6.7.9) to solve the Navier-Stokes equations for low-speed flows based on Large Eddy Simulations (LES) [28]. In order to visualize the results for postprocessing purposes, Smokeview (SMV) program is used [29]. Some scripts are also written in MATLAB and Python environments for better data visualization.

The basic governing equations in FDS, including mass, momentum, and energy conservation equations and Radiation Transport Equation (RTE), are coupled to solve the fire-mist problem. It should be noted that interested readers can refer to the FDS technical reference guide [30] for detailed mathematical modeling of FDS and formulations. In addition to the above-mentioned equations, the initial and boundary conditions must be defined, and the equation of state should be added to make the complete set of equations. It is worth mentioning that in FDS, air, fuel, and products consisting of primitive species are considered lumped species to formulate combustion. Moreover, a pressure zone is defined in this study by considering a bigger computational domain, and the external boundaries are selected as 'OPEN'. For air extraction, the 'EXHAUST' vent is considered as the volumetric flow of the air blower is known. The general steps of the numerical procedure of this study are depicted in Figure 2.

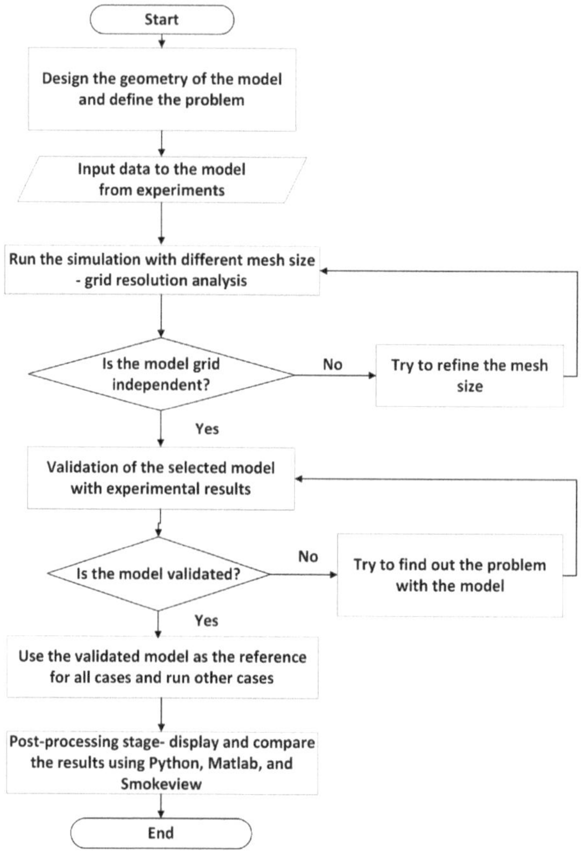

Figure 2. Flow chart of the numerical procedure in FDS for the present study.

According to [28], there are three basic principles for simulating the water suppression phenomenon in FDS, namely, transporting the water droplets through the air, tracking the water along the solid surface, and predicting the reduction of the burning rate. The local mass loss rate of the fuel and, consequently, HRR in the suppression model of FDS is formulated as follows [28]:

$$\dot{m}''_f(t) = \dot{m}''_{f,0}e^{-\int k(t)dt} \tag{1}$$

where $\dot{m}''_{f,0}$ is the mass loss rate per unit area in case of dry test, and $k(t)$ is a function of the water mass per unit area ($\dot{m}''_w$) and is defined as follows:

$$k(t) = \alpha\dot{m}''_w(t) \tag{2}$$

The coefficient α is obtained experimentally or through sensitivity analysis. In FDS, an 'e_coefficient' can be introduced to see the reduction in the burning rate in HRR curves after nozzle activation.

The extinction and evaporation models in FDS and their formulations are demonstrated here and can also be found in [28,30]. In order to simulate the flame extinction using water mist droplets, two extinction models, namely 'EXTINCTION 1' and 'EXTINCTION 2' are introduced in FDS. The 'EXTINCTION 1' model is based on oxygen concentration, whereas the 'EXTINCTION 2' model is based on both fuel and oxygen concentration. In the current work, 'EXTINCTION 2' is implemented into the model. In this model, the

Critical Flame Temperature (CFT—T_{OI}) is of importance, and if the temperature of a cell goes below the CFT, the fire is suppressed. The CFT can be calculated as follows:

$$T_{OI} = T_0 + X_{OI}\frac{\Delta H_c/r}{n\overline{c_p}} \tag{3}$$

where T_0, X_{OI}, $\Delta H_c/r$, n, and $\overline{c_p}$ are the initial temperature of the fuel/air mixture, the limiting oxygen volume fraction, the heat of combustion per mole of oxygen consumed, the number of moles of products of combustion per mole of fuel/air mixture, and the average heat capacity of products of combustion, respectively [28].

Additionally, in order to capture the evaporation effect of water mist droplets on the fire suppression process, it is critical to understand and define the evaporation model in FDS. The heat and mass transfer in this model can be estimated as follows [31]:

$$\frac{dm_p}{dt} = -A_{p,s}h_m\rho_f(Y_{\alpha,l} - Y_{\alpha,g}) \tag{4}$$

$$\rho_g V\frac{dY_{\alpha,g}}{dt} = -(1 - Y_{\alpha,g})\frac{dm_p}{dt} \tag{5}$$

$$\frac{dT_p}{dt} = \frac{1}{m_p c_p}[\dot{q}_r + A_{p,s}h_g(T_g - T_p) + A_{p,s}h_w(T_w - T_p) + \frac{dm_p}{dt}h_v] \tag{6}$$

$$\frac{dT_g}{dt} = \frac{1}{m_g c_g}[A_{p,s}h_g(T_p - T_g) + \frac{dm_p}{dt}(h_{\alpha,p} - h_{\alpha,g})] \tag{7}$$

$$\frac{dT_w}{dt} = \frac{A_{p,s}h_w}{m_w c_w}(T_w - T_p) \tag{8}$$

where the subscripts α, g, and p refer to the gas species, the average of the quantity in the cell occupied by the droplet, and the liquid droplet, respectively. $Y_{\alpha,l}$ is the liquid equilibrium vapor mass fraction, $Y_{\alpha,g}$ represents the local gas phase vapor mass fraction, T_p stands for the droplet temperature, T_g is the local gas temperature, m_p displays the droplet mass, c_p stands for the specific heat of the liquid, $A_{p,s}$ is the surface area of the liquid droplet, h_m shows the mass transfer coefficient, $\dot{q}_r$ is the droplet radiative heating rate, ρ_g represents the gas density, $h_{\alpha,g}$ is the vapor-specific enthalpy, c_g shows the gas-specific heat, h_g is the heat transfer coefficient between the droplet and the gas, $Y_{\alpha,g}$ is the vapor mass fraction of the gas, m_g is the mass of the local gas, ρ_f denotes the density of the particle film, m_w, h_w, and c_w represent the mass of the first node of the solid, the heat transfer coefficient between the droplet and the solid, and the solid specific heat, respectively. It should be noted that Equation (8) is employed when the droplets hit a wall surface.

After demonstrating the mathematical and physical explanations of the model, the geometry and computational domain of the model are defined. The details about the simulation process are presented in the next section.

2.2. Simulation Model

The dimensions and the properties of the compartment, and the fuel and material characteristics are chosen according to the data presented by Jenft et al. [32]. The size of the room is $4.20 \times 4.30 \times 3.05$ m^3, and the material properties are given in Table 1. The diesel pool fire is defined by introducing a steel pan in the middle of the compartment with the size of $30 \times 30 \times 10$ cm^3. The properties of the diesel are mentioned in Table 2. The pyrolysis is introduced to the model by defining a specific heat release rate per unit area (HRRPUA—kW/m^2) and a peak HRR value of 75 kW. The HRR evolution is also extracted from [32], and it is implemented into the model using a time ramp via the 'RAMP_Q' function in FDS. The geometry of the FDS model representing the walls, the nozzle, the pool fire, and the obstacle is displayed in Figure 3a, and the cross-section of the model showing the grid can be seen in Figure 3b.

Table 1. Material properties.

Material	Conductivity (k) W/mk	Specific Heat (c_p) j/kgk	Density (ρ) kg/m^3
Concrete	1.575	1000	2100
glass	1	750	2500
steel	50	450	7800
wood	0.13	1600	500

Table 2. Fuel properties.

	Common Formula	Heat of Combustion (ΔH_c) kj/kg	Soot Yield kg/kg
Diesel	$C_{12}H_{23}$	42,200	0.059

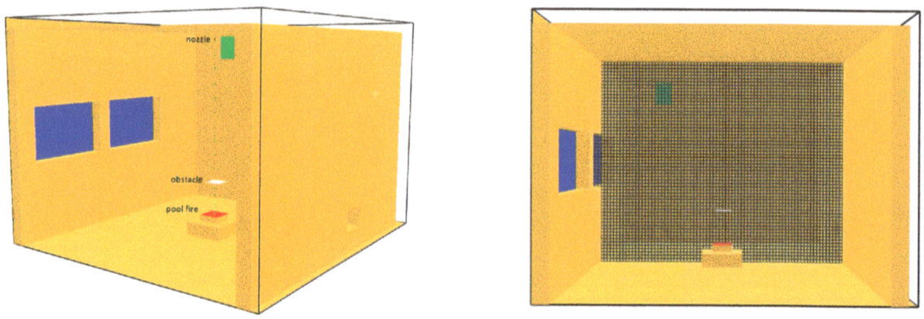

(a) (b)

Figure 3. (**a**) Snapshot of the model geometry in FDS and the (**b**) XZ plane of the grid.

Three single-orifice water mist systems with different nozzle characteristics, including the operating pressure, the droplet size distribution, and the cone angle, were used in this study. In order to analyze the effectiveness of water mist systems to control and extinguish the shielded diesel fire, a variety of input and output parameters were considered. The variables include the size of the obstacle, the vertical distance between the obstacle and the nozzle, and the characteristics of the nozzle. The characteristics of the nozzles can be seen in Table 3. In this paper, a total number of 23 different cases have been defined, as shown in Table 4. It is worth mentioning that a useful way to predict the movement of liquid droplets on horizontal surfaces, like shielded fire cases, is to employ the 'ALLOW_UNDERSIDE_PARTICLES=T' function in FDS [28]. The obstacle sizes include 25 cm × 25 cm (Obstacle 1), 50 cm × 50 cm (Obstacle 2), and 1 m × 1 m (Obstacle 3) at two heights: 800 mm (H1) and 1500 mm (H2) above the floor. In Table 4, different categories are defined with respect to the nozzle and height numbers; for instance, N1H1 refers to cases in which nozzle 1 is used, and the obstacles are located at a height of 800 mm above the floor. It should be noted that the HRR value of all cases is 75 kW.

Table 3. Nozzle characteristics of water mist systems.

Nozzle 1	Nozzle 2	Nozzle 3
D = 46 μm Operating pressure = 100 bar Flow rate = 11.9 L/min Velocity = 10 m/s Cone angle = 0–48° K factor = 1.9 L/min/bar$^{1/2}$	D = 124.6 μm Operating pressure = 10 bar Flow rate = 22.8 L/min Velocity = 42.5 m/s Cone angle = 0–90° K factor = 7.2 L/min/bar$^{1/2}$	D = 112 μm Operating pressure: 10 bar Flow rate: 6.3 L/min Nozzle exit diameter: 0.0008 m half injection angle: 65° K factor: 1.9 L/min/bar$^{1/2}$

Table 4. Defined simulation cases.

Category	Case No.	Obstacle No.	Distance between Obstacle and Floor	Nozzle Number	Activation Time
Validation	I—dry validation	-	-	-	-
	II—wet validation	-	-	3	540 s
	III	-	-	1	75 s
N1H1	IV	1	H1	1	75 s
	V	2	H1	1	75 s
	VI	3	H1	1	75 s
	VII	-	-	2	75 s
N2H1	VIII	1	H1	2	75 s
	IX	2	H1	2	75 s
	X	3	H1	2	75 s
	XI	-	-	3	75 s
N3H1	XII	1	H1	3	75 s
	XIII	2	H1	3	75 s
	XIV	3	H1	3	75 s
N1H2	XV	1	H2	1	75 s
	XVI	2	H2	1	75 s
	XVII	3	H2	1	75 s
N2H2	XVIII	1	H2	2	75 s
	XIX	2	H2	2	75 s
	XX	3	H2	2	75 s
N3H2	XXI	1	H2	3	75 s
	XXII	2	H2	3	75 s
	XXIII	3	H2	3	75 s

3. Results and Discussion

3.1. Grid Sensitivity Analysis

The aim of the grid resolution study is to find out a trade-off between the accuracy of the model and the computational cost in which the cell size does not affect the results significantly. The characteristic fire diameter is calculated as the following equation, suggested by the FDS user's guide [28]:

$$D^* = \left(\frac{\dot{Q}}{\rho_\infty c_p T_\infty \sqrt{g}}\right)^{2/5} \tag{9}$$

where $\dot{Q}$ is HRR, and ρ_∞, c_p, *and* T_∞ are the density, the specific heat, and the temperature of the ambient, respectively. The nondimensional quantity $\frac{D^*}{\delta x}$ is then applied to define the proper grid size for buoyant plume simulations, where δx is the nominal size of a mesh cell. In the current study, the sensitivity analysis was carried out considering three cases with the same characteristics but different mesh sizes, including coarse, moderate, and fine, with the total number of grid cells varying from about 300,000 to almost 1,700,000. The average temperature distribution along a vertical line that is lateral to the fire (from the floor to the ceiling) between 50 s and 150 s after ignition is illustrated in Figure 4. It should be noted that all three sets of meshes are selected based on an acceptable range of cell numbers for this study.

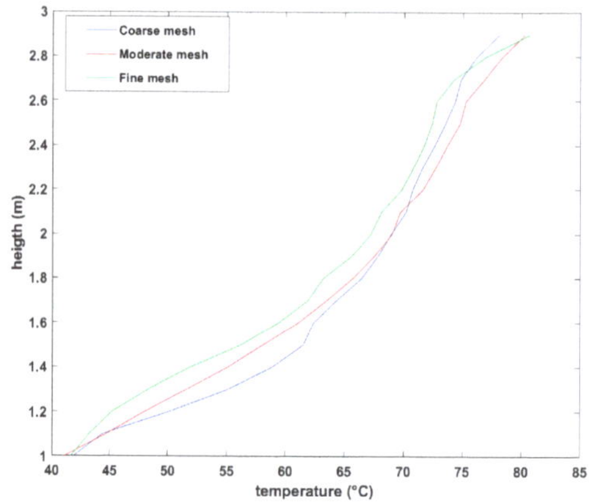

Figure 4. Temperature comparison between three cases for sensitivity study.

The temperature deviation between the moderate and the fine meshes, which follow the same trend, does not go beyond 7%. However, this percentage for the coarse mesh is 13% compared to the fine mesh. The deviation of the quantities between the moderate mesh and the fine mesh is not significant. Therefore, the moderate mesh was selected for further studies, which has reasonable accuracy and an acceptable computational time. The selected model has a single mesh with a total number of approximately 500,000 cells.

3.2. Model Validation

3.2.1. Dry Test Validation

The simulation results of the basic case (case I—no water mist application) were compared with the experimental data of Reference [32] for the diesel pool fire dry test validation with a peak HRR value of 75 kW. The comparison of the results for O_2 concentration near the exhaust fan is displayed in Figure 5a. The average difference between the results is below 1%, and the simulation outputs are perfectly matched with the experimental data for this quantity. Moreover, the temperature was recorded at the heights of 50 cm, 150 cm, 250 cm, and 290 cm on a thermocouple tree with 140 cm offset from the central axis of the room in the corner. From Figure 5b, it can be seen that the simulation results are in good agreement with the data from the experiments; however, the FDS model generally overpredicts the temperature values in the corner axis at the height of 50 cm by a maximum difference of 20%. This percentage decreases at a higher altitude, from 150 cm to 290 cm, by a maximum deviation of 10%. The maximum temperature discrepancy between the results does not exceed 18 °C.

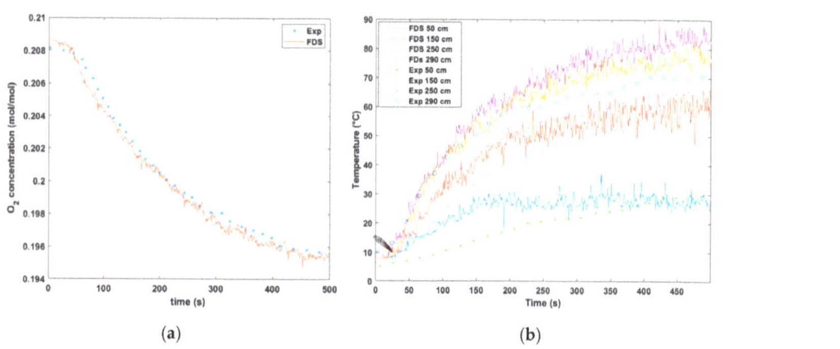

Figure 5. Comparison of (**a**) O_2 concentration and (**b**) temperature evolution between FDS and experimental data [32] (dry test).

3.2.2. Wet Test Validation

In addition to the dry test validation, the wet test validation (using a water mist system) was also carried out using experimental data. In this regard, the water mist system, with the same characteristics as mentioned in Reference [32] (see Table 3—nozzle 3), was introduced into the model, and the outputs were compared with the experimental data.

For the wet test validation, the simulation time was set to 700 s, and the nozzle activation time was 540 s after ignition (case II). As can be seen in Figure 6, the comparisons can be analyzed in two parts: before and after nozzle activation. Before nozzle activation, the deviation in the results for both temperature and oxygen concentration values is small. However, the difference between the results is higher after nozzle activation. The FDS model showed an immediate increase in O_2 concentration after mist activation. However, the maximum deviation does not go beyond 5%, with an average difference of below 3%. This shows a good match between the results. Qualitatively speaking, the temperature decreases for both the experimental and simulation measurements follow the same trend after nozzle activation in the wet test, but this reduction is sharper for FDS modeling. Generally, FDS overestimates the impact of the mist droplets on the reduction in temperature. The same behavior and observation (for both dry and wet validations) were seen in the FDS model of Reference [32], and the complete explanation was also reported in that article. Considering the delay in data logging and experimental bias, the validation results are satisfactory for the wet test. After validation, the current model can be used to further study the performance of water mist systems and the shielded fire suppression process.

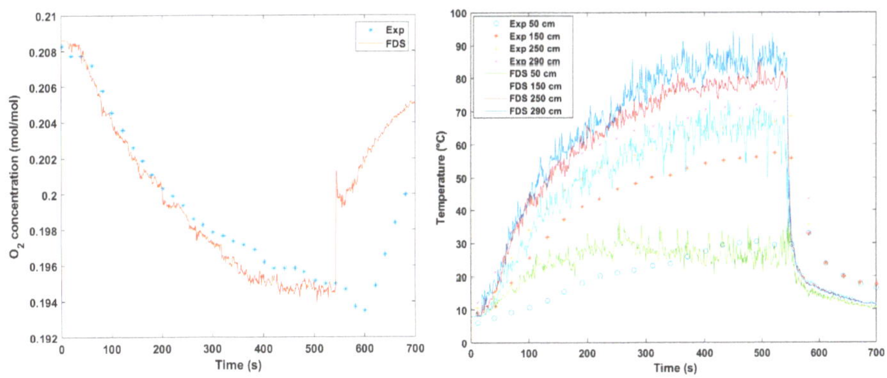

(a) (b)

Figure 6. Comparison of (**a**) O_2 concentration and (**b**) temperature between FDS and experimental data [32] (wet test).

3.3. HRR and Temperature Fields

The HRR evolution and the designed fire curve are defined according to [32] in FDS. For the early application of the water mist, the nozzles are activated 75 s after ignition (the HRR at the time of activation is around 50 kW—the fire is not fully developed). For cases with different obstacle scenarios, the HRR and temperature curves are plotted to compare the evolution after nozzle activation. In this study, the criterion for comparing the extinguishing time for any case is the time when HRR becomes zero.

In Figure 7a, the HRR comparison for cases III to VI (nozzle 1) is illustrated. The only case where the water mist system was unable to suppress the pool fire completely is case VI, in which the obstacle is 1 m × 1 m in size and is located 800 mm (H1) above the floor. In this case, the pool fire is completely covered by the obstacle, and the droplets are unable to reach the fire plume or the fuel surface. In case IV, the water mist system was able to extinguish the fire around 7 s after nozzle activation. Cases III and IV represent almost the same behavior, where the burning rate decreases to zero sharply after activation. In case IV, the obstacle is the smallest, and the mist droplets could reach the flame and the fuel surface. The longest extinguishing time among the suppressed cases belongs to case V, in which the mist nozzle extinguished the fire after almost 37 s. In this case, although the fire was blocked by the obstacle, some droplets could bypass the obstacle and reach the flame.

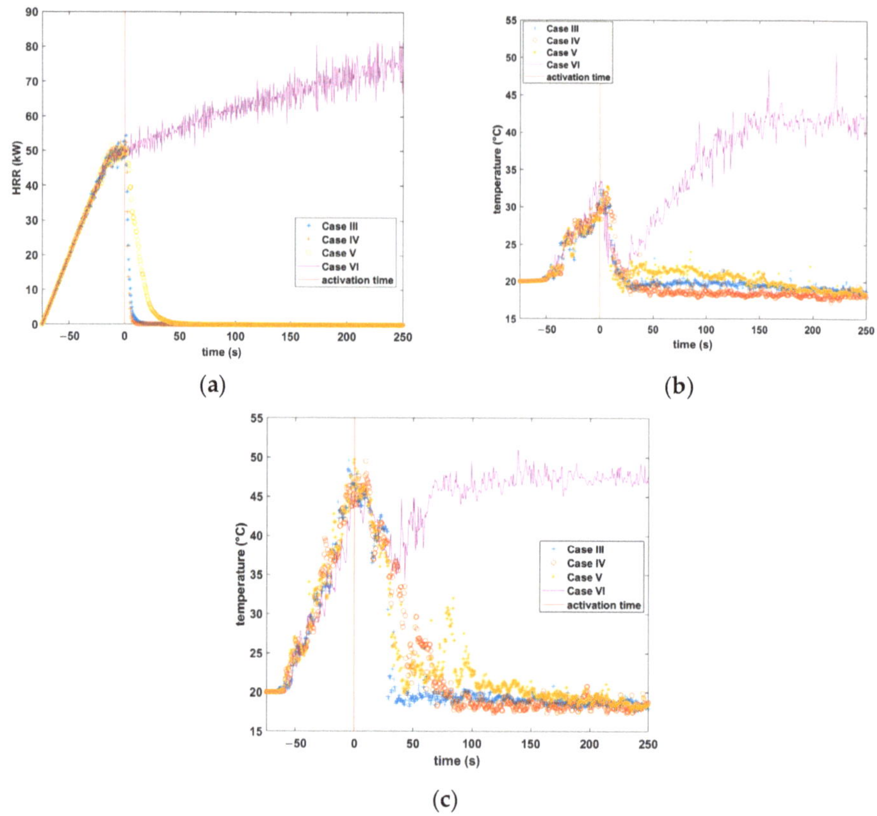

Figure 7. (**a**) HRR evolution for cases III to VI; (**b**) temperature evolution in the corner at the height of 50 cm for cases III to VI; (**c**) temperature evolution in the corner at the height of 290 cm for cases III to VI.

The temperature evolution of different cases is also illustrated. The temperature values were obtained in three thermocouple trees: one in the center axis of the room above the fire and the other two in the corners, 1 m away from the central axis. Only the temperature evolution in the corners is shown here. The temperatures were measured at different heights, from 50 cm above the floor up to 290 cm. Figure 7b,c show the temperature evolution for cases using the early application of nozzle 1. The temperature first decreased with nozzle activation in case VI (the largest obstacle) but then increased rapidly up to 45 °C. In other cases, the temperature went down sharply and stayed at a constant value until the end of the simulation. The temperature was controlled successfully for the extinguished cases after the activation of nozzle 1. The flame temperature plots are not shown here; however, as the fire was suppressed quickly in cases III to V, the flame temperature at the thermocouple tree above the fire (from 30 cm to 50 cm above the fire) dropped to the ambient temperature after activation. On the other hand, the near flame temperature for case VI (failed suppression case) did not change significantly and kept fluctuating up to 800 °C for the closest thermocouples to the fuel surface.

Figure 8a shows the HRR evolution for four cases using the second nozzle. As can be seen, the water mist system was unsuccessful in suppressing the fire where the obstacle was the largest (case X). However, the mist system could extinguish the shielded fire in other cases. The extinguishing time for cases VII and VIII is below 10 s, but this time increased to around 45 s when the size of the obstacle increased to 50 cm × 50 cm. Figure 8b,c represent the temperature evolution for cases VII to X at 50 cm and 290 cm above the floor in the corner of the compartment. In the failed extinguishment case, the temperature in the corner first decreased after nozzle activation, then increased after a few seconds. The temperature was completely controlled in the other cases.

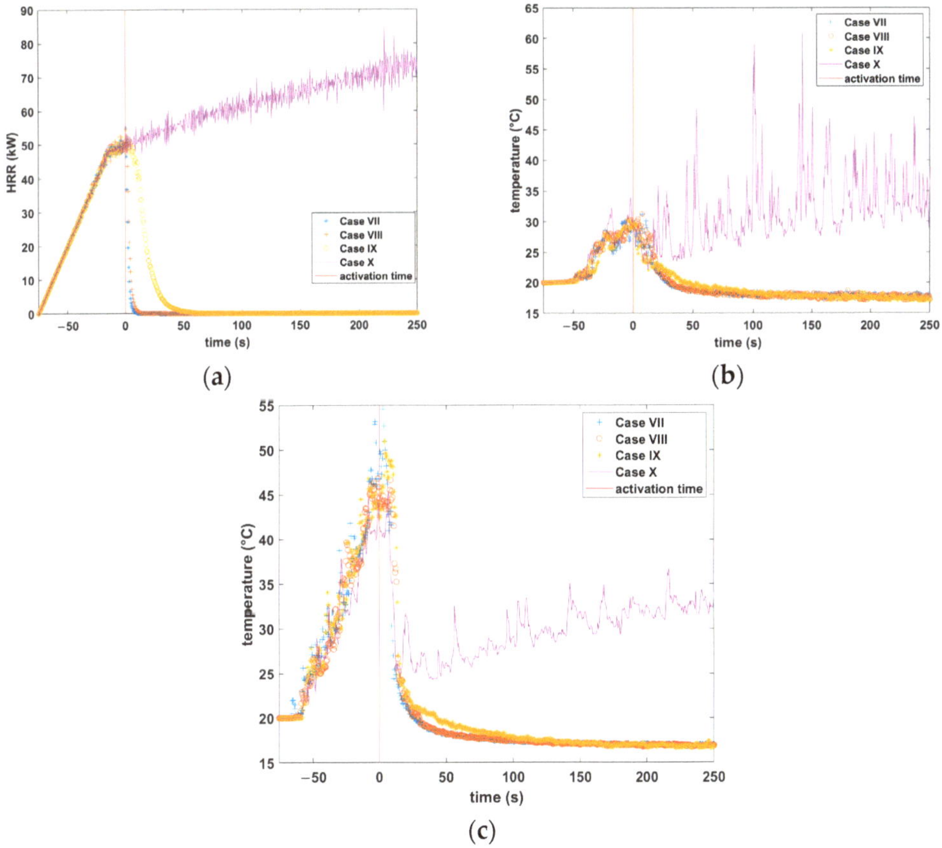

Figure 8. (**a**) HRR evolution for cases VII to X; (**b**) temperature evolution on the corner at the height of 50 cm for cases VII to X; (**c**) temperature evolution on the corner at the height of 290 cm for cases VII to X.

The same visualizations are illustrated for cases XI to XIV using nozzle 3 in Figure 9. For this nozzle, the HRR and temperature distributions also follow the same trends as nozzles 1 and 2. However, the extinguishing time is almost 5 to 10 times the extinguishing time of those cases using the first nozzle with the same k-factor. Nozzle 3 was capable of controlling far-field temperatures at the corners, which is almost the same as the other two nozzles, thanks to its high cone angle (130°). The nozzle acts as a mist curtain around the fire leading to a temperature drop, although the suppression process is slow compared to nozzles 1 and 2. Although the working pressure and the droplet size of this nozzle are almost the same as nozzle 2, the suppression time is longer due to the lower flow rate.

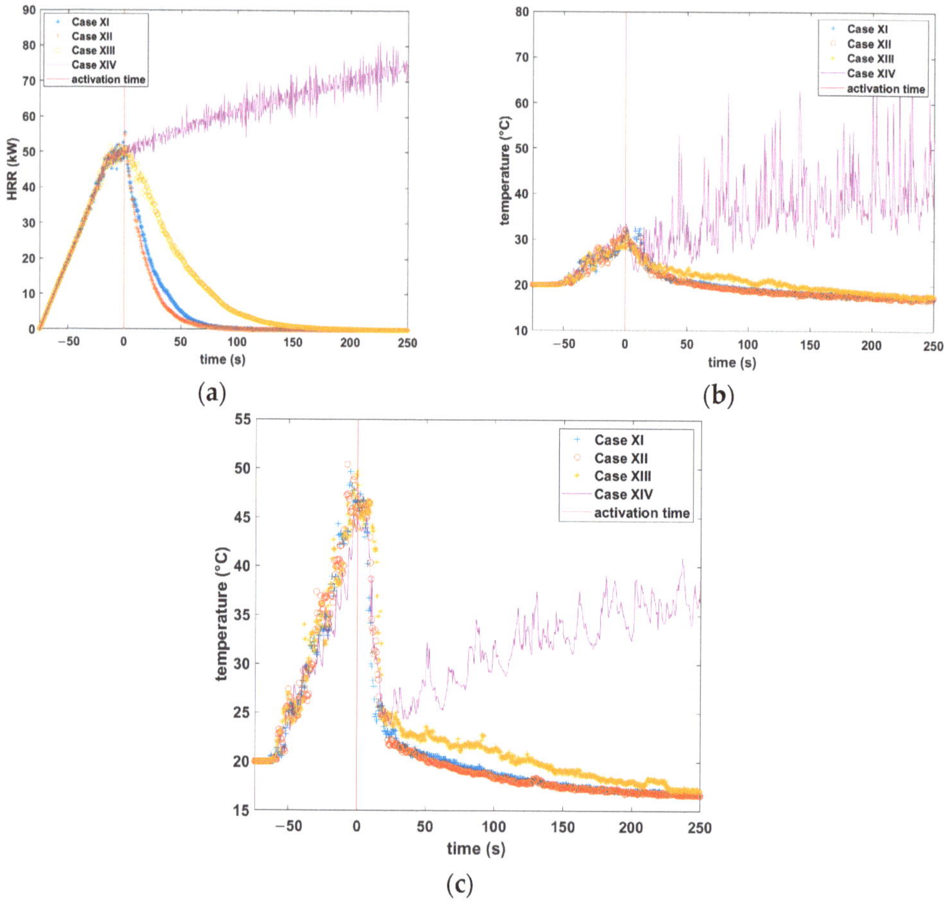

Figure 9. (**a**) HRR evolution for cases XI to XIV; (**b**) temperature evolution on the corner at the height of 50 cm for cases XI to XIV; (**c**) temperature evolution on the corner at the height of 290 cm for cases XI to XIV.

The results for the HRR values of cases XV to XXIII are shown in Figure 10. In order to avoid repetition, only the HRR evolutions are displayed in this figure. In cases XV to XXIII, the obstacle is placed at a height of 1500 mm (H2) above the floor. The water mist system successfully suppressed the fire almost immediately for cases XV and XVI (Figure 10a). For case XVII, nozzle 1 failed to control the fire. Similar to the previous cases, the simulations were carried out for cases XVIII to XX using the second nozzle. As can be seen, nozzle 2 was also unable to suppress the shielded fire when the obstacle was the largest (case XX—Figure 10b). With the use of nozzle 3 (case XXI–XXIII), the same suppression behavior can be seen (Figure 10c). When the obstacle is placed at a height of 1500 mm, the extinguishing time is generally shorter compared to the H1 cases.

In order to compare the nozzles more clearly in terms of the extinguishing time with different shielding conditions, the suppression time of the successful cases is shown in Figure 11. The extinguishing time in case IX is almost 8 s longer than the time in case IV (considering the time that the HRR becomes almost zero). In this regard, the first nozzle with finer droplets and higher pressure excelled over nozzle 2 in the shielded fire application, with an obstacle size of 50 cm × 50 cm. In cases VII and VIII, the extinguishing time is less than 10 s. Nozzles 1 and 2 performed almost the same for those cases in which there was no obstacle and when the obstacle size was the smallest (25 cm × 25 cm).

However, the extinguishing time is almost 3 to 11 times longer than nozzles 1 and 2. When the obstacle is closer to the nozzle (H2), the droplets have a better chance to bypass the obstacle, therefore reaching the flames. It should be noted that if the obstacle is very close to the fuel surface or to the nozzle, the fire can be fully covered, preventing mist droplets from reaching the flames or bypassing the horizontal surface. Even though in such scenarios, water mist droplets, especially finer droplets, can occupy the space around the shielded fire, the spray momentum is not sufficient to overcome the buoyant fire plume momentum and, consequently, there will be no penetration to the fire plume and flames.

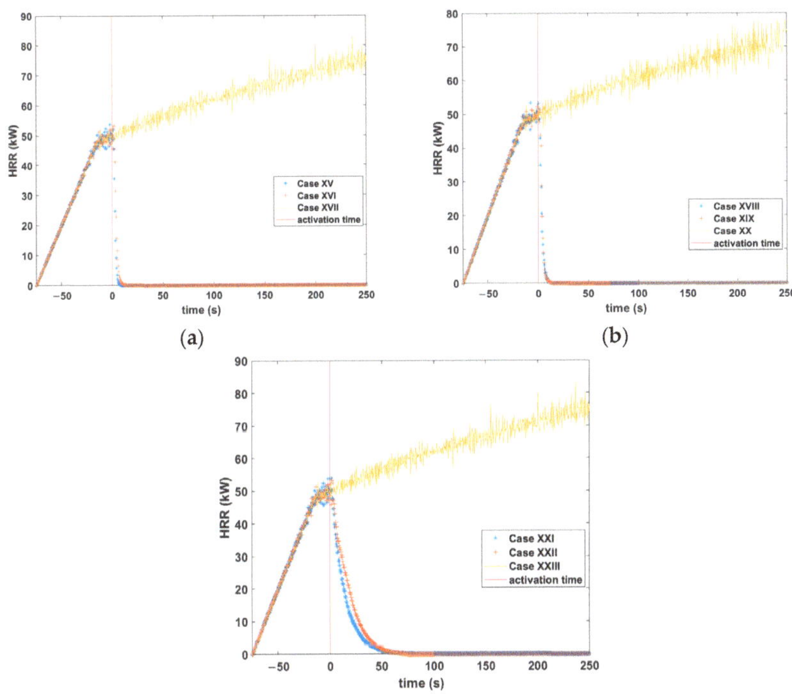

(c)

Figure 10. (**a**) HRR evolution for cases XV to XVII; (**b**) HRR evolution for cases XVIII to XX; (**c**) HRR evolution for cases XXI to XXIII.

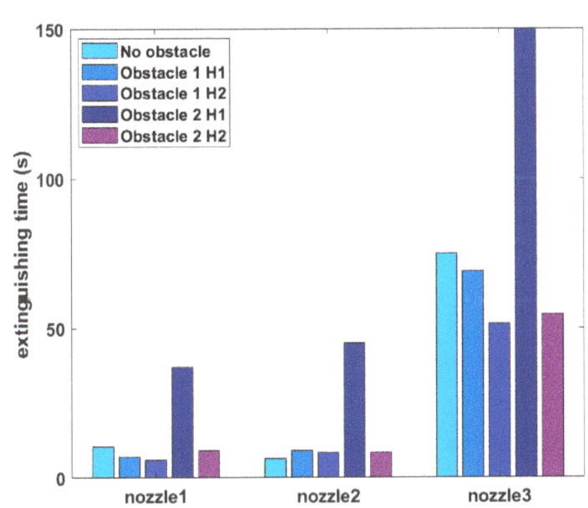

Figure 11. Comparison of extinguishing times in the successfully suppressed cases.

As depicted before (Figure 1), there are different fire-extinguishing mechanisms involved in the suppression process facilitated by the water mist systems. In shielded fire suppression, the dominant extinguishing mechanisms can be diverse with respect to the shielding condition. When the fire is fully blocked by an obstacle (obstacle 3 in this study) even large droplets cannot penetrate the fuel surface or flames; therefore, the fire can only be controlled through thermal radiation attenuation or oxygen displacement. As mentioned before, all three nozzles failed to extinguish the fire in the existence of the largest obstacle however, the temperature in the corners of the enclosure decreased. In the shielded fire suppression regarding obstacles 1 and 2, the fire is not fully covered, and larger droplets have a chance to penetrate the fire plume and extract heat from the flames so that the smaller droplets can bypass the obstacle, leading to endothermic cooling, oxygen displacement and radiation attenuation.

Figure 12a,b show the velocity contour plots at the plane y = 2.15 m in the middle of the compartment for case V, as an example. For the velocity contours, one plot displays 1 s before water mist activation (Figure 12a), and the other one shows 1 s after activation (Figure 12b) in order to compare the effect of the water droplets on the velocity field. As can be observed clearly, the velocity is the highest in the middle, close to the nozzle after activation. The obstacle above the fire prevents most of the droplets from directly reaching the fuel's surface, as shown in the velocity contour plot after activation.

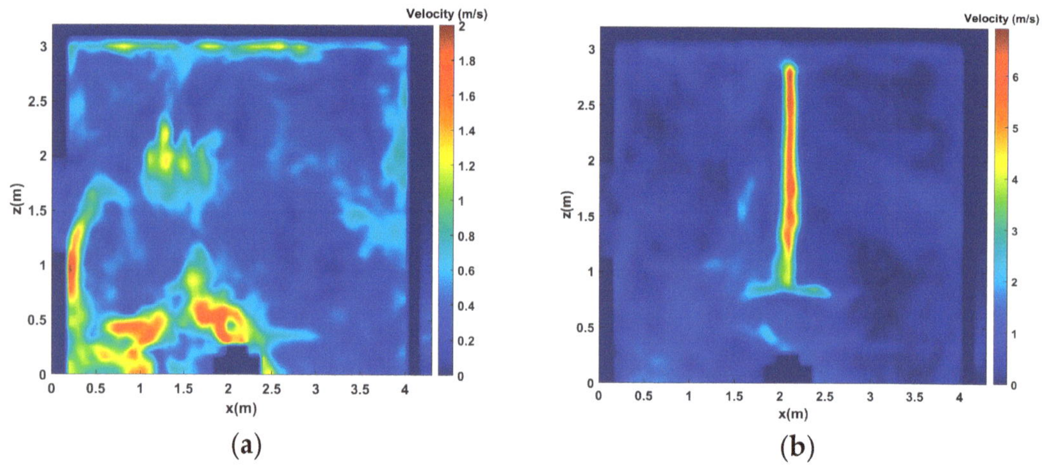

Figure 12. Velocity contour plots for case V (**a**) at t = 75 s, just before activation, and (**b**) at t = 76 s, 1 s after activation.

For the temperature contours, six different plots at the plane y = 2.15 m are shown for case V at different times, including 1 s before activation (Figure 13a) and 1 s (b), 5 s (c), 10 s (d), 25 s (e), and 100 s (f) after activation. It can be seen that, for this fully extinguished case, the flame almost disappeared 25 s after spray activation (the predicted extinguishing time for this case is around 37 s), and the temperature in the upper layers close to the ceiling was higher after suppression. Additionally, the maximum temperature dropped from above 800 °C to around 70 °C 25 s after nozzle activation.

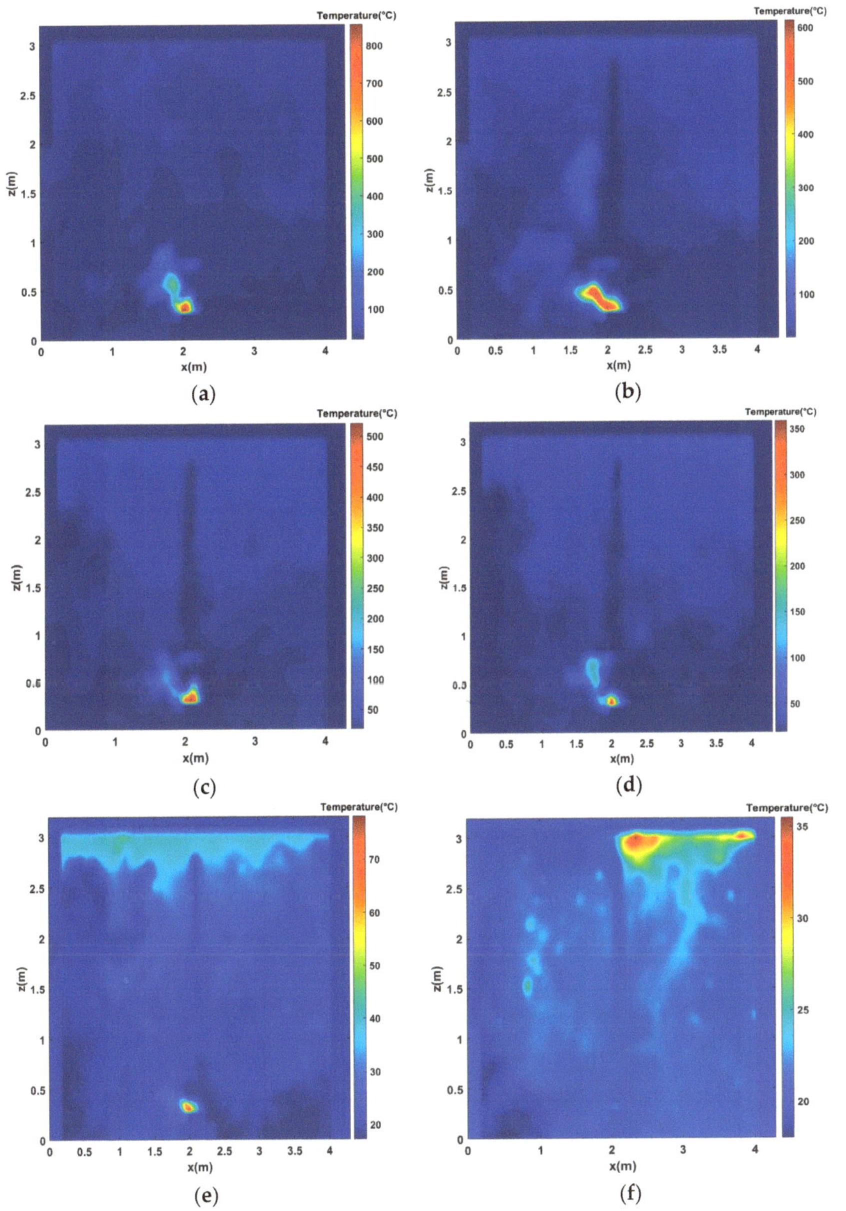

Figure 13. Temperature contour plots for case V (**a**) at t = 75 s, just before activation, (**b**) at t = 76 s, just after activation, (**c**) at t = 80 s, 5 s after activation, (**d**) at t = 85 s, 10 s after activation, (**e**) at t = 100 s, 25 s after activation, and (**f**) at t = 175 s, 100 s after activation.

4. Conclusions

In this study, the performance of three single-orifice water mist systems with different characteristics on extinguishing shielded fires in an enclosure was investigated using FDS simulations. The impact of the shielding condition, including three obstacle sizes and two vertical distances from the nozzle, on the shielded fire suppression behavior was thoroughly assessed. A 4.20 m × 4.30 m × 3.05 m enclosure was modeled in FDS, and 23 different cases were defined for comparisons. The diesel pool fire with an HRR peak value of 75 kW was placed in the middle of the compartment. The following conclusions can be drawn:

- The maximum deviation in O_2 concentration between the FDS predictions and the experimental measurements was 1% for the dry test and less than 5% for the wet test. Moreover, the discrepancy in the temperature values did not exceed about 18 °C in both cases;
- The first two nozzles (nozzles 1 and 2) were able to suppress the fire completely in the absence of obstacles in a very short time. However, all three nozzles failed to suppress the shielded fire where the obstacle size was 1 m × 1 m, located at any distance from the nozzle;
- The high-pressure spray system performed better in terms of the extinguishing time of the shielded fire compared to nozzles 2 and 3, where the obstacle sizes were 25 cm × 25 cm and 50 cm × 50 cm. However, the suppression time using nozzle 2 (low-pressure) was close to nozzle 1 (high-pressure) due to the higher flow rate;
- Although nozzles 2 and 3 had the same pressure (10 bar) and almost the same droplet size, nozzle 3 had a longer extinguishing time due to a lower flow rate;
- In the successful cases of extinguishment, the temperature inside the enclosure decreased sharply after nozzle activation;
- The obstacle size, its distance from the nozzle, and the nozzle characteristics are critical parameters in the study of shielded fire suppression;
- When the fire is not fully covered by the obstacle, a portion of droplets can penetrate the flame by overcoming the plume thrust and bypassing the obstacle.

The authors recommend that researchers perform more experimental tests and simulations to investigate different water mist systems in the case of shielded fires and to analyze the fire extinguishing mechanisms. Moreover, the nozzle application time (early or late application) as an effective parameter can be taken into consideration.

Our future work will focus on experimental shielded fire suppression tests with different shielding conditions to further develop the FDS models.

Author Contributions: Conceptualization, V.V. and R.B.; methodology, A.H. and V.V.; software, A.H.; validation, A.H.; formal analysis, A.H.; writing—original draft preparation, A.H.; writing—review and editing, V.V. and R.B.; visualization, A.H.; supervision, V.V. and R.B.; project administration, V.V. and R.B.; funding acquisition, V.V. and R.B.; All authors have read and agreed to the published version of the manuscript.

Funding: This research received no external funding.

Institutional Review Board Statement: Not applicable.

Informed Consent Statement: Not applicable.

Data Availability Statement: Not applicable.

Acknowledgments: Computational resources were provided by HPC@POLITO, a project of Academic Computing within the Department of Control and Computer Engineering at Politecnico di Torino (http://www.hpc.polito.it).

Conflicts of Interest: The authors declare no potential conflict of interest.

References

1. *NFPA 750*; Standard on Water Mist Fire Protection Systems. National Fire Protection Association: Quincy, MA, USA, 2006.
2. Li, Q.; Tang, Z.; Fang, Z.; Yuan, J.; Wang, J. Experimental study of the effectiveness of a water mist segment system in blocking fire-induced smoke and heat in mid-scale tunnel tests. *Tunn. Undergr. Space Technol.* **2019**, *88*, 237–249. [CrossRef]
3. Gupta, M.; Pasi, A.; Ray, A.; Kale, S.R. An experimental study of the effects of water mist characteristics on pool fire suppression. *Exp. Therm. Fluid Sci.* **2013**, *44*, 768–778. [CrossRef]
4. Tian, G.; Li, H.; Xu, H.; Li, Y.; Raj, S.M. Spray characteristics study of DMF using phase doppler particle analyzer. *SAE Int. J. Passeng. Cars Mech. Syst.* **2010**, *3*, 948–958. [CrossRef]
5. Wang, Z.; Wang, X.; Huang, Y.; Tao, C.; Zhang, H. Experimental study on fire smoke control using water mist curtain in channel. *J. Hazard. Mater.* **2018**, *342*, 231–241. [CrossRef]

6. Jeong, C.S.; Lee, C.Y. Experimental investigation on spray characteristics of twin-fluid nozzle for water mist and its heptane pool fire extinguishing performance. *Process Saf. Environ. Prot.* **2021**, *148*, 724–736. [CrossRef]
7. Husted, B.P.; Petersson, P.; Lund, I.; Holmstedt, G. Comparison of PIV and PDA droplet velocity measurement techniques on two high-pressure water mist nozzles. *Fire Saf. J.* **2009**, *44*, 1030–1045. [CrossRef]
8. Ditch, B.; Yu, H.-Z. Characterization of Water Mist Sprays Using a Phase-Doppler-Particle-Analyzer and an Iso-Kinetic Sampling Probe. In Proceedings of the ASME 2004 Heat Transfer/Fluids Engineering Summer Conference, Charlotte, NC, USA, 11–15 July 2004; pp. 221–230. [CrossRef]
9. Qin, J.; Chow, W.K. Experimental Data on Water Mist Suppression. *Procedia Eng.* **2013**, *62*, 868–877. [CrossRef]
10. Santangelo, P.E.; Tarozzi, L.; Tartarini, P. Full-Scale Experiments of Water-Mist Systems for Control and Suppression of Sauna Fires. *Fire* **2022**, *5*, 214. [CrossRef]
11. Yang, P.; Liu, T.; Qin, X. Experimental and numerical study on water mist suppression system on room fire. *Build. Environ.* **2010**, *45*, 2309–2316. [CrossRef]
12. Zhou, Y.; Bu, R.; Zhang, X.; Fan, C.; Gong, J. Performance evaluation of water mist fire suppression: A clean and sustainable fire-fighting technique in mechanically-ventilated place. *J. Clean. Prod.* **2019**, *209*, 1319–1331. [CrossRef]
13. Yinshui, L.; Zhuo, J.; Dan, W.; Xiaohui, L. Experimental research on the water mist fire suppression performance in an enclosed space by changing the characteristics of nozzles. *Exp. Therm. Fluid Sci.* **2014**, *52*, 174–181. [CrossRef]
14. Fan, C.; Bu, R.; Xie, X.; Zhou, Y. Full-scale experimental study on water mist fire suppression in a railway tunnel rescue station: Temperature distribution characteristics. *Process Saf. Environ. Prot.* **2021**, *146*, 396–411. [CrossRef]
15. Ferng, Y.-M.; Liu, C.-H. Numerically investigating fire suppression mechanisms for the water mist with various droplet sizes through FDS code. *Nucl. Eng. Des.* **2011**, *241*, 3142–3148. [CrossRef]
16. Wang, Z.; Wang, W.; Wang, Q. Optimization of water mist droplet size by using CFD modeling for fire suppressions. *J. Loss Prev. Process Ind.* **2016**, *44*, 626–632. [CrossRef]
17. Sikanen, T.; Vaari, J.; Hostikka, S.; Paajanen, A. Modeling and Simulation of High Pressure Water Mist Systems. *Fire Technol.* **2014**, *50*, 483–504. [CrossRef]
18. Liang, Q.; Li, Y.; Li, J.; Xu, H.; Li, K. Numerical studies on the smoke control by water mist screens with transverse ventilation in tunnel fires. *Tunn. Undergr. Space Technol.* **2017**, *64*, 177–183. [CrossRef]
19. Wang, J.; Nie, Q.; Fang, Z.; Tang, Z. CFD Simulations of the Interaction of the Water Mist Zone and Tunnel Fire Smoke in Reduced-scale Experiments. *Procedia Eng.* **2018**, *211*, 726–735. [CrossRef]
20. Magdolenová, P. CFD Modelling of High-Pressure Water Mist System in Road Tunnels. *Transp. Res. Procedia* **2021**, *55*, 1163–1170. [CrossRef]
21. De Cachinho Cordeiro, I.M.; Liu, H.; Yuen, A.C.; Chen, T.B.; Li, A.; Wang, C.; Cao, R.; Yeoh, G.H. On the Large Eddy Simulation Modelling of Water Suppression Systems Droplet Impact and Coverage Area. *Fire* **2022**, *5*, 165. [CrossRef]
22. Kim, S.C.; Ryou, H.S. An experimental and numerical study on fire suppression using a water mist in an enclosure. *Build. Environ.* **2003**, *38*, 1309–1316. [CrossRef]
23. Mawhinney, J.R.; Back, G.G. Water mist fire suppression systems. In *SFPE Handbook of Fire Protection Engineering*; Springer: Berlin/Heidelberg, Germany, 2016; pp. 1587–1645.
24. Liu, Y.; Wang, X.; Liu, T.; Ma, J.; Li, G.; Zhao, Z. Preliminary study on extinguishing shielded fire with water mist. *Process Saf. Environ. Prot.* **2020**, *141*, 344–354. [CrossRef]
25. Liu, Y.; Fang, Z.; Tang, Z.; Beji, T.; Merci, B. The combined effect of a water mist system and longitudinal ventilation on the fire and smoke dynamics in a tunnel. *Fire Saf. J.* **2021**, *122*, 103351. [CrossRef]
26. Beihua, C.; Guangxuan, L.; Zhen, H. Extinction Limit of Diesel Pool Fires Suppressed by Water Mist. *J. Fire Sci.* **2009**, *27*, 5–26. [CrossRef]
27. Alpert, R.L. Numerical modeling of the interaction between automatic sprinkler sprays and fire plumes. *Fire Saf. J.* **1985**, *9*, 157–163. [CrossRef]
28. McGrattan, K.; Hostikka, S.; McDermott, R.; Floyd, J.; Weinschenk, C.; Overhold, K. *Fire Dynamics Simulator User 's Guide (FDS)*, 6th ed.; National Institute of Standards and Technology: Gaithersburg, MD, USA, 2020.
29. Forney, G.P. *Smokeview, A Tool for Visualizing Fire Dynamics Simulation Data, Volume I: User's Guide*; National Institute of Standards and Technology: Gaithersburg, MD, USA, 2015; pp. 1011–1017.
30. McGrattan, K.; Hostikka, S.; Floyd, J.; McDermott, R.; Vanella, M. Fire dynamics simulator technical reference guide volume 1: Mathematical model. *NIST Spec. Publ.* **2020**, *1018*, 175.
31. Cheremisinoff, N.P. *Encyclopedia of Fluid Mechanics. Vol. 3: Gas-Liquid Flows*; Gulf Publishing Company: Houston, TX, USA, 1986.
32. Jenft, A.; Collin, A.; Boulet, P.; Pianet, G.; Breton, A.; Muller, A. Experimental and numerical study of pool fire suppression using water mist. *Fire Saf. J.* **2014**, *67*, 1–12. [CrossRef]

fire

MDPI

Article

Containment and Suppression of Class A Fires Using CO_2 Hydrate

Olga Gaidukova [1], Vladimir Morozov [2], Roman Volkov [1] and Pavel Strizhak [1,*]

1 Heat and Mass Transfer Laboratory, National Research Tomsk Polytechnic University, Tomsk 634050, Russia
2 Kutateladze Institute of Thermophysics, Novosibirsk 630090, Russia
* Correspondence: pavelspa@tpu.ru

Abstract: This paper presents the experimental findings on fire containment and suppression by dropping CO_2 hydrate granules and tablets on burning solid materials. We used the combustible materials typical of compartment fires—wood, linoleum, and cardboard—to determine the volume and mass of gas hydrate powder necessary for the effective fire suppression. Gaseous emissions were recorded from the combustion with and without fire suppression using hydrates. Conditions were specified in which a fire can be extinguished with minimum air pollution. We also identified the conditions for effective fire containment and suppression using hydrates as compared to water spray, snow, and ice. The necessary volume of hydrate was determined for effective fire suppression in a compartment filled with various materials. Experimental data show that the impact of temperature on the CO_2 hydrate decomposition is highly nonlinear. The carbon dioxide hydrate exhibited a much better fire suppression performance than water spray in the course of total flooding of solid combustible materials. It was established that fine water spray failed to reach the lower levels of multi-tier crib fires. Finally, key patterns of total flooding with CO_2 hydrate powder were identified when applied to fires.

Keywords: CO_2 hydrate; powder and tablet; fire containment; extinguishing; anthropogenic gaseous emissions; experiment

Citation: Gaidukova, O.; Morozov, V.; Volkov, R.; Strizhak, P. Containment and Suppression of Class A Fires Using CO_2 Hydrate. *Fire* **2023**, *6*, 82. https://doi.org/10.3390/fire6030082

Academic Editor: Grant Williamson

Received: 11 January 2023
Revised: 10 February 2023
Accepted: 14 February 2023
Published: 21 February 2023

1. Introduction

Compartment fires are among the most severe ones in terms of fire safety [1–3] because buildings, structures and vehicles contain substances and materials with significantly different properties [4–6]. The thermal decomposition and combustion of these materials and substances produce toxic gases, making it difficult to evacuate people. Excessive use of water or other extinguishing agents on its basis leads to great property loss and again makes it difficult for people to escape the premises. Extinguishing agents are often available in enclosed spaces in very limited amounts [1,7]. Therefore, both time and extinguishing agents should be used efficiently to contain and suppress fire. Quite often, it is more efficient to evacuate people from fire-affected premises and suppress the fire quickly using extinguishing agents instead of sprinkling the adjacent compartments. The most widespread water, gas, and powder firefighting systems are adjusted for the conditions of potential fires to optimize the fire containment and suppression [8–11]. Each of such systems has its strengths and weaknesses. The strength of the firefighting systems based on water mist consists in their minimum negative impact on the environment [9,12]. However, when fires are suppressed by water mist, the flame is briefly intensified, which poses a threat to surrounding people including firefighters. Pei et al. evaluate the efficiency of a two-liquid N_2 water mist containing a KQ additive for the suppression of an ethanol pool fire [9]. This helped improve both physical and chemical effects of fire suppression. Droplet size was found to be one of the major factors influencing the water mist atomization efficiency during the suppression of multiple pool fires [12]. The minimum optimal droplet size was determined for water mist. Making the particle size smaller than this threshold value is not feasible [12]. To enhance the strengths of the corresponding systems and minimize their weaknesses, it is necessary to intensify three fire suppression mechanisms [8,13,14]: cooling the surface of the reacting material and gas–vapor mixture through heat transfer and phase transitions; displacing the oxidizer and thermal decomposition products from the high-temperature zone using inert gas (or gas–vapor mixture); filling the pores of the pyrolyzing material with an inert component—water. The analysis of the experience with extinguishing agents (in particular, emulsions, solutions, slurries, foams, etc.) shows that each of the three mechanisms or all of them at once can be activated at different stages of a fire (pyrolysis, growth, flame combustion, smoldering, and decay). Here, both the consumption of extinguishing agents and fire suppression time can be reduced by using multi-phase composite agents [15]. These include gas hydrates (ice and gas in a crystal lattice; when heated, a liquid film is formed on the hydrate surface) based on inert components, CO_2 in particular.

CO_2 hydrate has some unique benefits [16]. It is conventionally used for the capturing and storage of CO_2 [17,18], sea water desalination [19], production and cooling of carbonated solid foods [20], as well as replacement of CH_4 in natural gas hydrate deposits [21]. The use of gas hydrate powder for firefighting purposes has not received much attention [22]. Authors report on their experimental research into the pool fire extinction using carbon dioxide hydrates [23]. The results show that CO_2 hydrate can extinguish a pool flame using less water as compared to ice and cause lower CO_2 emission as compared to dry ice [23]. It is of great scientific interest to study different materials to find the optimal reacting substances to hydrate mass ratio as well as their contact areas. The relevance of this research also stems from the boost that the production of artificial gas hydrates [24] and recovery of natural gas hydrates [25,26] has received over the recent years. Both granulated powder and pressed tablets are used to transport different volumes of gas. As a result, new knowledge has been obtained on the mechanisms of hydrate dissociation [27,28], melting, evaporation and boiling [29,30] as well as ignition and combustion [31,32]. These papers have become a base for promoting the effective use of different gas hydrates (in particular, methane, ethane, propane, isopropanol, carbon dioxide, and their combinations) in the petrochemical and energy industries. A relevant task is to study compartment fire containment and suppression using CO_2 hydrates. This served as a motivation for this study.

The analysis of the literature quoted above indicates that firefighting is a global problem. The destructive force of fires often leads to injuries and loss of life. The currently known firefighting technologies are notable for some weaknesses that are under the radar of specialists who are busy developing new, more effective firefighting tools. Firefighting services are facing a multitude of challenges, while the number of fires is on the increase and so is the material damage they cause. Under these conditions, it is necessary to equip industrial and housing facilities with cutting-edge, highly effective extinguishing media based on brand-new technology. The research of new effective fire suppression agents is a major objective. In particular, it is a relevant task to study compartment fire containment and suppression using CO_2 hydrate as a new extinguishing component. The aim of this research was to experimentally identify the conditions for the effective suppression of compartment fires with typical solid combustibles involved using CO_2 hydrates in the form of pressed tablets and granulated powder. Here, we look into several key aspects of carbon dioxide hydrate as a component for class A fire suppression. We also investigate the patterns of CO_2 hydrate dissociation as one of the key processes characterizing the gas hydrate decomposition rate and time, which has a significant impact on the fire suppression parameters. A separate section deals with anthropogenic emissions from the combustion and suppression of flammable materials. The main section reports the necessary and sufficient conditions of effective class A fire containment and suppression using carbon dioxide hydrates as compared to water aerosol, snow, and ice.

2. Experimental Technique

2.1. Materials

The experiments involved three of the most widespread fire-hazardous solid materials typical of miscellaneous compartments including unoccupied ones (e.g., warehouses): wood, linoleum, and cardboard (Table 1). Wood is the most abundant combustible material found in compartments. The fire hazard of wood is defined by its thermal decomposition under external heat fluxes that start at above 110 °C [33]. Further heating causes the evaporation of free and bound moisture from timber. This process finishes at 180 °C when the least heat-resistant components decompose to release CO_2 and H_2O. At above 250 °C, timber pyrolyzes to emit the following gaseous products: CO, CH_2, H_2, CO_2, H_2O. The gaseous mixture is flammable and capable of igniting from an open flame, a spark, heated surfaces, a short circuit, etc. Higher temperatures accelerate the thermal decomposition of timber. One of the important factors determining the fire safety of timber is its ability to ignite and to stimulate the propagation of fire in an oxidizer environment [33]. Linoleum is one of the most affordable flooring materials. Being used everywhere from residential premises to industrial buildings, linoleum is manufactured with strict fire safety regulations. However, when exposed to fire, it decomposes to emit a large volume of toxic gases. According to the statistics [34,35], most fire fatalities are the result of smoke inhalation rather than burns. Apart from wood, paper is another widespread flammable material. Cardboard is often used in industrial premises. In actual fires, paper products often serve as the primary combustible medium leading to one of the most hazardous and fast spreading types of fire. Linoleum, rubber goods, particle board, and fiberboard are widely used as finishing materials. These materials fall into one group in terms of fire safety indicators. Linoleum is the most frequently used of them. Thus, the research findings on wood, cardboard, and linoleum can be representative of typical solid materials found indoors. Due to the high fire hazard of the materials under study, it is important to investigate the fire containment and suppression for these very materials.

Table 1. Characteristics of combustible materials used in the experiments.

Combustible Material	Specifications
Pine wood	Pine density is 520 kg/m^3 and moisture content is 12–15%. Total heating value of pine wood is 4.4 kW·h/kg. The combustion of pine wood produces water vapor, heat, carbon dioxide and carbon monoxide, aldehydes, acids and different gases.
Fabric backed linoleum	Linoleum backed with fabric is made of polyvinyl chloride with added plasticizers, fillers and dyes. A quality material does not support active combustion. The main combustion product of polyvinyl chloride is hydrogen chloride.
Corrugated cardboard	It consists largely of recycled materials (semicellulose, straw, waste paper, etc.). The rest is primary cellulose fibers.

The following extinguishing agents were chosen for the containment and suppression of fires: snow, water, ice, and carbon dioxide hydrate in the form of powder and tablets. Carbon dioxide is actively used in firefighting systems [8,21]. It provides fast fire containment and suppression because it does not contribute to fire propagation and displaces the fire-sustaining oxygen. Using CO_2 for firefighting has a number of benefits: it is easy to use, suitable for fires involving any materials, etc. The main weakness of using CO_2 is the potential hazard to building occupants due to the critical lack of oxygen during fire extinguishing with carbon dioxide. CO_2 hydrate can significantly reduce the impact of this weakness due to the content of water. The content of water in the hydrate also helps effective firefighting even if multi-phase reacting media are involved. To enhance the role of steam in the presence of CO_2, we used a hydrate with a limited content of gas in the crystal lattice. The CO_2 hydrate used in the experiments had the following component composition: 71–72% of water and 28–29% of CO_2 (powder); 73–75% of water and 25–27% of CO_2 (tablet). The CO_2 hydrate powder particles were 0.3–0.5 mm in diameter, which is a typical diameter of hydrate granules [36]. The CO_2 hydrate tablet had the following dimensions: 20 mm (diameter d_0) and 5 mm (height h_0). Experiments with alternative extinguishing agents involved tap water, natural snow and ice from a freezer in the form of 3–5-mm granules. The masses of extinguishing agents used in the experiments were preliminarily determined using an AJH-620CE balance with an accuracy of ±0.001 g. Table 2 presents the specifications of the extinguishing agents used in the experiments.

Table 2. Specifications of the extinguishing agents used in the experiments.

Extinguishing Agent	Specifications
Water	Tap water
Snow	Snow from natural precipitation particle size: 1–2 mm.
Ice	Frozen tap water ice particle (granule) size: 3–5 mm.
CO_2 hydrate powder	Component composition: 71–72% water, 28–29% CO_2. Powder particle size: 0.3–0.5 mm.
CO_2 hydrate tablet	Component composition: 73–75% water, 25–27% CO_2. Hydrate tablet dimensions: 20 mm (diameter), 5 mm (height).

2.2. Experimental Setup and Methods

In the experimental research into the thermal decomposition (dissociation) of a carbon dioxide hydrate tablet, it was exposed to radiant heating with heat fluxes comparable to those typical of compartment fires. The setup is schematically presented in Figure 1. The thermal decomposition of a CO_2 hydrate tablet occurred in a high-temperature air environment generated in a ceramic tube of an R 50/250/13 muffle furnace. The air temperature in the furnace ranged from 500 to 900 °C. It was measured using a type S thermocouple with a temperature range of –50 ... 1400 °C and accuracy of ±2.5 °C. A hydrate tablet was retrieved from a Dewar vessel filled with liquid nitrogen, weighed on an AJH-620CE electronic balance with an accuracy of ±0.001 g, and put into a fine-mesh metal holder. The time from retrieving the sample to when it was placed inside the tubular muffle furnace did not exceed 10 s. Using a positioning mechanism, we introduced the holder with a hydrate tablet into a hollow tube of the heated furnace. The thermal decomposition was recorded at 100 fps using a high-speed Phantom v411 camera with a resolution of 1280 × 800 pix, 12-bit depth, 20 μm pixel size, 1 μs minimum exposure, and image-based auto-trigger. The resulting video recordings were analyzed using the Tema Automotive software. The concentration of gases emitted from the decomposition was measured using a Test 1 gas analyzer (Bonair, Russia). The Test 1 gas analyzer probe was introduced into the central part of the heated furnace. The gas flowed through the sampling hose to the gas analyzer sensors. The resulting gas concentrations (CO_2, CO) were analyzed using the custom Test software. The average gas concentrations were calculated using the trapezoidal rule. The calculation procedure is described in more detail in [37].

The fire containment and suppression were experimentally studied on a setup schematically represented in Figure 2. A wood/linoleum/cardboard sample with a fixed initial mass was placed in a metal container and ignited by a gas burner (evenly across the entire surface). The fuel was exposed to the gas burner for 15–20 s (wood), 12–15 s (linoleum), and 8–10 s (cardboard). The research was carried out in a transparent, hollow parallelepiped (0.35 × 0.25 × 0.2 m) made of a fire-resistant material. The extinguishing agent was delivered to the burning material by a robotic arm. The mass of the fuel was preliminarily measured using an electronic balance. The temperature in the central part of fire (on the surface of the material and above it) was recorded by type K thermocouples (Owen) with a measurement range of 233–1573 K, accuracy of ±1.5 K at T = 233–573 K and ±0.004·T at T = 574–1573 K. The temperature readings were transferred to the PC. The gas concentrations emitted from the combustion of the material were measured by a Test 1 gas analyzer; its probe was introduced into the center of the cube through a designated orifice on top. The combustion processes were recorded by a high-speed Phantom MIRO C110 camera with a frame rate of up to 1000 fps at a resolution of 1280 × 1280 pix.

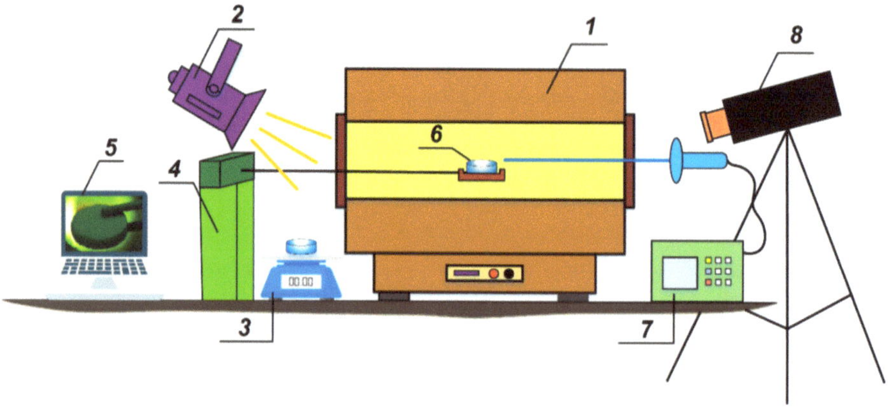

Figure 1. Scheme of the setup for studying the thermal decomposition of gas hydrate on radiant heating: 1—muffle furnace; 2—spotlight; 3—electronic balance; 4—positioning mechanism; 5—laptop; 6—metal holder with a hydrate tablet; 7—gas analyzer; 8—camera.

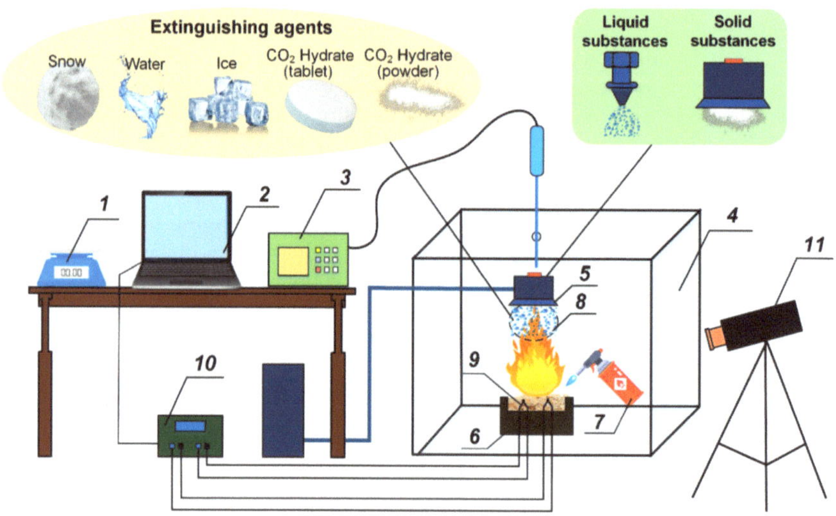

Figure 2. Scheme of the setup for studying the fire containment and suppression: 1—electronic balance; 2—laptop; 3—gas analyzer; 4—transparent cube; 5—extinguishing agent feeder; 6—container with burning material (wood/linoleum/cardboard); 7—gas burner; 8—extinguishing agent; 9—thermocouple; 10—temperature transmitter; 11—camera.

The distance from the extinguishing agent feeder to the surface of the controlled laboratory-scale fire was about 0.15 m. The combustion time of the fire approximated 7–10 s (from the moment when the fuel exposure to the gas burner stopped up until the moment of an extinguishing agent was supplied). Ice, snow, and CO_2 hydrate were delivered by opening a sliding gate on the extinguishing agent feeder 5 (Figure 2). The extinguishing agent traveled to the fire under gravity. Water was delivered using an FMT-100 spray nozzle [38]. The water pressure before the spray nozzle was 1.5 bar, the average droplet velocity was 3.6 m/s, and the size of the droplets ranged from 2 to 120 μm. The sizes and velocities of droplets generated by the nozzle were determined using the optical techniques of SP and PIV, such as in [38].

The proposed experimental method is versatile: it can be used for studying the suppression of fire involving combustibles with different initial masses using water, snow, ice, and gas hydrate powder. Preliminary experimental findings were used to calculate the integral characteristics of fire containment and suppression. Video recording and thermocouple measurements allowed us to keep track of the fire suppression stages even during active vaporization and smoke generation as well as thermal decomposition of materials producing a gas–vapor mixture that poorly transmitted radiation.

3. Results and Discussion

3.1. Patterns of CO_2 Hydrate Dissociation

Figure 3 shows the thermal decomposition times of a CO_2 hydrate tablet when varying the ambient gas temperature in the range of 500–900 °C. The error bars in the figure illustrate the random error in a series of measurements of thermal decomposition time (confidence intervals). To process the results (including gross error identification and elimination), standard approaches were used [39]. These involved calculating the mathematical expectation (Equation (1)), variance of a random variable (Equation (2)), and standard deviation for each series (Equation (3)). Then the width of the error bars was calculated (Equation (4)).

$$M_X = \frac{1}{n}\sum_{i=1}^{n} X_i \quad (1)$$

$$V = \frac{1}{n-1}\sum_{i=1}^{n} (X_i - M_X)^2 \quad (2)$$

$$\sigma = V^{1/2} \quad (3)$$

$$\Delta = t_{\alpha n} \cdot \sigma \quad (4)$$

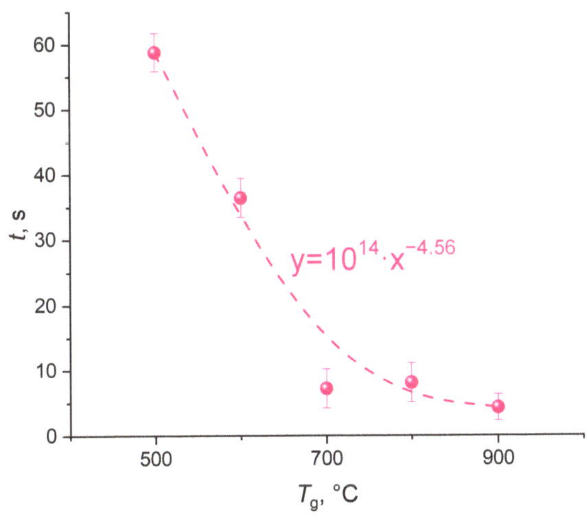

Figure 3. Average thermal decomposition times of a carbon dioxide hydrate tablet at varying temperatures in the muffle furnace.

The following nomenclature was taken in Equations (1)–(4): M_X—mathematical expectation; X_i—measurement result; n—number of measurements; V—variance; σ—standard deviation; Δ—width of error bar; $t_{\alpha n}$—Student's coefficient.

When choosing the values of $t_{\alpha n}$, the confidence coefficient was taken as equal to 0.95. In Section 3, all the main figures show the error bars that illustrate the range of possible values of the measured parameter (with a 95% probability). If error bars are not shown, this indicates that either their values are too small, or the results of continuous/instantaneous measurements of the corresponding parameter are presented (for example, droplet velocity fields, velocity profiles, droplet size distributions). The accuracy of the latter is described by systematic measurement errors of the corresponding parameters (Section 2).

The thermal decomposition time was found to decrease to about 1/9 of the initial value with an increase in the furnace temperature from 500 to 700 °C. A temperature increase from 700 °C to 900 °C caused a 35% time reduction (Figure 3). This result indicates that the gas–vapor area near the dissociating powder surface has a certain degree of saturation with gases. The gas hydrates used in the experiments are 70–75% water. This factor is the key to the patterns recognized in the experiments. Clearly, the higher the gas temperature, the more intense the ice melting and water evaporation. The intensification of these processes accelerates the self-preservation of hydrate granules, leading to the clogging of some pores in the near-surface layer. The release of carbon dioxide is inhibited. Thus, the dissociation rates reach a certain asymptotic value. This suggests that both low and extremely high ambient gas temperatures are not effective at catalyzing the hydrate dissociation. It is important to choose the right average temperature of the gas environment.

The ambient gas temperature (T) was found to affect the dissociation time (τ_{dis}) of carbon dioxide hydrate. The carbon dioxide hydrate was placed into a muffle furnace with a constant ambient gas temperature. The kinetic equation for the dissociation as well as kinetic constants are given in [40]. A model of gas hydrate dissociation at negative temperatures (beyond the self-preservation region) controlling for the dissociation kinetics and gas filtration through pores is considered in [41]:

$$3\left(Y^{1/3}-1\right)+\frac{3}{2}B\left(Y^{2/3}-1\right)-B(Y-1)=-K\tau_{dis} \tag{5}$$

where Y is the degree of carbon dioxide hydrate particle conversion to ice, parameter $B=\frac{R_0 k^R \mu}{k^F \rho}$, parameter $K=\frac{3k^R\left(p^{Eq}-p_0\right)}{b\rho_H D_0}$, R_0 is the radius of the sphere, k^{R} and k^{F} are the kinetic and filtration coefficients, ρ_H is the CO_2 hydrate density, b is the initial carbon dioxide concentration C_0, μ is the dynamic viscosity of gas, p^{Eq} is the equilibrium pressure in the CO_2 hydrate, D_0 is particle diameter and p^0 is the ambient pressure. Modeling also involved the calculation of thermal balance controlling for the dissociation, ice melting, and water evaporation [41]. The results of predictive calculations are given in Figure 4. The curve is nonlinear. At above 700 °C, the impact of temperature (slope of the curve) is much lower than in the temperature range under 600 °C. The computational findings are in acceptable agreement with experimental data given in Figure 3.

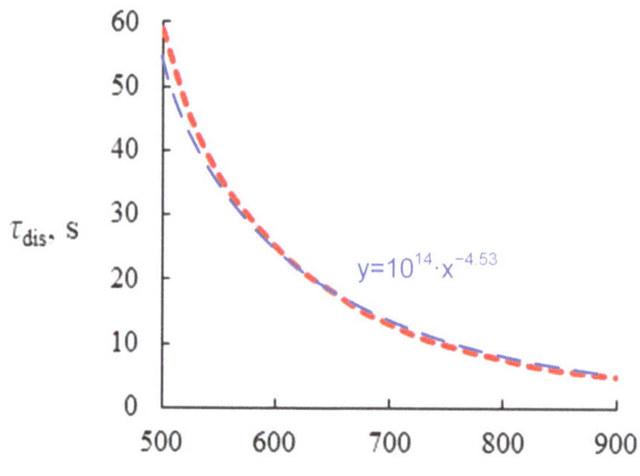

Figure 4. Calculated CO_2 hydrate dissociation times at varying ambient gas temperatures (the red curve shows the calculated data; the blue curve is an approximation based on the calculated data).

When analyzing Figures 3 and 4, we singled out important patterns of gas dissociation from the hydrate. First, the relationship of hydrate dissociation rate and full dissociation time (gas release) versus temperature is exponential. This aspect can predict the gas release time for different applications. In particular, gas–vapor mixtures are produced in reactors at an ambient gas temperature of less than 600 °C. The relationships obtained in this research show that the gas mixing times (i.e., the preparation times of gas–vapor mixtures) should be several dozens of seconds. In this case, all of the gas will leave the hydrate, and complete miscibility will be achieved. The temperature range of 500 °C to 700 °C corresponds to the typical technologies of composite fuel pyrolysis and gasification as well as the co-combustion of several components, in particular, hydrocarbons, coal and oil processing wastes, biomass, municipal wastes, etc. Carbon dioxide and water vapor are commonly regarded as promising gas environments for effective pyrolysis and gasification of composite fuels. Thermal conversion of composite fuels in carbon dioxide and water vapor proceeds with an intense release of carbon monoxide, methane, and hydrogen. The concentrations of sulfur and nitrogen oxides are minimized due to minimum oxygen concentrations. The systems of the so-called low-temperature fuel combustion and high-temperature gasification are triggered at ambient gas temperatures of over 700 °C. The CO_2 added to the gas–vapor mixture provides control of oxidation reactions, and other reactions also become more controllable under oxygen deficiency. As a result, the concentrations of unspecified gas emissions decrease. The curves of hydrate dissociation time against T show that the durations of the typical processes remain practically the same (several seconds) at ambient gas temperatures of over 800 °C. Thus, it is advisable to set the limit at 800 °C for the commercial implementation of the processes in reactors and chambers. Small-size hydrate heating units would suffice to generate the entire volume of gas when preparing a gas–vapor mixture with the required CO_2 concentration.

Another important pattern identified during the analysis of Figures 3 and 4 is that the times of complete hydrate dissociation determined in a series of experiments were quite well reproducible. This is crucial for the use of the data obtained for predicting the duration of commercial production processes. The experimental data were obtained for fixed sizes of hydrate tablets. As these samples are similar to natural hydrate layers in shape and structure, and these layers dissociate consecutively, it is possible to predict the complete hydrate dissociation times at identical temperatures with different layer thicknesses.

The third pattern is related to a more intense hydrate dissociation in an experimental chamber as compared to typical commercial systems in which hydrates are delivered in layers. Such layers are heated in production units in the same way as in a tubular muffle furnace in the experiments—all around—but substrates for samples have different structures. In the experiments, the hydrate samples were placed on a perforated mesh to make it similar to tablets suspended in a chamber or free-falling granules. Such meshes are often used in commercial production processes. The heat is supplied to the hydrate sample surface through this mesh, and water drains from the surface of a tablet through the mesh as well. With a non-perforated substrate, the heat supply to the lower surface of the hydrate tablet proceeds in a different way, and there is no drainage. In this case, the complete hydrate dissociation will last longer than it was established in this research. It is more efficient to heat hydrate granules and tablets from all sides and let water drain from their surface. This will minimize the self-preservation of hydrate pores, thus accelerating hydrate decomposition and gas dissociation.

Figures 5–7 present the images showing the thermal decomposition of a carbon dioxide hydrate tablet in a muffle furnace at $T_g \approx$ 500–900 °C. Gas and vapor release from the surface was recorded in the form of the outflow of a gas–vapor mixture with vortices. These processes had interesting and distinct stages. At first, we recorded a gas release in the form of almost transparent tracks from the hydrate surface. This stage was relatively short. The higher the ambient gas temperature was, the faster this stage finished. At the second stage, a water film was formed on the hydrate surface. The vapor outflow was difficult to identify on the images at this stage because the formation of a water film on the surface of the hydrate particle triggered the self-preservation of pores, so gases passed through pores only partially. After that, the water film served as a membrane of sorts and only let a small fraction through. After a short time, bubbles started to form on the surface of the water film. The surface of the hydrate particle became uneven. When the critical pressure was exceeded, bubbles imploded, and the gas–vapor mixture was released from the hydrate surface. These processes were not monotonous. It was only when the deep layers were heated that the final stage started, notable for the irreversible gas and vapor release from the hydrate surface causing the dispersion of the near-surface layer and increase in the hydrate surface area. The hydrate sample dispersion led to the expansion of the channels of gas release from the depth of the hydrate. The physical dispersion mechanism is related to water boiling, water film retaining vapor and gas, and the third phase in the form of ice reinforcing the vapor-water frame. Under such conditions, the vapor and gas pressure in the deep layers increased rapidly, but their release from the surface was suppressed by the reinforced frame. Thus, when the critical pressure was exceeded, the vapor and gas release caused the ice particles to break off with a new portion of gas inside. The release of this gas from the newly formed hydrate fragments (commonly known as secondary or child fragments) accelerated after that. This is how the cascade dissociation was triggered for the primary gas hydrate sample and its secondary fragments.

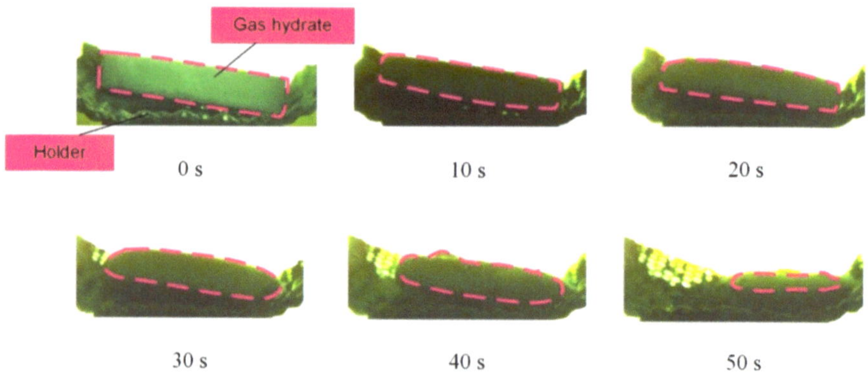

Figure 5. Images of thermal decomposition of a carbon dioxide hydrate tablet in a muffle furnace at $T_g \approx 500$ °C.

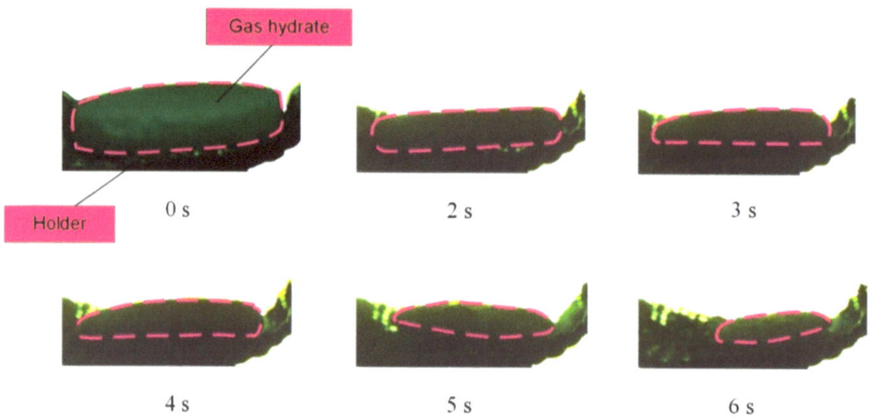

Figure 6. Images of thermal decomposition of a carbon dioxide hydrate tablet in a muffle furnace at $T_g \approx 700$ °C.

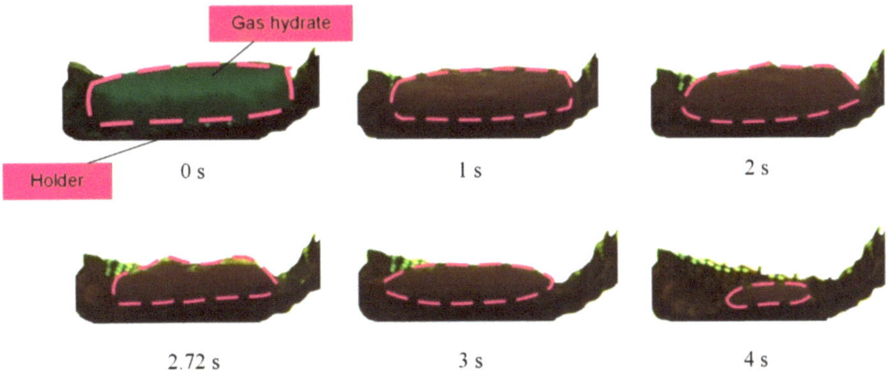

Figure 7. Images of thermal decomposition of a carbon dioxide hydrate tablet in a muffle furnace at $T_g \approx 900$ °C.

3.2. Anthropogenic Emissions from the Decomposition of Gas Hydrates

Water vapor is known to reduce the harmful emissions from the combustion of a wide range of materials and substances, which is important from the environmental perspective. Compartment fires involve various materials and produce a large amount of anthropogenic emissions that are harmful to respiratory organs. That is why it is important to evaluate how water and temperature of the gas environment temperature affect the concentration of emissions. Water is formed when gas hydrate decomposes and ice melts at a high ambient temperature. Figures 8 and 9 show the measured concentrations of CO_2 and CO emitted during the thermal decomposition of carbon dioxide hydrate. The error bars given in Figures 8 and 9 (on the right) were calculated according to the procedure described in Ref. [39] and illustrate the random error in the average gas component concentration (confidence intervals) in a series of five measurements. As the temperature increases, the gas concentration drops. Both the maximum concentration and the gas release time decrease. An increase in the concentration with a higher ambient temperature stems from a more intense interaction between the decomposition products and water vapor (i.e., radicals responsible for the reactions with carbon are formed more rapidly). A slight increase in the CH_4 concentrations was recorded at high temperatures in the furnace ($T_g \geq 700$ °C). In particular, the methane concentration was 0.01% at $T_g \approx 700$ °C and 0.02% at $T_g \approx 900$ °C.

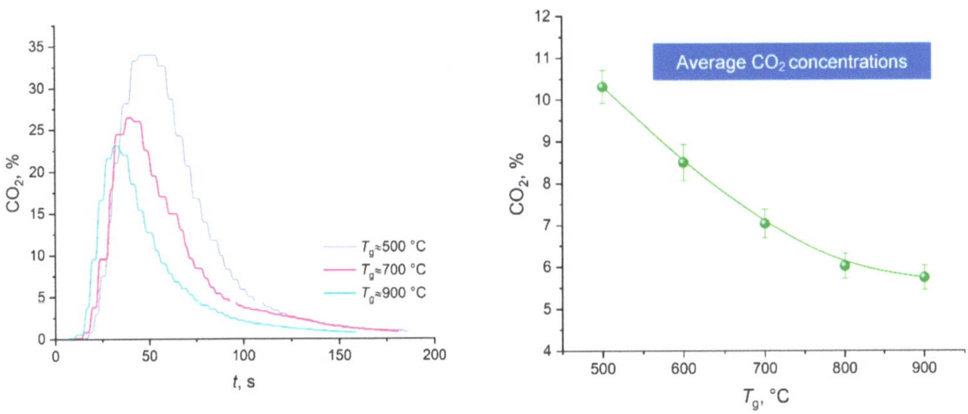

Figure 8. CO_2 concentrations from the thermal decomposition of carbon dioxide hydrate tablet at varying temperatures in the muffle furnace: changes in the concentration over time (on the **left**); average concentrations (on the **right**).

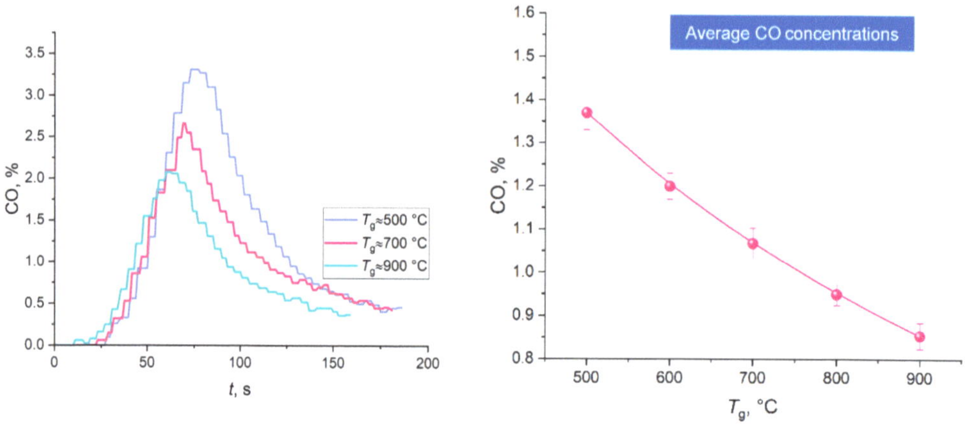

Figure 9. CO concentrations from the thermal decomposition of carbon dioxide hydrate tablet at varying temperatures in the muffle furnace: changes in the concentration over time (on the **left**); average concentrations (on the **right**).

Table 3 presents the maximum concentrations of CO_2 and CO from the combustion and extinguishing of a fire involving flammable materials with varying extinguishing agents.

Table 3. Maximum concentrations of CO_2 and CO from the combustion and extinguishing of a fire involving flammable materials with varying extinguishing agents.

	Wood		Cardboard		Linoleum	
	CO_2, %	CO, %	CO_2, %	CO, %	CO_2, %	CO, %
No extinguishing	5.88	1.4	5.21	2.3	2.56	0.8
Water aerosol	1.05	0.4	2.66	1.9	0.57	0.3
Snow	3.41	1.1	1.9	1.4	0.38	0.2
Ice	2.37	0.8	3.7	2.7	1.52	0.6
CO_2 hydrate powder	1.14	0.4	1.43	0.7	0.67	0.3
CO_2 hydrate tablet	2.94	1.1	4.17	1.7	2.37	0.8

Figure 10 presents the concentrations of the main gases emitted from the combustion and extinguishing of a fire involving wood with the help of several extinguishing agents. The fire was deemed extinguished when the thermocouple measuring the temperature in the internal layer of the material (at a depth of 2–3 mm from the upper free surface) read$\leq$ 100 °C (minimum pyrolysis temperature of the combustible materials under study). The gas component concentrations were no longer recorded at this temperature. According to the data obtained, the maximum CO_2 concentrations were observed from the combustion of wood without extinguishing and equaled 5.9 vol%. The use of any of the extinguishing agents provided a 42% to 82% reduction in CO_2 emissions into the atmosphere. Water and CO_2 hydrate showed the highest suppression performance for wood. Hydrated CO_2 powder triggered two main mechanisms necessary for fire containment. The temperature in the gas phase decreased due to the gas hydrate dissociation, ice shell melting, and water evaporation. The decrease led to the deceleration of chain-branching oxidation reactions with a high activation energy. The inert gas—CO_2—rapidly released from the hydrate displaced the oxygen from the combustion zone, thus inhibiting the oxidation of wood-based material. Fire is suppressed by a gas hydrate in an air–vapor environment due to ice crust melting and water evaporation. A certain fraction of carbon monoxide is spent in the water gas shift reaction: OH + CO→H + CO_2. The lower production rate of H and OH radicals in the reaction zone leads to a significant deceleration of the combustion front propagation and flame quenching. The experimental findings obtained for snow and ice confirm the positive effect of CO_2 hydrate powder. Snow and ice provide a significant decrease in the flame temperature, but the fire containment takes longer because the gas hydrate dissociation requires some time as well. The delay of the carbon dioxide hydrate decomposition is determined by the filtration and kinetic resistance of the hydrate [41], as well as the heat of dissociation, which reduces the velocity of the thermal front within the powder particle. Due to longer extinguishing time, more material burns out, increasing the CO_2 emissions into the atmosphere. However, this drawback (longer extinguishing time) is typical of fire suppression on a quasi-flat surface of the fuel. In real-life conditions, however, materials are located at different levels, hence the high altitude of flame propagation. Further, we will prove that the fire suppression performance of gas hydrate powder is significantly higher than that of water spray.

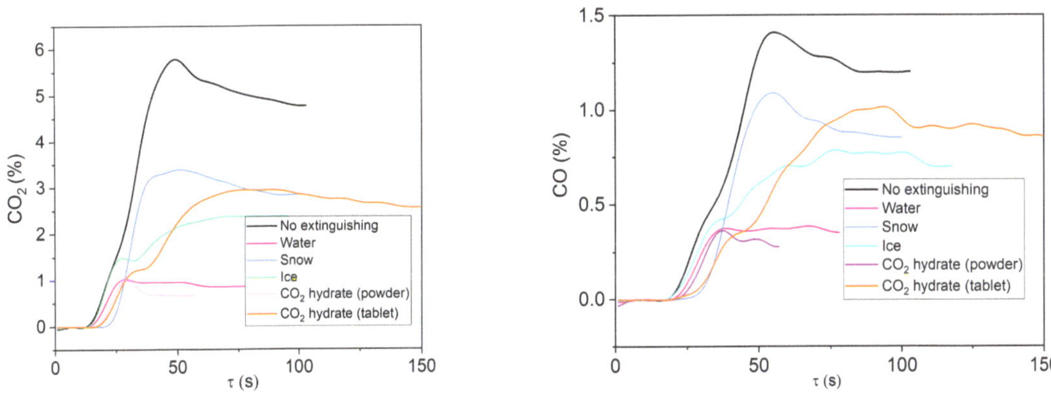

(a) (b)

Figure 10. Concentrations of CO_2 (**a**) and CO (**b**) emitted during wood fire suppression.

All the trends with the recorded component concentrations of a gas–vapor–air mixture show significantly different durations. This effect stems from the different durations of flame combustion and thermal decomposition of materials exposed to different extinguishing agents. When research findings are presented in this format, it is possible to analyze how fast the corresponding physicochemical processes slow down in the depth of the material and in the close vicinity of its surface. Especially valuable are the recorded extrema on the trends as well as their number, because they reflect the cyclic and cascade manner of the heat exchange processes and chemical reactions. With the physical mechanisms of fire suppression described above, it is possible to reliably predict the amount and type of extinguishing agent required for effective firefighting. On average, it took 50 to 200 s for the concentrations of all the components detected in the gas–vapor–air mixture to go down to zero. The time depended on the mass of the pyrolyzing material, type of extinguishing agent, and operation of the exhaust system. If a gas hydrate powder remained on the surface of the material by the time the concentration of pyrolysis products started to decrease, this decrease remained monotonic. However, if the film of the extinguishing agent based on water, ice, or hydrate granules became thinner and had gaps, the trends of the decrease in the concentration of pyrolysis products were not fully monotonic. This stems from the unstable heat exchange conditions across the material surface.

The combustion of wood without extinguishing produced the maximum CO concentrations. The use of extinguishing agents reduced the CO emission by 3.5 times. The extinguishing agents under study were ranked as follows in terms of the efficiency of carbon monoxide emission decrease (from least effective to most effective): snow, CO_2 hydrate tablet, ice, water, and CO_2 hydrate powder. Taking the CO_2 emissions into account, the hydrate powder turned out to be the most effective agent under the conditions of the CO concentration decrease, mainly because the rate of the reaction OH + CO→H + CO_2 decreased. The reaction was inhibited by the release of more CO_2 from the hydrate, which led to a decrease in the CO and CO_2 production.

The comparison of experimental findings on hydrate powder and tablets gave an interesting result. Wood fire suppression using a hydrate tablet took a longer time and produced higher CO_2 and CO concentrations. This can be explained by the differences in the surface areas of the hydrate and burning solid material. When hydrate was applied in the powder form, it spread evenly across the wood surface, so the combustion reaction was also suppressed evenly across the entire surface of the material. The size of the hydrate tablet (in particular, its diameter of 20 mm) prevented it from covering the entire free surface of the reacting material. As a result, the thermal decomposition of the sample was not suppressed evenly: pyrolysis and combustion continued on the edges. The central part, however, covered by the hydrate tablet, stopped burning almost instantaneously after the suppression started. Thus, it was experimentally proven that the class A fire suppression efficiency using CO_2 hydrate is largely defined by its dissociation rate and free surface area (hydrate/fire contact area).

The trends obtained for cardboard (Figure 11) agree overall agree with the data on wood fire suppression. The minimum emissions of carbon oxides were produced during the suppression of a cardboard fire with CO_2 hydrate powder. The difference in the threshold concentrations of CO_2^{max} and CO^{max} from the combustion of cardboard without suppression and with suppression using CO_2 powder was 70%. The maximum gas release time and rather high concentrations of carbon oxides indicate that ice is the least effective extinguishing material of those considered in this research. When used to extinguish burning cardboard, water exhibited lower efficiency in decreasing the CO_2 and CO emissions as compared to wood fire suppression (Figure 10). Water quickly evaporated from the cardboard surface and got absorbed into cardboard without penetrating into the depth of the porous cardboard layer, several centimeters high. As a result, pyrolysis continued in the depth, and local combustion continued over the surface of the material.

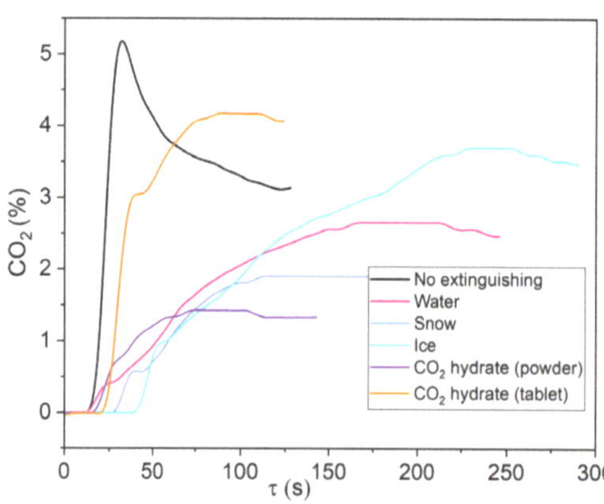

(a)

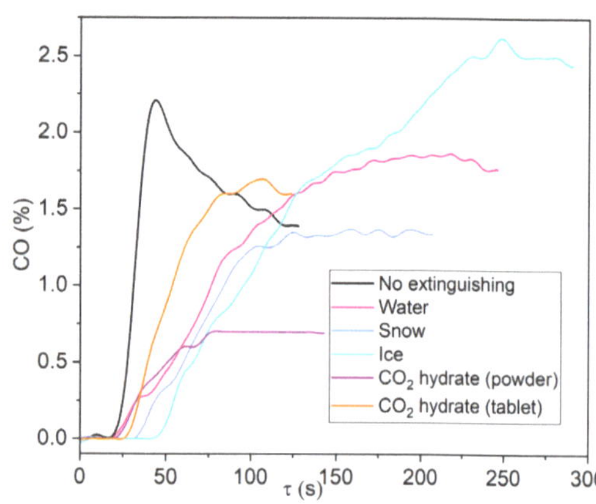

(b)

Figure 11. Concentrations of CO_2 (**a**) and CO (**b**) emitted during cardboard fire suppression.

Figure 12 presents the concentrations of the main gaseous emissions from the combustion of linoleum with and without extinguishing. The minimum concentrations of carbon oxides were recorded when water was used for extinguishing. The concentrations of CO_2

and CO emitted during the suppression with carbon dioxide hydrate are somewhat higher than those for water. The volume of CO_2 hydrate contained in one tablet turned out to be insufficient for the complete containment of a fire involving linoleum.

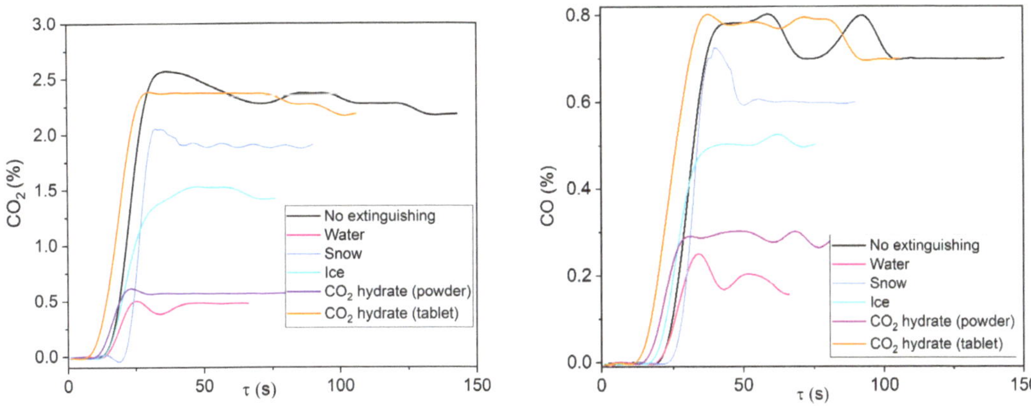

(**a**) (**b**)

Figure 12. Concentrations of CO_2 (**a**) and CO (**b**) emitted during linoleum fire suppression.

Experimental research into the suppression of combustible materials must identify the minimum (critical) masses of extinguishing agents. When the ratio of the combustible material mass to the extinguishing agent mass changes, the concentration of pyrolysis and combustion products changes as well. Figure 13 shows the data on CO_2 and CO concentrations as a function of the CO_2 hydrate mass. The functions are nonlinear. When the mass increases from 7 g to 15 g, the emission decreases by three times for CO and by eight times for CO_2. Further changes in the mass do not affect the volume and maximum concentrations of emitted gases that much. The minimum extremum is observed for CO_2 and CO concentrations. The concentrations of CO_2 and CO increase when a certain mass of carbon dioxide hydrate is exceeded. This nonlinear behavior is associated not only with the combustion temperature but also with the water vapor concentration in the combustion and pyrolysis zones. The water temperature and concentration govern both the elementary reactions and the oxidation rate.

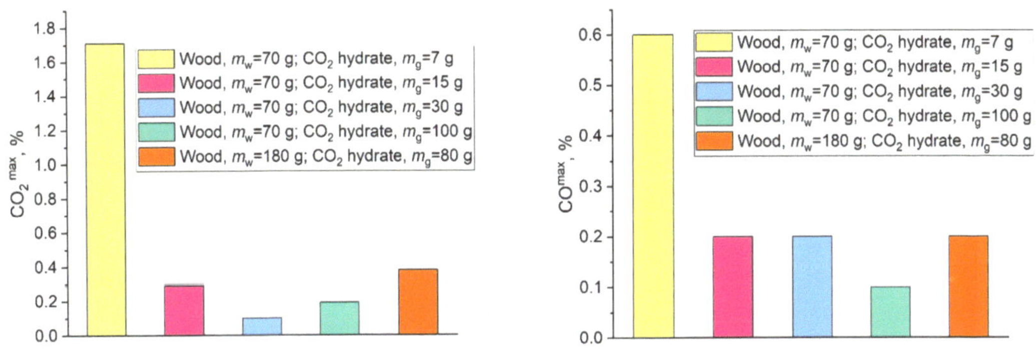

(**a**) (**b**)

Figure 13. Concentrations of CO_2 (**a**) and CO (**b**) emitted during wood fire suppression with varying masses of the extinguishing material and the seat of the fire.

Figure 14a,b shows the curve of the time when the CO_2 and CO emissions started to decrease versus the mass of hydrate (m_g = 7–30 g) with the mass of wood remaining constant (m_w = 70 g). It is clear that the time until the gas concentrations begin to decrease shortens with an increase in the mass of the extinguishing hydrate. A change in the mass of carbon dioxide hydrate from 7 g to 15 g reduces the time until the CO_2 and CO concentrations start to go down by 71% and 28%, respectively. An increase in the hydrate mass from 15 g to 30 g leads to a negligible change (up to 10%) in the time before the concentrations begin to decrease. Figure 14a,b presents the approximation curves of the exponential (CO_2) and polynomial (CO) nature with the mathematical expressions describing these curves. The resulting mathematical expressions allow the extrapolation to greater masses of extinguishing agents and larger fire areas.

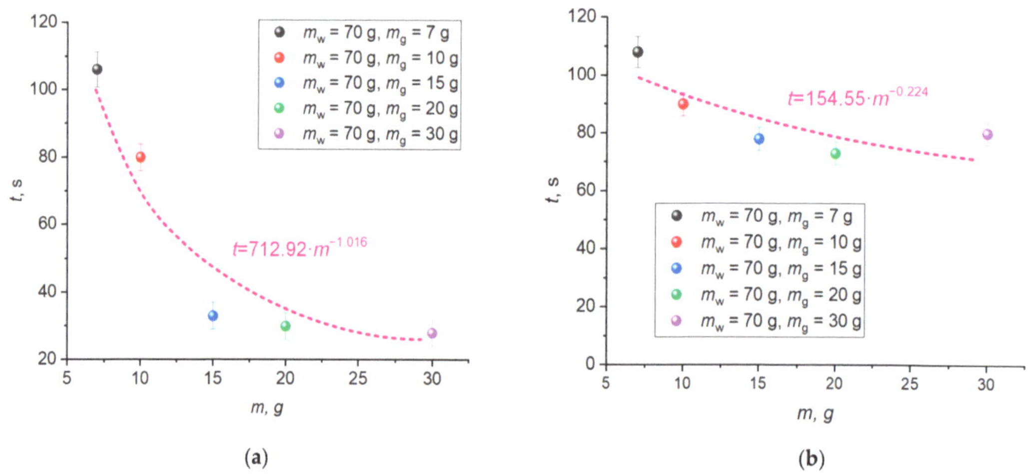

Figure 14. Averaged starting times of decrease in CO_2 (**a**) and CO (**b**) concentrations with varying masses of gas hydrate m_g used to suppress the combustion of wood with the constant mass m_w.

3.3. Conditions for the Effective Fire Containment and Suppression Using Gas Hydrates

Figure 15 presents the typical images showing a fire comprised of pine rods being extinguished with carbon dioxide hydrate powder. The mass of timber remained constant (m_w = 70 g), and the mass of gas hydrate was varied. The footage of wood fire suppression using hydrate powder with a mass of 15 g and 30 g is given. According to the data obtained, no more than 15 g of hydrate powder is needed to fully extinguish a laboratory-scale wood fire of this mass. A further increase in the hydrate mass is impractical (superfluous). Thus, 5/1 is the optimal ratio of the burning material mass to the CO_2 hydrate powder mass. However, experiments with a sample of greater mass (180 g) show that more hydrate is needed to extinguish the fire and achieve the same CO_2 and CO concentrations as with a small sample. In this case, the mass ratio was equal to 2.25/1 (sample mass/hydrate mass). This result indicates that the relationship between the size of the sample and the mass of the gas hydrate used for fire suppression is nonlinear. Therefore, this nonlinearity needs to be considered when extending the results to larger fires.

In real conditions of compartment fire suppression, combustible materials are located in ties, cascades, and layers. In this case, it is necessary to switch to total flooding. There are quite large air gaps between the layers of burning material, so a different mechanism is required for the suppression. When upper layers are extinguished, middle and lower layers continue to burn because air (oxidizer) can still access them. Moreover, extinguishing agents may fail to reach lower layers. We performed preliminary experiments with multi-tier structures (wood pieces were arranged in several layers) and found that most of the extinguishing agent failed to reach lower layers with this type of structures. Just a small volume of sprayed water reached the lower layers as it drained down the wood pieces. Isolated zones were formed where water interacted with the lower layers of the fire. As part of the experiments, we extinguished a fire comprised of several tiers of wood (Figure 16). The fire suppression footage is presented. Pine rods were arranged in several tiers. The rods were 150–200 mm long and 5–7 mm thick; the total height of the tiers was 50–70 mm. Water spray and carbon dioxide hydrate were used for fire suppression. The mass of the extinguishing agent was varied. The minimum critical mass of the carbon dioxide hydrate powder required for fire suppression was 80 g. The wood to gas hydrate mass ratio was 2/1.

It is a known fact that water spray provides effective fire suppression due to the large surface area of droplets. This factor leads to a large vapor flow and a vapor cloud forming. The temperature in the flame combustion zone drops sharply, and the access of the oxidizer to the pyrolysis and combustion products becomes limited. The experiments have shown that water spray only suppresses the fire in the upper tier, which has a vapor cloud over it. Water droplets evaporate too fast to reach the lower tiers of wood pieces. Water vapor goes up due to gravitational convection, which blocks the access to the lower part of the combustion zone not only for vapor but also for small water droplets. Carbon dioxide hydrate powder, however, shows high efficiency in suppression of flame combustion and temperature reduction below the pyrolysis temperature. The dissociation of gas hydrate and its melting to form water took 0.1–1 s, and this time was enough for most of the powder granules to fall on different tiers and reach the base of the fire. The fall took less than 0.1 s. As a result, a high concentration of the inert mixture components—carbon dioxide and water vapors—was provided in the entire wood crib. Due to the heat of the gas hydrate dissociation, ice melting, and water evaporation, as well as due to the low initial temperature of the granules (about −30 °C), the temperature inside the fire crib quickly fell lower than the wood pyrolysis temperature. The main factors providing the suppression of flame combustion using the carbon dioxide hydrate are as follows: (i) solid particles falling to the bottom of the burning wood crib, (ii) phase transitions leading to a dramatic decrease in the temperature of the material and gas–vapor mixture, (iii) a large amount of carbon dioxide and vapor released from the hydrate preventing the access of the oxidizer. In addition, the density of carbon dioxide is much higher than that of air or vapor, which slows down the diffusion and convection of carbon dioxide from the combustion zone.

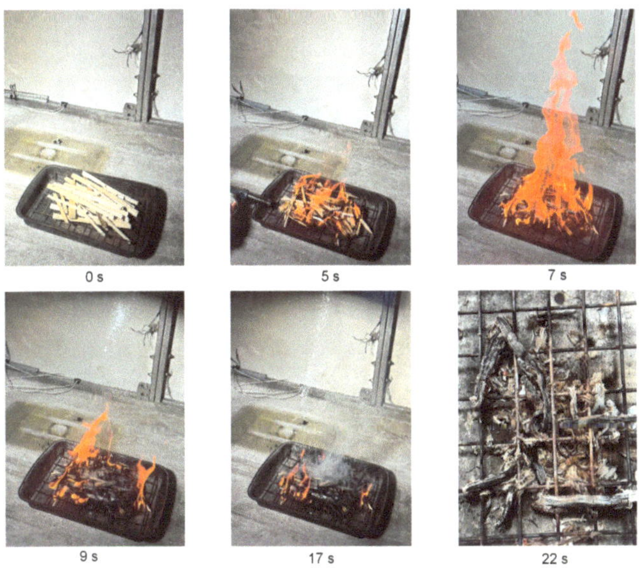

(a)

Figure 15. *Cont.*

(**b**)

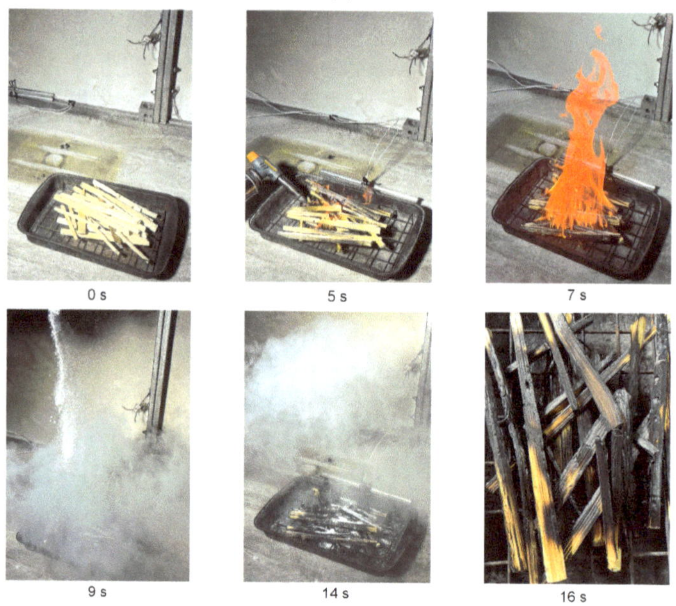

(**c**)

Figure 15. *Cont.*

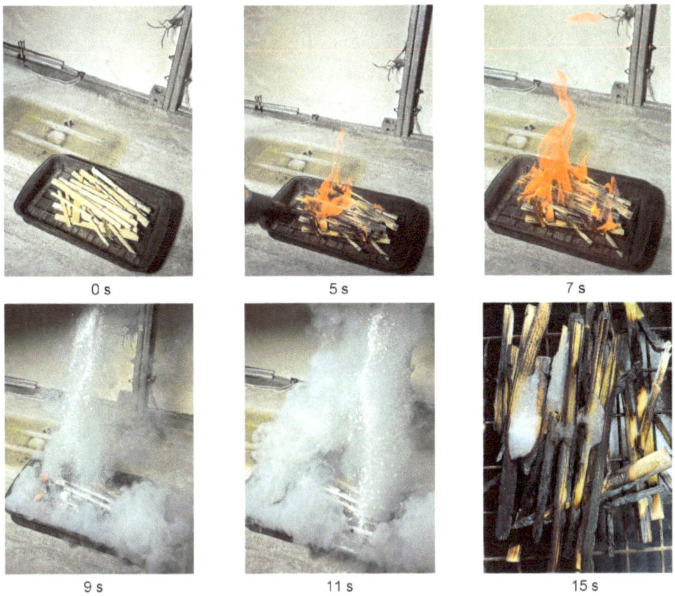

(**d**)

Figure 15. Snapshots of wood fire ($m_w \approx 70$ g) being suppressed using carbon dioxide hydrate powder: (**a**) $m_g \approx 7$ g; (**b**) $m_g \approx 15$ g; (**c**) $m_g \approx 30$ g; (**d**) $m_g \approx 100$ g.

Figure 16. Snapshots of the suppression of a model wood crib fire with several tiers using carbon dioxide hydrate powder: $m_g \approx 80$ g.

It is also important to note that total flooding with unsprayed water requires a large volume of the extinguishing agent and does not provide fire suppression. In this case, it is impossible to provide the even supply of water to the tiers of wood pieces in a wood crib. Local combustion zones are formed that water cannot reach. However, it is possible to distribute carbon dioxide hydrate powder (when it falls by gravity) evenly throughout the wood crib.

Figure 17 shows the images of materials ($m_w \approx 6$ g) used in the experiments with and without extinguishing using water, snow, ice, CO_2 hydrate powder, and CO_2 hydrate tablet.

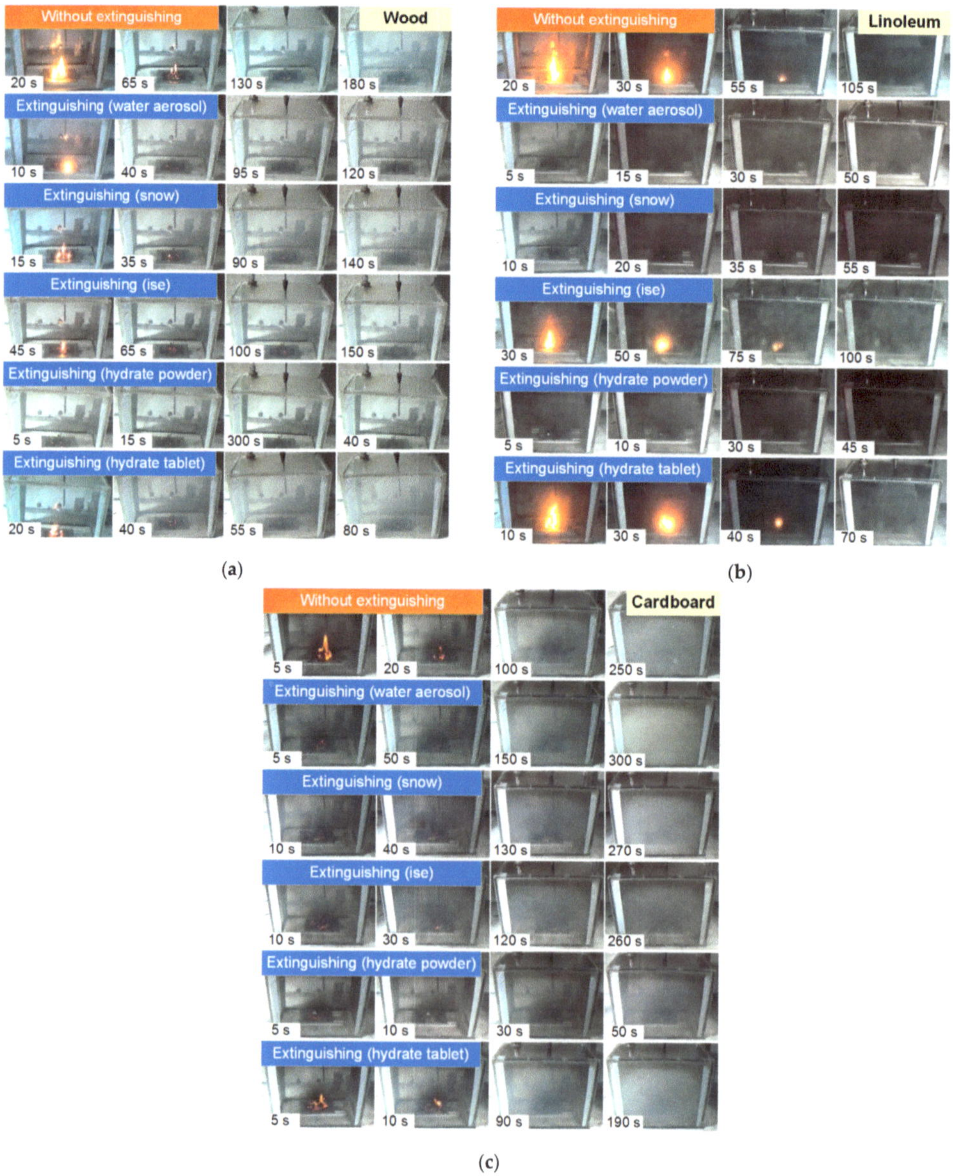

Figure 17. Images of fires ($m_w \approx 6$ g) based on wood (**a**), linoleum (**b**), and cardboard (**c**) burning without suppression by extinguishing agents as well as after contact with extinguishing agents (water, snow, ice, CO_2 hydrate).

Figure 18 shows the temperature trends obtained using thermocouples in the center of fires with and without being exposed to extinguishing agents (water spray, snow, ice, and CO_2 hydrate tablet). There are distinct intervals of intense pyrolysis and combustion of materials, supply of extinguishing agents, as well as the fire containment and suppression. The non-monotonic sectors on the trends illustrate the highly unsteady nature of the combustion after ignition and suppression using a wide range of agents. The conditions of fire containment and suppression were provided in all the experiments: a temperature decrease below 100 °C in the near-surface and deep layers. The analysis of Figure 18 shows that CO_2 hydrate powder provided the shortest time between the supply of the extinguishing agent and the moment when the flame combustion of the material stopped for all the laboratory-scale fires under study (1–5 s). Water aerosol was the second best extinguishing agent with 2–19 s. The worst result was obtained using ice: here, the flame quenching time was comparable to the one without extinguishing.

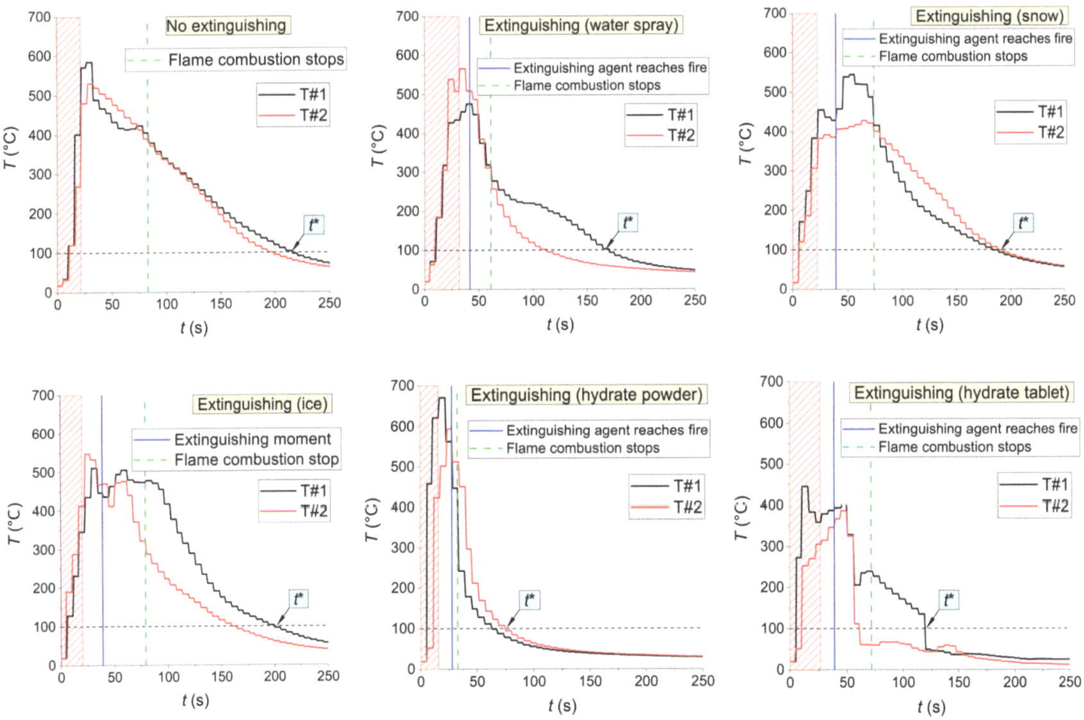

(a)

Figure 18. *Cont.*

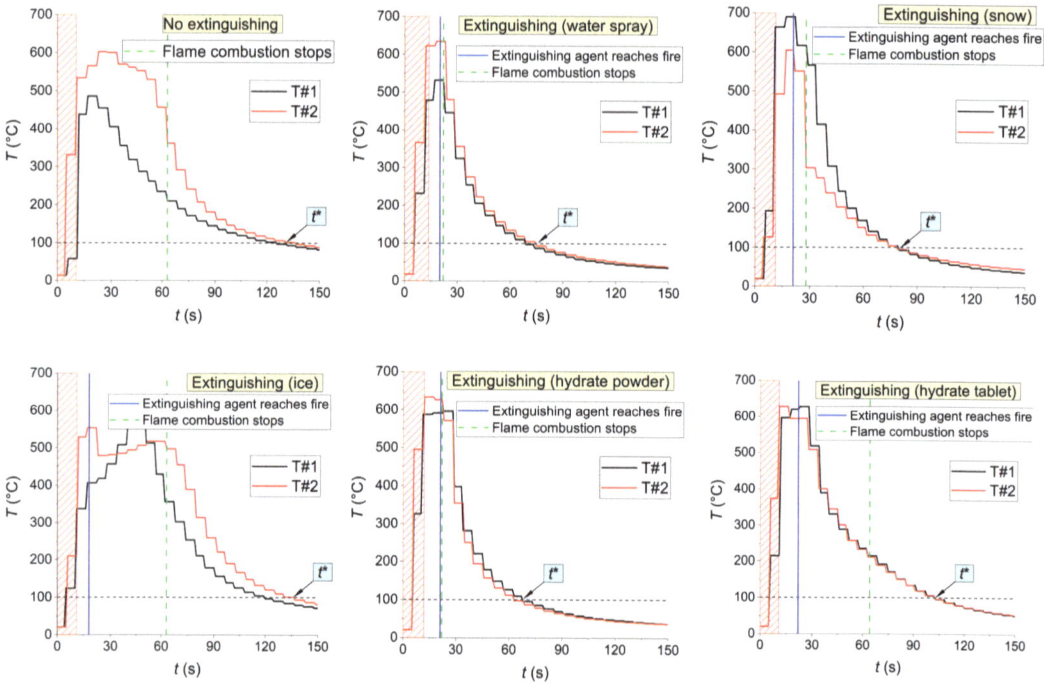

(**b**)

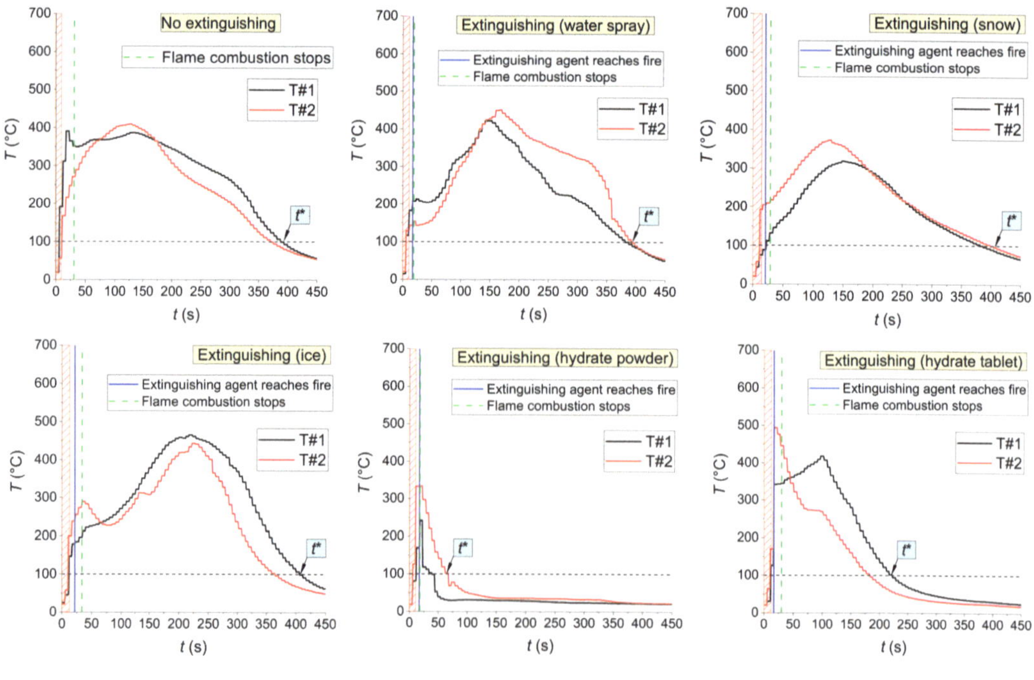

(**c**)

Figure 18. Dynamics of temperature variation in the center of fires for wood (**a**), linoleum (**b**), and cardboard (**c**) with and without suppression (the crosshatched (red) area is the interval when the seat of the fire was heated by a gas burner; t* is the time of complete burnout/decay of the fire).

The gas burner application time was determined by the type of fuel and the specific nature of its flaming. For instance, laboratory-scale fires made of wood were exposed to the burner until the temperature inside it reached 450–500 °C as recorded by at least one thermocouple (for about 25–30 s on average) (Figure 18). For fires made of linoleum, these values were 500–600 °C and 10–15 s, and for those made of cardboard, 200–250 °C and 15–20 s, respectively. Figure 18 reports the thermocouple measurements inside the fuel layer. The flame temperature does not play a major role in terms of laboratory-scale fire suppression because pyrolysis proceeds in the depth of the fuel even when no flame combustion is observed. Table 4 presents the data illustrating the action of extinguishing agents on the laboratory-scale fires based on the above-mentioned fuels. The experimental findings (Figure 18, Table 4) demonstrate the dynamics of the physical and chemical processes occurring in the structure of the pyrolyzing fuel. The analysis of dynamic processes is what makes it possible to confirm the guaranteed conditions of the full suppression of a chemical reaction. Fires exposed to suppression systems often reignite because fuel continues to pyrolyze even when flame combustion is contained. The experimental data show principal differences between the physics and chemistry of pyrolysis and fire containment using snow, ice, water, and inert gas hydrates in the form of powder and tablets. Differences in the temperature variation trends between different layers of pyrolyzing fuel exposed to extinguishing agents are the basis for extending the research findings to thicker samples. According to our conclusion, the thermal decomposition of materials arranged in a thick layer cannot be suppressed effectively only using the high heat capacity of the extinguishing agent. Phase transitions play an important part as they provide fast heat removal. The intense release of the inert gas during the hydrate dissociation improves the efficiency of flame fire suppression through diluting the gas–air mixture and displacing the oxidizer.

Table 4. Action of extinguishing agents on the fire: times until the flame combustion stops (t_b) and times until the fire is fully extinguished (t*).

	Wood		Cardboard		Linoleum	
	t_b (s)	t* (s)	t_b (s)	t* (s)	t_b (s)	t* (s)
No extinguishing	88	193	20	377	53	127
Water aerosol	19	129	2	376	2	55
Snow	35	149	8	378	7	59
Ice	70	161	12	384	45	119
CO_2 hydrate powder	5	47	1	51	1	46
CO_2 hydrate tablet	33	81	13	206	42	81

Snow and ice were selected as alternative extinguishing agents to determine how the extinguishing agent temperature and phase condition affected the suppression process. The use of snow and ice allowed us to evaluate the effect of dissociation (CO_2 formation). According to the experimental findings (Table 4), when ice and snow are used as extinguishing agents, the time until flame combustion stops (t_b) is 2–4 times longer compared to fire suppression with water aerosol, while extinction times (t) are comparable. Flame quenching time depends more on the contact area between the extinguishing agent and the pyrolyzing material, while complete fire suppression time largely depends on the volume of the extinguishing agent (the volumes of all the agents were comparable in the experiments). The temperature of the extinguishing agent has a negligible effect on the times t_b and t (Table 4), as well as on the temperatures of the laboratory-scale fires (Figure 18). With the phase condition and free surface temperatures being similar, the flame quenching time is on average 85% shorter when hydrate powder is used as an extinguishing agent compared to snow (irrespective of the type of fuel), and the complete fire suppression time is 15–70% shorter (depending on the type of fuel) (Table 4). A similar comparison of suppression by ice and a hydrate tablet shows overall similar values of t_b. However, t^* is on average 50% shorter for hydrate tablets than for ice. The results indicate that hydrate dissociation (CO_2 formation and oxidizer displacement from the combustion zone) has the decisive influence on suppression characteristics when CO_2 hydrates are used as extinguishing agents (in the form of both powder and tablets), and all the other conditions are similar.

Figure 19 gives the times of complete burnout/decay of the fires used in the experiments. We considered the conditions of their combustion without suppression by extinguishing agents and with suppression using water spray, snow, ice, as well as CO_2 hydrate tablet and powder. The principal physical differences in the conditions of fire suppression using different extinguishing agents were recorded quite consistently. In particular, the interaction of water spray with the surface of almost all the fires was notable for the bounce of some of the droplets and their repeated collision with burning and pyrolyzing fragments of the material. In addition, the ascending flue gases entrained small droplets of water. The fire suppression using water spray took quite a long time in the case of burst injection. Burst injection of liquid into the combustion zone reduced the consumption of the extinguishing agent but increased the fire containment and suppression time. Unlike water droplets, the interaction of gas hydrate granules with the burning surface was notable for their sticking to the surface, so the agglomeration of granules and material was the dominating regime. This enhanced the important mechanism of fire containment and suppression based on cooling the surface. Hydrate granules formed a film on the surface of the burning material as part of the heat exchange with it. Due to the high heat capacity and vaporization heat of water, there was a significant heat sink from the surface to the depth of the hydrate. This enhanced the CO_2 release. The hydrate layer became heterogeneous: it formed a composition with ice particles, water, as well as vapor and gas bubbles. Such structures are more effective than homogeneous ones during heat exchange. This is because the heat sink was intensified not only during the heating of liquids but also during their evaporation and boiling. The drop of snow and ice on the surface of the reacting materials in the form of powder also provided a denser coverage of the surface, but the presence of vaporization centers in hydrates in the form of ice particles, water droplets, and gas bubbles ensured better cooling of the surfaces of the burning samples. The principal differences between the mechanisms of fire containment and suppression used in the experiments were especially noticeable at high temperature in the reaction zone. An important benefit of gas hydrates used as extinguishing agents is the rapid displacement of the oxidizer from the combustion and pyrolysis zone. This does not only inhibit the oxidation reactions but also triggers the reactions stifling the growth of carbon monoxide, hydrogen, and methane concentrations typical of a certain stage of fire suppression by water.

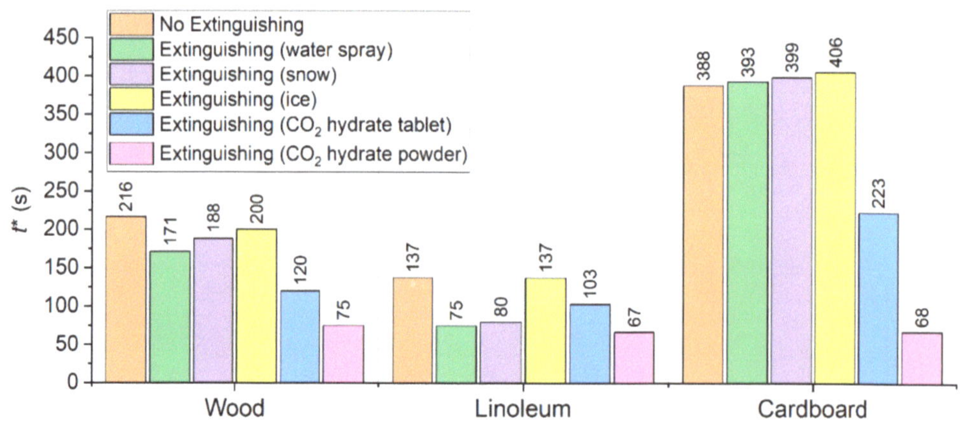

Figure 19. Times of complete burnout/decay of the fires used in the experiments with and without suppression with the extinguishing agents.

Interesting physical differences in the fire containment and suppression have been revealed between the materials used in this study: wood, linoleum, and cardboard. Wood burns with a stable and rather large flame zone due to the active release of pyrolysis gases and gas-phase reactions. It is impossible to contain wood fire by replacing oxygen with inert gas alone. Fire suppression can only be effective if several mechanisms are used, in particular, cooling the gas phase and wood surface as well as preventing the pyrolysis products, oxidizer, and combustion products from mixing. Cardboard and paper burn out very fast in the gas phase to form a fragile solid frame. The fire containment and suppression in this case is possible by cooling and destroying this frame but in a reserved manner. In particular, if the water droplets, ice and snow particles, hydrate tablets and powder granules did not cause the formation of firebrands, (i.e., actively pyrolyzing fragments of material), the combustion stopped rather quickly. However, if the burning material broke up after contact with the extinguishing agent, the fire area inevitably grew. The supply of hydrate tablets and granules caused the minimum fragmentation of the reacting material. The intense diffusion of gas from the hydrate surface prevents the particles of burning material from breaking off and covering long distances. Linoleum only has a flame combustion zone during the uninterrupted supply of the oxidizer and energy to its surface. Therefore, it is possible to effectively suppress the combustion of this material by blocking these two flows.

The acceptable repeatability of the experimental results suggests that they can be extended to other materials and systems as a whole because the mechanisms of combustion and, hence, of fire containment and suppression for most materials are the same as those considered here. For instance, rubber goods, particle board, and fiberboard are close to wood, and plastics are close to linoleum. Therefore, the interaction patterns of hydrate tablets and granules with the materials under study will mostly be identical for the most common materials, substances, and systems found in compartments.

The analysis of experimental footage allowed us to hypothesize the feasibility of using a set of hydrated gases and implementing alternative approaches to fire suppression. In particular, it is possible to effectively control the composition of pyrolysis gases by using specialized gaseous environments in reactors and chambers. We have established experimentally that the composition of pyrolysis and combustion products may vary in a wide range when different gasifying agents are used. Combustible gases in hydrates can be used to intensify the so-called back fire that is initiated in front of an active fire front. When these fronts meet, the fire gets contained (this solution is often used when combating wildfires). The main limitations of using multiple gases as part of hydrates stem from different critical (threshold) volumes of gases that a crystal lattice of a hydrate can hold. As a rule, the maximum concentrations in the hydrate structure range from 15 to 40% for most gases. This should be considered when using gas hydrates for fire containment and suppression. For instance, a hydrate lattice can hold up to 40% of carbon dioxide, but no more than 20% of methane or propane. Thus, double or triple hydrates can be effective in terms of controlling certain reactions in the pyrolysis, gasification, and flame combustion zone, but the amount of gases is limited by the hydrate structure. The obtained concentrations of gases in the pyrolysis and gasification products as well as gas–vapor mixtures in the reaction zone with and without suppression show the impact of the initial gas content within the material and hydrate, as well as the thermal conditions. These findings can be used to predict the necessary amount (volume of gas and mass of water) and type of hydrate for fire suppression.

The experimental findings (Figures 18 and 19) show that the CO_2 hydrate powder is the most effective extinguishing agent in terms of fire suppression time when the mass of the extinguishing agent approximates 1.5 g (Figure 19). The experiments established the necessary and sufficient volumes of water to extinguish the combustion of the materials under study. When analyzing the research findings, we took into account the discharge density specified in [42] that a firefighting system must be able to provide. Thus, according to [42], the required discharge density for the first and second groups of facilities is in the range of 0.08–0.12 $l/(m^2s)$ with a maximum duration of 30–60 min. Thus, with the above parameters, the maximum specified discharge density reaches 144–432 l/m^2. For the spray nozzle used in the experiments with a water droplet size ranging from 5 to 120 μm (with the specific discharge density set at 0.3 $l/(m^2s)$ and spraying time of 4–10 s), the specific water volume per unit area of the fire required to extinguish a fire is 8.7 l/m^2 for wood, 7.2 l/m^2 for cardboard, 3.6 l/m^2 for paper, and 0.9 l/m^2 for linoleum. The specific water volume required to extinguish a fire can increase by 1.5–2 times if the average droplet radius is increased to 250–300 μm. These values would be 17.4 l/m^2 for wood, 14.4 l/m^2 for cardboard, 7.2 l/m^2 for paper, and 1.8 l/m^2 for linoleum. These values are considerably lower than the required 144–432 l/m^2 [42]. This confirms the efficiency of the approach used in this research. With gas hydrates as extinguishing agents, a much smaller amount is necessary for fire containment and suppression. The calculations show that wood requires more CO_2 hydrate than cardboard, paper, or linoleum do due to the nuances of thermal decomposition and flame combustion described above. In particular, the minimum (threshold) mass of the hydrate powder is 20 g/m^2 for wood, 10 g/m^2 for cardboard, 5 g/m^2 for linoleum, and 4 g/m^2 for paper. The conditions of hydrate powder distribution over the surface of pyrolyzing material play an important role. Tablets effectively contain and suppress combustion only when the reacting materials have a small area and there is no inflow of oxygen. Granulated powder is more versatile in this respect. It can effectively cover quite a large surface area of the material. In this case, the gas release from the hydrate will be quite considerable and fast. The fire energy is also extensively spent on heating ice and water as well as their crystallization and evaporation, respectively. In the case of high temperatures in the combustion zone, water boiling begins to play an important part as well. Moreover, due to the multi-phase frame and heterogeneous structure of hydrates, we observe film and bubble boiling. Any application needs reliable experimental data on the dissociation rates and complete decomposition times to describe these patterns. Such data as well as the approximations and formulas obtained in this research can be used to extend the results to different sizes of fires, reacting materials, and temperatures of the gaseous environment in the flame combustion zone and intense pyrolysis area.

As gas hydrates are a multi-phase system, it is quite difficult to describe their heating, gas decomposition, changes in the phase structure and component proportions (gas, vapor, water, and ice) using the known dimensionless similarity criteria. Clearly, that would require large-scale and long-term research involving leading teams of specialists from different countries to obtain the criterial proportions that could be used to reliably predict the fire suppression characteristics for a wide range of materials and substances as well as single, double, triple, and multi-component hydrates with different compositions and gas types. The experimental data made it possible to analyze the patterns of physical and chemical processes as well as phase transitions during the interaction of gas hydrate samples and burning materials with a variable free surface area.

The numerical data were scaled to the areas of compartment fires. The necessary and sufficient masses of gas hydrate were determined for extinguishing fires on different areas In particular, burning materials with an area ranging from 10 to 20 m^2 can be extinguished using 0.2–0.4 kg of gas hydrate powder, which is approximately 45 wt% less than water spray. Moreover, the fire containment and suppression time is 11–82% shorter for hydrate powder than for water spray. Gas dissociation, ice melting, as well as water evaporation and boiling play a major role in the effective fire suppression and containment. That is why it is important to know the relationships between these characteristics and temperature in the combustion and pyrolysis zone as well as consider the complete gas dissociation and liquid evaporation times. The latter can be used to calculate the relative spraying densities for the surface of reacting materials. The experiments involved much smaller fire areas than those observed in real-life conditions. In a simplified statement, an increase in the fire surface area was proportional to an increase in the mass of gas hydrate required for fire containment and suppression. A more precise simulation would significantly complicate the prediction model. There is good reason to do that in the future as an independent work

Simple prediction of the necessary amount of gas hydrate is necessary to provide a quick response to fire outbreaks. For a detailed description of the estimation model, its limitations and deviations from the experiments, please see [40,41,43]. These models describe the dissociation of gas hydrate with varying heat exchange conditions (for instance, ambient temperature). The models do not describe the pyrolysis or chemical reactions during gas-phase combustion. Such a simulation would require further research. A simulation that would factor in the gas hydrate dissociation time depending on the heat exchange conditions is also a major scientific objective. The results of simplified modeling here can be used for scaling—selecting the mass of an extinguishing agent (carbon dioxide hydrate)—with varying fire surface areas.

The key barrier to the large-scale use of gas fire suppression systems is the complicated evacuation of people. As a rule, gas fire suppression is launched when no living organisms are present in the fire-affected compartments, buildings, and structures. Due to their controllable dissociation rate, gas hydrates coupled with water vapor injection make it possible to evacuate people even in the course of fire containment and suppression. One can estimate the time for human evacuation from compartments given the known hydrate dissociation time, gas-to-water ratio in hydrates, as well as the gas and steam concentrations as functions of the interaction time with the reacting materials. At the same time, if the temperature in the combustion zone, pyrolysis gas concentration, and type of burning material are known, the fire containment and suppression can be optimized in terms of suppression time, volume of the extinguishing agent involved, and conditions of the agent supply. In particular, systems of CO_2 hydrate granules with different initial temperatures and, hence, different initial dissociation rates can be regarded as especially promising. It is possible to vary not only the initial concentrations of the hydrated gas and water vapors but also their rheological characteristics (fluidity, viscosity, and structure of layers). This will allow firefighters to employ each of the three fire suppression mechanisms separately or together: cooling the surface and gas environment, displacement of oxidizer and pyrolysis products, and intensification of endothermic phase transitions to control chemical reactions.

We propose that future research should focus on using gas hydrate as an extinguishing agent for liquid and composite systems. Fires involving such systems are difficult to combat due to their high reactivity, and composite systems are even more challenging because they can react both in the gas phase and on the surface (heterogeneous combustion). However, where conventional water, gas, and foam firefighting may struggle or fail, gas hydrates bring new benefits to the table. According to the experiments, these benefits improve the fire-extinguishing performance for a wide range of materials and substances. It is important to adapt the proposed engineering solutions based on gas hydrates to challenging operating conditions.

Carbon dioxide hydrates have not yet been used for fire containment and suppression. Here, we have, for the first time, shown the experimental findings explaining the physical effects emerging during the suppression of class A fires using carbon dioxide hydrates. The data obtained are extremely important as they confirm that hydrates with inert gases can be used for fire suppression. The data will also be helpful for developing physical and mathematical models to predict the effective conditions of fire containment by exposure to a gas hydrate. The physicochemical processes behind this are extremely complex, as the research has shown. Therefore, we concentrated on performing experiments with small-size fires. Due to the small size and limited mass of the gas hydrate powder and tablets, we managed to determine the critical ratios between the mass of pyrolyzing and burning materials and gas hydrate. We also determined the factors, processes, and effects exerting the most significant influence on the conditions of combustion front propagation. When processing the experimental findings, we obtained enough data to extend them to large-scale fires. These aspects define the scientific novelty and practical value of the research findings. The new knowledge on hydrate dissociation under the conditions considered in this study can serve as a basis for the development of the theory of changes in the hydrate structure and component composition of the gas–vapor–air mixture during their dissociation under limited energy supply. Using this new knowledge, balance models can be constructed that will quickly estimate the required ratios of the hydrate mass and volume of gases to limit the growth of heat fluxes not just for firefighting but also in heat exchange and cooling systems, which are widely used in direct-contact selective technologies. Using the data obtained, it is advisable to develop a summarized model simulating the heat exchange of inert gas hydrate with chemically active materials and substances.

4. Conclusions

In this paper, we conducted experiments on fire containment and suppression by dropping CO_2 hydrate granules and tablets, ice, snow, water spray on burning solid materials. We used the combustible materials typical of compartment fires—wood, linoleum, and cardboard—to determine the volume and mass of gas hydrate powder necessary for the effective fire suppression.

(i) We have experimentally determined the optimal ratios of the CO_2 hydrate mass to surface area for fire containment and suppression. The laboratory-scale fires used in the experiments contained wood, linoleum, and cardboard, most commonly involved in compartment fires. The mass of CO_2 hydrate sufficient for the complete fire suppression is shown to be much lower than that of water, snow, or ice. In particular, minimum (critical) masses of hydrate powder required for fire containment and suppression were as follows: 20 g/m^2 for wood, 5 g/m^2 for linoleum, 4 g/m^2 for paper, and 10 g/m^2 for cardboard. The extinguishing agents under study triggered the main mechanisms of fire suppression: accumulation of the heat from the fire for heating the agent and for subsequent phase transitions; blocking the access of the oxidizer to thermal decomposition and combustion products; cooling the gas–air environment around the fire.

(ii) The main anthropogenic emissions were evaluated from the combustion and suppression of a fire involving wood, cardboard, and linoleum using variable extinguishing agents. Carbon dioxide hydrate powder was proved the most effective under decreasing CO and CO_2 concentrations compared to other extinguishing agents. The difference in the maximum concentrations of CO_2 and CO emitted from the combustion of cardboard, wood, and linoleum without suppression and with suppression using CO_2 powder was up to 70%.

(iii) The experimental findings formed a database for developing the practices of fast compartment fire suppression using gas hydrate tablets and granulated powder. The research findings make it possible to use specialized gas mixtures in hydrates as well as additives to water that will alter the fire suppression and containment parameters in a wide range. If such conditions are provided, this will increase the efficiency of firefighting management against fires involving widespread solid combustible materials in industrial and residential facilities in line with the fire parameters and firefighting resources available.

Author Contributions: Conceptualization, data curation, funding acquisition, P.S.; investigation, P.S., R.V., V.M. and O.G.; writing—original draft, P.S. and O.G.; writing—review and editing, R.V., P.S., O.G. All authors have read and agreed to the published version of the manuscript.

Funding: The research was supported by the program of the National Research Tomsk Polytechnic University (Priority-2030-NIP/EB-006-0000-2023).

Conflicts of Interest: The authors declare no conflict of interest.

References

1. Guo, X.; Zhang, H.; Pan, X.; Zhang, L.; Hua, M.; Zhang, C.; Zhou, J.; Yan, C.; Jiang, J. Experimental and Numerical Simulation Research on Fire Suppression Efficiency of Dry Powder Mediums Containing Molybdenum Flame Retardant Additive. *Process Saf. Environ. Prot.* **2022**, *159*, 294–308. [CrossRef]
2. Bi, Y.; Yang, Z.; Cong, H.; Bi, M.; Gao, W. Experimental and Theoretical Investigation on the Effect of Inclined Surface on Pool Fire Behavior. *Process Saf. Environ. Prot.* **2022**, *162*, 328–336. [CrossRef]
3. Ma, Q.; Guo, Y.; Zhong, M.; You, J.; He, Y.; Chen, J.; Zhang, Z. Theoretical Prediction Model of the Explosion Limits for Multi-Component Gases (Multiple Combustible Gases Mixed with Inert Gases) under Different Temperatures. *Fire* **2022**, *5*, 143. [CrossRef]
4. Ma, Y.; Zhang, Z.; Du, H.; Li, K.; Wu, Q.; Wang, S.; Wang, X. Experimental Study on Combustion Characteristics of Mixed Sodium Fire in a Well-Ventilated Environment Compared with Confined Space. *Ann. Nucl. Energy* **2022**, *172*, 109069. [CrossRef]
5. Wang, K.; Yuan, Y.; Chen, M.; Lou, Z.; Zhu, Z.; Li, R. A Study of Fire Drone Extinguishing System in High-Rise Buildings. *Fire* **2022**, *5*, 75. [CrossRef]
6. Liu, R.; Yuan, C.; Ma, W.; Liu, S.; Lu, S.; Zhang, H.; Gong, J. Simulation Study on Aircraft Fire Extinguishing Pipeline with Different Filling Conditions and Pipeline Characteristics. *Fire* **2022**, *5*, 86. [CrossRef]
7. Liu, R.; Shi, H.; Zhou, Q.; Ma, W.; Wang, T.; Lu, S. Simulation of Fire Extinguishing Agent Transport and Dispersion in Aircraft Engine Nacelle. *Fire* **2022**, *5*, 97. [CrossRef]
8. Lv, D.; Tan, W.; Zhu, G.; Liu, L. Gasoline Fire Extinguishing by 0.7 MPa Water Mist with Multicomponent Additives Driven by CO_2. *Process Saf. Environ. Prot.* **2019**, *129*, 168–175. [CrossRef]
9. Pei, B.; Zhu, Z.; Yang, S.; Wei, S.; Pan, R.; Yu, M.; Chen, L. Evaluation of the Suppression Effect on the Flame Intensification of Ethanol Fire by N2 Twin-Fluid Water Mist Containing KQ Compound Additive. *Process Saf. Environ. Prot.* **2021**, *149*, 289–298. [CrossRef]
10. Cui, Y.; Liu, J. Research Progress of Water Mist Fire Extinguishing Technology and Its Application in Battery Fires. *Process Saf. Environ. Prot.* **2021**, *149*, 559–574. [CrossRef]
11. Xing, H.; Lu, S.; Yang, H.; Zhang, H. Review on Research Progress of $C_6F_{12}O$ as a Fire Extinguishing Agent. *Fire* **2022**, *5*, 50. [CrossRef]
12. Dasgotra, A.; Rangarajan, G.; Tauseef, S.M. CFD-Based Study and Analysis on the Effectiveness of Water Mist in Interacting Pool Fire Suppression. *Process Saf. Environ. Prot.* **2021**, *152*, 614–629. [CrossRef]
13. Wang, W.; He, S.; He, T.; You, T.; Parker, T.; Wang, Q. Suppression Behavior of Water Mist Containing Compound Additives on Lithium-Ion Batteries Fire. *Process Saf. Environ. Prot.* **2022**, *161*, 476–487. [CrossRef]
14. Liu, M.; Liu, Y.; Sun, H.; Hu, J.; Wang, X. Experimental Study on the Interaction of Water Mist Spray with Two Buoyant Non-Premixed Flames. *Process Saf. Environ. Prot.* **2022**, *161*, 1–12. [CrossRef]
15. Wu, Y.; Yu, X.; Wang, Z.; Jin, H.; Zhao, Y.; Wang, C.; Shen, Z.; Liu, Y.; Wang, W. The Flame Mitigation Effect of N_2 and CO_2 on the Hydrogen Jet Fire. *Process Saf. Environ. Prot.* **2022**, *165*, 658–670. [CrossRef]
16. Liu, F.P.; Li, A.R.; Qing, S.L.; Luo, Z.D.; Ma, Y.L. Formation Kinetics, Mechanism of CO_2 Hydrate and Its Applications. *Renew. Sustain. Energy Rev.* **2022**, *159*, 112221. [CrossRef]
17. Riestenberg, D.E.; Tsouris, C.; Brewer, P.G.; Peltzer, E.T.; Walz, P.; Chow, A.C.; Adams, E.E. Field Studies on the Formation of Sinking CO_2 Particles for Ocean Carbon Sequestration: Effects of Injector Geometry on Particle Density and Dissolution Rate and Model Simulation of Plume Behavior. *Environ. Sci. Technol.* **2005**, *39*, 7287–7293. [CrossRef]
18. Liu, N.; Meng, F.; Chen, L.; Yang, L.; Liu, D. Investigating the Effects of MWCNT-HB on Gas Storage Performance of CO_2 Hydrate. *Fuel* **2022**, *316*, 123289. [CrossRef]
19. Kang, K.C.; Linga, P.; Park, K.n.; Choi, S.J.; Lee, J.D. Seawater Desalination by Gas Hydrate Process and Removal Characteristics of Dissolved Ions (Na^+, K^+, Mg^{2+}, Ca^{2+}, B^{3+}, Cl^-, $SO_4{}^{2-}$). *Desalination* **2014**, *353*, 84–90. [CrossRef]
20. Sato, T.; Takeya, S.; Nagashima, H.D.; Ohmura, R. Preservation of Carbon Dioxide Clathrate Hydrate Coexisting with Sucrose under Domestic Freezer Conditions. *J. Food Eng.* **2014**, *120*, 69–74. [CrossRef]
21. Tupsakhare, S.S.; Castaldi, M.J. Efficiency Enhancements in Methane Recovery from Natural Gas Hydrates Using Injection of CO2/N2 Gas Mixture Simulating in-Situ Combustion. *Appl. Energy* **2019**, *236*, 825–836. [CrossRef]
22. Sugahara, A.; Nakano, H.; Yokomori, T.; Ohmura, R.; Ueda, T. Effect of Fuel Boiling Point of Pool Flame for the Flame Extinction by CO_2 Hydrate. In Proceedings of the 10th Asia-Pacific Conference on Combustion, ASPACC 2015, Beijing, China, 19–22 July 2015.
23. Hatakeyama, T.; Aida, E.; Yokomori, T.; Ohmura, R.; Ueda, T. Fire Extinction Using Carbon Dioxide Hydrate. *Ind. Eng. Chem. Res.* **2009**, *48*, 4083–4087. [CrossRef]
24. Hu, W.; Chen, C.; Sun, J.; Zhang, N.; Zhao, J.; Liu, Y.; Ling, Z.; Li, W.; Liu, W.; Song, Y. Three-Body Aggregation of Guest Molecules as a Key Step in Methane Hydrate Nucleation and Growth. *Commun. Chem.* **2022**, *5*, 33. [CrossRef] [PubMed]
25. Zhang, P.; Liu, B.; Hu, L.; Meegoda, J.N. Coupled Multiphase Flow and Pore Compression Computational Model for Extraction of Offshore Gas Hydrates. *Comput. Geotech.* **2022**, *145*, 104671. [CrossRef]
26. Chen, H.; Du, H.; Shi, B.; Shan, W.; Hou, J. Mechanical Properties and Strength Criterion of Clayey Sand Reservoirs during Natural Gas Hydrate Extraction. *Energy* **2022**, *242*, 122526. [CrossRef]
27. Liu, H.; Li, H.; Yao, D.; Guo, P.; Wen, L. The Research on the Natural Gas Hydrate Dissociation Kinetic from Hydrate-Sediments/Seawater Slurries. *Chem. Eng. J.* **2022**, *435*, 135127. [CrossRef]

28. Li, X.Y.; Wang, Y.; Li, X.S.; Zhang, Y.; Chen, Z.Y. Experimental Study of Methane Hydrate Dissociation in Porous Media with Different Thermal Conductivities. *Int. J. Heat Mass Transf.* **2019**, *144*, 118528. [CrossRef]
29. Song, Y.; Tian, M.; Zheng, J.; Yang, M. Thermodynamics Analysis and Ice Behavior during the Depressurization Process of Methane Hydrate Reservoir. *Energy* **2022**, *250*, 123801. [CrossRef]
30. Wei, R.; Xia, Y.; Qu, A.; Lv, X.; Fan, Q.; Zhang, L.; Zhang, Y.; Zhao, J.; Yang, L. Dependence of Thermal Conductivity on the Phase Transition of Gas Hydrate in Clay Sediments. *Fuel* **2022**, *317*, 123565. [CrossRef]
31. Cui, G.; Dong, Z.; Xie, K.; Wang, S.; Guo, T.; Liu, J.; Xing, X.; Li, Z. Effects of Gas Content and Ambient Temperature on Combustion Characteristics of Methane Hydrate Spheres. *J. Nat. Gas Sci. Eng.* **2021**, *88*, 103842. [CrossRef]
32. Cui, G.; Wang, S.; Dong, Z.; Xing, X.; Shan, T.; Li, Z. Effects of the Diameter and the Initial Center Temperature on the Combustion Characteristics of Methane Hydrate Spheres. *Appl. Energy* **2020**, *257*, 114058. [CrossRef]
33. Lin, S.; Huang, X.; Gao, J.; Ji, J. Extinction of Wood Fire: A Near-Limit Blue Flame Above Hot Smoldering Surface. *Fire Technol.* **2022**, *58*, 415–434. [CrossRef]
34. Ruøycka, M.; Giebułtowicz, J.; Wroczy-Ski, P. Present-Day Toxicity of Smoke Inhalation. *Acta Pol. Pharm.-Drug Res.* **2018**, *75*, 281–295.
35. Giebułtowicz, J.; Rużycka, M.; Wroczyński, P.; Purser, D.A.; Stec, A.A. Analysis of Fire Deaths in Poland and Influence of Smoke Toxicity. *Forensic Sci. Int.* **2017**, *277*, 77–87. [CrossRef] [PubMed]
36. Misyura, S.Y.; Manakov, A.Y.; Morozov, V.S.; Nyashina, G.S.; Gaidukova, O.S.; Skiba, S.S.; Volkov, R.S.; Voytkov, I.S. The Influence of Key Parameters on Combustion of Double Gas Hydrate. *J. Nat. Gas Sci. Eng.* **2020**, *80*, 103396. [CrossRef]
37. Dorokhov, V.V.; Kuznetsov, G.V.; Nyashina, G.S.; Strizhak, P.A. Composition of a Gas and Ash Mixture Formed during the Pyrolysis and Combustion of Coal-Water Slurries Containing Petrochemicals. *Environ. Pollut.* **2021**, *285*, 117390. [CrossRef] [PubMed]
38. Chvanov, S.V.; Kuznetsov, G.V.; Strizhak, P.A.; Volkov, R.S. The Necessary Water Discharge Density to Suppress Fires in Premises. *Powder Technol.* **2022**, *408*, 117707. [CrossRef]
39. Rabinovich, S.G. Measurement Errors and Uncertainties. In *Theory and Practice*; Springer Science and Media, Inc.: New York, NY, USA, 2005; ISBN 0387253580.
40. Misyura, S.Y.; Donskoy, I.G. Dissociation Kinetics of Methane Hydrate and CO_2 Hydrate for Different Granular Composition. *Fuel* **2020**, *262*, 116614. [CrossRef]
41. Misyura, S.Y.; Donskoy, I.G. Dissociation of a Powder Layer of Methane Gas Hydrate in a Wide Range of Temperatures and Heat Fluxes. *Powder Technol.* **2022**, *397*, 117017. [CrossRef]
42. Fire Protection Systems. Automatic Fire Extinguishing Installations. SP 485.1311500.2020; Russia, 2020.
43. Misyura, S.Y.; Donskoy, I.G. Dissociation of Gas Hydrate for a Single Particle and for a Thick Layer of Particles: The Effect of Self-Preservation on the Dissociation Kinetics of the Gas Hydrate Layer. *Fuel* **2022**, *314*, 122759. [CrossRef]

fire

Article

Synthesis of High-Efficiency, Eco-Friendly, and Synergistic Flame Retardant for Epoxy Resin

Jiaxiang Gao, Hanguang Wu *, Yang Xiao, Wenjing Ma, Fei Xu, Rui Wang and Zhiguo Zhu *

School of Materials Design and Engineering, Beijing Institute of Fashion Technology, Beijing 100029, China
* Correspondence: hanguangwu@bift.edu.cn (H.W.); clyzzg@bift.edu.cn (Z.Z.)

Abstract: It remains a challenge to prepare flame-retardant composites via the addition of a low content of flame retardant. In this work, a novel DOPO-functionalized reduced graphene oxide hybrid (DOPO-M-rGO) flame-retardant system was synthesized for epoxy resin (EP). The phosphorus-nitrogen-reduced graphene oxide ternary synergistic effect provided DOPO-M-rGO with high flame-resistance efficiency in EP; thus, the EP-based composite exhibited superior fire-resistant performance at extremely low DOPO-M-rGO loading. The limiting oxygen index (LOI) of the EP-based composite was increased from 25% to 32% with only 1.5 wt% DOPO-M-rGO addition, and the peak heat release rate (pHRR), total heat release (THR), and total smoke production (TSP) were significantly decreased by 55%, 30%, and 20%, respectively. In addition, as a halogen-free flame-retardant system, DOPO-M-rGO presents great application potential as an eco-friendly additive for the flame-resistance improvement of thermosetting polymer materials.

Keywords: flame retardant; epoxy resin; synergistic effect; high efficiency

Citation: Gao, J.; Wu, H.; Xiao, Y.; Ma, W.; Xu, F.; Wang, R.; Zhu, Z. Synthesis of High-Efficiency, Eco-Friendly, and Synergistic Flame Retardant for Epoxy Resin. *Fire* **2023**, *6*, 14. https://doi.org/10.3390/fire6010014

Academic Editors: Song Lu, Changcheng Liu, Guohui Li and Paweł Wolny

Received: 11 December 2022
Revised: 31 December 2022
Accepted: 1 January 2023
Published: 4 January 2023

1. Introduction

As a polymer material with excellent properties and low cost, epoxy resin (EP) is widely used in various fields, such as adhesives, coatings, electronics, and aerospace industries [1–8]. Nevertheless, the inherent flammability of EP might pose a serious threat to the environment, people's life and property, and severely limits its further application in automobile, household, and aerospace fields. Therefore, the flame-resistance modification for EP has attracted much attention [9–19].

Graphene oxide (GO) is one of the most promising materials as the flame retardant for EP because it presents high thermal stability and an excellent barrier effect during combustion [20–24]. In addition, taking advantage of the abundant functional groups, GO is easily doped by metal ions or chemically reduced with other molecules, leading to the formation of a variety of flame-retardant systems with further improved fire-resistant performances, in which the reduced graphene oxide (rGO) component functions as the multi-layered skeleton [25–30]. In Wang's work, a flame retardant for EP was prepared by functionalizing the rGO with a sheet-like metal-organic zinc N, N′-piperazine (bismethylene phosphonate) [31]. The addition of 5 wt% of flame retardant promoted the EP-based composite to reach UL-94 V-0 rating, and resulted in 38.7%, 30.5%, and 33.7% lowered peak heat release rate (pHRR), total heat release (THR), and total smoke production (TSP) values, respectively, than that of pure EP. Zhu synthesized an iron hexamethylenediaminetetrakis-(methylenephosphonate) (Fe-HDTMP)-rGO hybrid, and used it in EP to improve the fire resistance. Compared to the pure EP, the EP modified by 5 wt% (Fe-HDTMP)-rGO exhibited 68.2% lower TSP, 54.5% lower peak CO production rate, 66.3% lower THR, and 47.7% reduced pHRR [32]. Through grafting silane coupling agent (KH550) as a bridge to connect GO and Octa (propyl glycidyl ether) POSS, Qu synthesized an organosilane-functionalized GO (FGO) as the flame retardant of EP, and the modification of EP-based composite with only 0.7 wt% FGO could reduce pHRR, THR, and TSP values by 49.7%,

34.3%, and 41.5%, respectively [33]. Xiao produced a flame retardant for EP by using melamine and GO, which is called AGO@COF [34]. The limiting oxygen index (LOI) value of the EP-based composite increased from 24% to 25.5% when 2 wt% AGO@COF was added, and the pHRR value was reduced by 43.6%. However, the flame retardancy performances of the graphene-based flame retardants, especially the UL-94 rating, need to be further improved (see Table S1).

9,10-Dihydro-9-oxa-10-phosphaphenanthrene-10-oxide (DOPO) is an H-phosphinate compound with active hydrogen, which is also widely used as the flame retardant for EP in previous studies (see Table S2) [35–41]. However, the DOPO-based flame retardants face the challenge of low flame-retardant efficiency, which can be solved by developing the hybrid flame-retardant systems of DOPO and rGO (see Table S3) [42–47]. Through the reaction between the active P-H from DOPO and the epoxy groups on GO, DOPO can be directly grafted onto the surface of GO to develop a DOPO-functionalized rGO, which plays an effective role in improving the flame resistance of the EP-based composites [42,43]. Zhi synthesized a functionalized GO grafted by DOPO and vinyltriethoxysilane (VTMS), which obviously increased the thermal stability and flame-retardant properties of EP [44]. In Ji's work, a functionalized rGO decorated with bi-DOPO groups (f-GO) was fabricated by using a covalent modification method, which endows EP with superior fire-resistant performances [45]. The extremely low loading (1 wt%) of the f-GO increased the LOI value of the composite from 19.9% to 30.8%, and the THR and the pHRR values were decreased by 44.0% and 55.5%, respectively. Furthermore, strategies have been developed to fabricate the DOPO-functionalized rGO hybrids through grafting DOPO onto the rGO skeleton by using other components as the bridge. Feng used glycidyl methacrylate as the bridge to synthesize a DOPO-functionalized rGO hybrid (GP-DOPO), and used it as the flame retardant for EP [46]. The addition of 2 wt% GO-DOPO decreased the pHRR value of the EP-based composite by 27%, the THR value is decreased by 32%, and the TSP value is decreased by 31%. In Qian's work, DOPO was firstly reacted with VTMS to form an intermediate, which was then grafted onto rGO by using (3-isocyanatopropyl) triethoxysilane as the bridge, and obtained a product which is highly effective in reducing the fire hazards of EP [47]. Although the DOPO-functionalized rGO flame retardants present higher flame-retardant efficiency, the vertical combustion grades of most DOPO-functionalized rGO flame retardants are required to be improved to achieve V-0.

As a nitrogen-containing compound, melamine is widely used as a typical eco-friendly flame retardant for EP, and its flame-resistance effect can be further improved when used cooperatively with other phosphorus-based flame retardants [48–52]. In this work, a novel DOPO-functionalized rGO hybrid (DOPO-M-rGO) was synthesized by using the synthesized methyl vinyl dichlorosilane/melamine polymeric intermediate as the bridge, which was used as a graphene-based phosphorus/nitrogen-containing flame retardant for EP. This study provides a facile approach for creating an extraordinarily effective flame retardant for EP, and aims to investigate the synergistic flame-resistance effect of DOPO, melamine, and the rGO skeleton components of DOPO-M-rGO.

2. Experimental Section

2.1. Materials

GO was prepared according to Hummer's method [53]. Melamine, methyl vinyl dichlorosilane, 1-ethyl-3-(3-dimethylaminopropyl) carbonized diimine hydrochloride (EDCI), and 4-dimethylaminopyridine (DMAP) were provided by Shanghai Maclin Biochemical Technology Co., Ltd. (Shanghai, China). DOPO, azo-diisobutyronitrile (AIBN), EP precursor (E51), and polyamide agent (650) were provided by Shanghai Aladdin Reagent Co., Ltd. (Shanghai, China). Tetrahydrofuran (THF) was provided by Beijing Tongguang Fine Chemical Co., Ltd. (Beijing, China).

2.2. Synthesis of DOPO-M-rGO

DOPO-M-rGO was synthesized by a two-step method, and the corresponding synthetic route is illustrated in Figure 1. First, melamine and methyl vinyl dichlorosilane (with the mass ratio of 1:1) was reacted in the THF suspension under mild stirring for 4 h to form a polymeric intermediate bearing -NH_2 groups and C=C bonds (see Figure 1a). Then, DOPO was added into the obtained system, and AIBN was used as the catalyst. The mass ratio of the melamine/methyl vinyl dichlorosilane polymeric intermediate, DOPO, and AIBN was controlled at 1:1:0.005. After reaction at 70 °C for 20 h, the white solid precipitate in the system was collected, washed, and dried, which is named DOPO-M in the following text (Figure 1b). Afterward, the obtained DOPO-M sample was added into THF along with GO to make a suspension, in which the mass ratio of DOPO-M and GO was 2:5. A total of 10 wt% EDCI and 1 wt% DMAP were added into the suspension as the catalyst. After 10 h reaction at 70 °C in the N_2 atmosphere, the ultimate suspension was filtered to remove the impurities, and the obtained black solid was abbreviated as DOPO-M-rGO in the following text (Figure 1b), which was used as the flame retardant of EP. The schematic for the preparation of DOPO-M and DOPO-M-rGO is shown in Figure 2a.

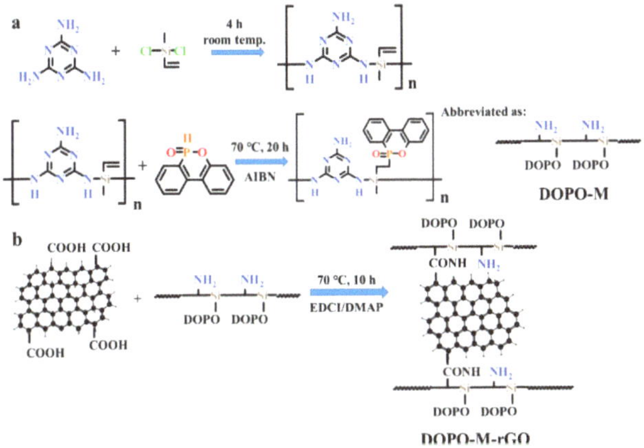

Figure 1. Synthetic route of (**a**) DOPO-M and (**b**) DOPO-M-rGO.

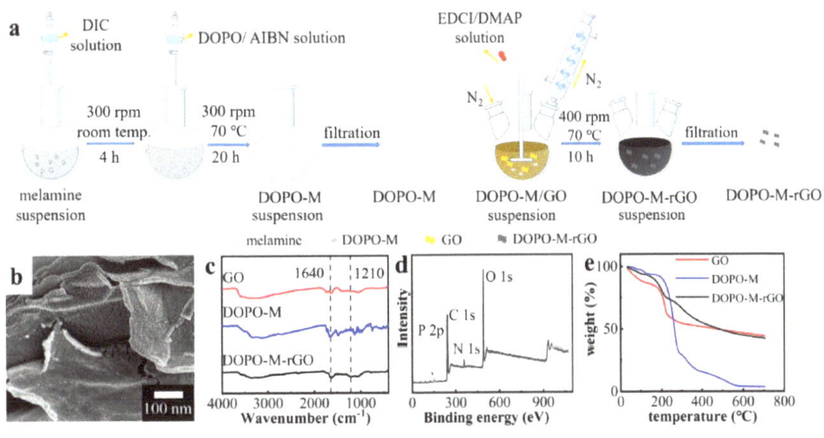

Figure 2. (**a**) Schematic representation for the preparation of DOPO-M-rGO; (**b**) SEM images of DOPO-M-rGO; (**c**) the FTIR spectra of GO, DOPO-M, and DOPO-M-rGO; (**d**) the XPS spectra for DOPO-M-rGO; (**e**) the TGA curves of GO, DOPO-M, and DOPO-M-rGO.

2.3. Preparation of DOPO-M-rGO/EP Composite

The prepared DOPO-M-rGO (0.5, 1, 1.5, 2, 3, and 5 wt%) was slowly added into the EP precursor (E51) and tenderly stirred at 60 °C for 10 min, and the polyamide agent (650) (45 wt%) was added into the system whilst continuously stirring as the curing agent of EP. Then, the obtained uniform mixture was casted and cured at 80 °C for 2 h, and the DOPO-M-rGO/EP composites with different DOPO-M-rGO contents were prepared.

The rGO/EP composite with 1.5 wt% rGO and the DOPO-M/EP composite with 1.5 wt% DOPO-M were also prepared for comparative study.

2.4. Characterization

Fourier transform infrared spectrometry (FTIR) was conducted by a Nicolet 6700 FTIR spectrometer (Thermo Fisher Scientific, Waltham, MA, USA) over the range of 400–4000 cm^{-1}. The X-ray photoelectron spectroscopy (XPS) was performed on an X-ray photoelectron spectrometer (AXIS, Kratos Analytical Ltd., Stretford, UK); the emission current and voltage were controlled at 5 mA and 10 kV, respectively. Raman spectroscopy was conducted by a SPEX-1403 laser Raman spectrometer (Renishaw in Via, London, UK) at an excitation wavelength of 532 nm. The morphologies of the samples were observed by using a scanning electron microscope (JSM-7500F, JEOL, Tokyo, Japan). The mechanical performances were measured by using an Instron universal material strength tester (4302, Instron Corporation, High Wycombe, UK) at a speed of 10 $mm \cdot min^{-1}$ according to the ASTM D638 standard. The width of the specimen was 6.0 ± 0.1 mm and the thickness was 2.0 ± 0.1 mm. All the samples were tested five times and the average values were selected as the final results.

Thermogravimetric analysis (TGA) was used to characterize the thermal behavior of the samples. The TGA was conducted by using a TG/DTA 6300 (Seiko, Japan), and the samples were heated from 30 °C to 700 °C at a rate of 10 °C min^{-1}. The flame-retardant properties of the pure EP and the EP-based composite were investigated by using LOI, vertical burning testing (UL-94), and cone calorimeter test (CCT). The LOI tests were conducted on a Dynisco LOI test instrument according to ASTMD 2863-97 standard. The size of the specimen was 80 mm × 6.5 mm × 3 mm. The UL-94 tests were carried out on a vertical burning instrument (CFZ-2, Jiangning Analytical Instrument Factory Co., Ltd., Nanjing, China) according to ASTM D3801 standard. The size of the specimen was 130 mm × 13 mm × 3 mm. CCT was conducted by using a calorimeter (iCone, Fire Testing Technology Co., Ltd., East Grinstead, UK) according to the ISO 5660-1 standard under a heat flux of 35 $kW \cdot m^{-2}$, and the sample with the dimensions of 100 mm × 100 mm × 3 mm was wrapped with aluminum foil.

3. Results and Discussion

3.1. Fabrication of DOPO-M-rGO

The fabrication process of DOPO-M-rGO is shown in Figure 2a. The chemical reaction between methyl vinyl dichlorosilane and melamine was firstly executed to synthesize a polymeric intermediate (see Figure 1a), which was then reacted with DOPO via addition reaction to form the DOPO-M. Afterward, the flame retardant DOPO-M-rGO was synthesized via the amidation reaction between the obtained DOPO-M and GO, accompanied by the reduction of GO (see Figure 1b). Therefore, by using the synthesized methyl vinyl dichlorosilane/melamine polymeric intermediate as the bridge, a DOPO-functionalized rGO hybrid flame retardant was successfully prepared, which was used as the flame retardant of EP. Figure 2b shows the SEM image of DOPO-M-rGO, in which a multilayered structure is exhibited thanks to the rGO component functioning as the skeleton. In addition, abundant clusters are attached on the rGO layers, which are formed by the grafting of DOPO-M.

The FTIR spectra of the melamine, DOPO, and GO monomers, as well as the synthesized DOPO-M and DOPO-M-rGO products are shown in Figures 2c and S1. It can be seen from the spectra of DOPO, melamine, and DOPO-M (as shown in Figure S1) that the peak at 835 cm^{-1} assigned to the Si-N bond shows up in the spectrum of DOPO-M, but not in the spectrum of DOPO or melamine, indicating the successful chemical reaction between methyl vinyl dichlorosilane and melamine and the formation of the Si-N bond in the methyl vinyl dichlorosilane/melamine polymeric intermediate. In addition, the absorption peak at 2440 cm^{-1} attributed to the P-H bond of DOPO disappears in the spectrum of DOPO-M, indicating the subsequent reaction between the methyl vinyl dichlorosilane/melamine intermediate oligomer and DOPO. Therefore, DOPO-M is successfully synthesized in our study. The peaks at 1640 cm^{-1} and 1210 cm^{-1} assigned to the amide bond show up in the spectrum of DOPO-M-rGO, indicating that the successful amidation reaction between DOPO-M and GO, and the successful synthesis of DOPO-M-rGO. In the XPS spectrum of synthesized DOPO-M-rGO (Figures 2d and S2), the peaks at 155.2 eV, 284.5 eV, 365.7 eV, and 490.1 eV are associated with the P atoms, C atoms, N atoms, and O atoms, respectively. In the high-resolution XPS spectrum for C 1s (Figure S2a), the main peaks centered at 284.3 eV and 286.5 eV are attributed to the C-C and C-O-C/C-O-P, whereas the additional component centered at 288.2 eV and 285.7 eV are assigned to C=O and C-N, respectively. The peaks in the XPS spectrum for N 1s (Figure S2b) are attributed to N-Si (398.5 eV), C-N-C (397.9 eV), and N-H (395.7 eV), respectively. In addition, the XPS spectrum for Si 2p (Figure S2c) is deconvoluted into two peaks at 101.9 eV and 101.3 eV, which correspond to Si-C and Si-N, respectively. The XPS results indicate the formation of the Si-N bonds and the amide bonds in DOPO-M-rGO, which is inconsistent with the FTIR results.

The thermal degradation behaviors of the GO, DOPO-M, and DOPO-M-rGO are shown by their TGA curves (see Figure 2e), in which the temperature of the maximum weight loss rate can be obtained. It can be observed that the thermal degradation of DOPO-M-rGO mainly happens at a wide temperature range, which combines the thermal degradation temperature ranges of GO and DOPO-M, further indicating the simultaneous existing of GO and DOPO-M segments in DOPO-M-rGO. In addition, the residual weight of DOPO-M-rGO is much higher than DOPO-M, indicating the promotion effect of rGO in DOPO-M-rGO on the char formation during thermal degradation.

3.2. Fabrication of DOPO-M-rGO/EP Composite

The synthesized DOPO-M-rGO was used as flame retardant in EP, and the fabrication process of the DOPO-M-rGO/EP composite is shown in Figure 3a. The DOPO-M-rGO/EP composites were fabricated through the crosslinking of the EP monomer in a DOPO-M-rGO/EP suspension. In order to study the effect of DOPO-M-rGO content on the performances of the EP-based composite and then optimize the amount of DOPO-M-rGO as the flame retardant, different amounts of DOPO-M-rGO (0.5 wt%, 1.0 wt%, 1.5 wt%, 2.0 wt%, 3.0 wt%, 5.0 wt%) were added to form the DOPO-M-rGO/EP composites for the subsequent characterization. Figure 3b shows the stress–strain curves of the pure EP and the DOPO-M-rGO/EP composites with different DOPO-M-rGO contents, demonstrating the improvement effect of DOPO-M-rGO content on the mechanical performances of EP composites. It can be seen that the addition of DOPO-M-rGO/EP improves the tensile strength and modulus of EP when the DOPO-M-rGO content is below 2 wt%, which is mainly attributed to the good dispersion of a small content of DOPO-M-rGO and the integrity of the internal structure of the composite (see Figures 3c and S4a–d). When the amount of the added DOPO-M-rGO is above 3 wt%, the DOPO-M-rGO shows poor dispersion and a great deal of agglomeration in the composites (see Figure S4e,f), leading to the significantly deteriorated mechanical performance of the composites. Therefore, the DOPO-M-rGO content should be controlled at below 2 wt% to satisfy the practical application requirements of the composite. In addition, the DOPO-M-rGO/EP composite shows a higher modulus compared with the DOPO-M/EP composite (see Figure S3), which is mainly caused by the supporting effect of the rGO skeleton in DOPO-M-rGO.

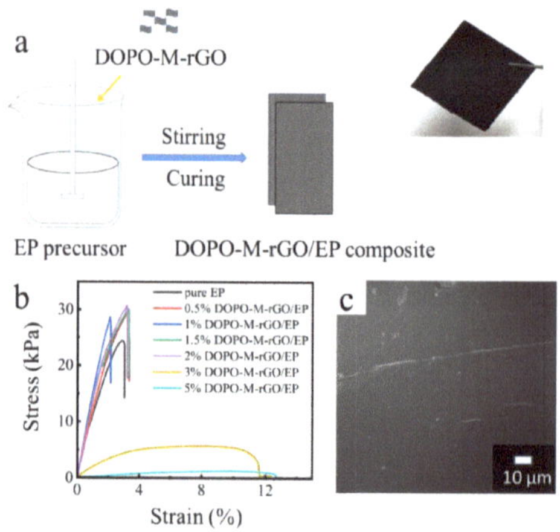

Figure 3. (**a**) Schematic representation for the preparation of DOPO-M-rGO/EP composite and the digital photographs of 1.5% DOPO-M-rGO/EP; (**b**) stress–stain curves of pure EP and the DOPO-M-rGO/EP composites with different DOPO-M-rGO contents; (**c**) SEM images of 1.5% DOPO-M-rGO/EP composite.

3.3. Flame-Resistant Effect of DOPO-M-rGO on EP

The flame-retardant properties of the pure EP and the EP-based composites were investigated by using LOI, UL-94, and CCT.

For the pure EP, the LOI value is 25%, and it could not pass any grade in the UL-94 test. The addition of DOPO-M-rGO endows the EP-based composite with a significant improved LOI value (see Figure 4). When 1.5 wt% DOPO-M-rGO is added into the EP, the LOI value is increased from 25% to 32%, while 1.5 wt% rGO or 1.5 wt% DOPO-M has no effect on the LOI value of the EP-based composite, indicating that DOPO-M and rGO components in DOPO-M-rGO provide a synergistic flame-resistance effect on EP. In addition, when the addition of DOPO-M-rGO is increased to 1.5 wt%, the DOPO-M-rGO/EP composite reaches V-0 rating in the UL-94 test (see Table S4). Figure 5 shows the digital photographs of the pure EP, rGO/EP composite, DOPO-M/EP composite, and DOPO-M-rGO/EP composite. It can be observed that pure EP, rGO/EP composite, and DOPO-M/EP composite is easily ignited with fast flame propagation, and then fiercely burn up to the clamp within just 50 s. Actually, although the incorporation of rGO or DOPO-M into EP matrix displays no rating in the UL-94 tests, both the rGO/EP composite and DOPO-M/EP composite show a lengthened combustion time compared with the pure EP. These results demonstrate that rGO and DOPO-M can reduce the combustion speed of the EP composites, and thus endow the composites with better flame retardancy. The addition of 1.5% DOPO-M-rGO makes the EP composite reach V-0 level in the UL-94 test. The flame extinguishes itself within 5 s, which reflects the high flame retardancy of DOPO-M-rGO in EP.

CCT is widely used to evaluate the combustion performances of polymers under a forced-flaming fire scenario. In this study, we used CCT to record the thermal parameters (including HRR, THR, TSP values, and char residue) of the EP-based composite samples to assess their fire and smoke risk, and the detailed data are shown in Table 1 and Figure 6. The DOPO-M-rGO/EP composite shows the optimized flame-retardant and smoke-suppression performances when 1.5 wt% DOPO-M-rGO is added (see Figure S5 and Table S5). When the addition of DOPO-M-rGO is further increased to 2 wt%, the pHRR value, THR value, and TSP value are all significantly enhanced, which should be attributed to the incombustible gas-inhibiting effect of the hyperdense char residue formed by the DOPO-M-rGO/EP composite with high flame-retardant content [41]. In addition, the LOI value reached the highest with the addition of 1.5% DOPO-M-rGO, and the UL-94 rating reached V-0 as well.

Therefore, the DOPO-M-rGO/EP composite material with 1.5% DOPO-M-rGO was taken as the example to illustrate the flame-retardant effect of DOPO-M-rGO.

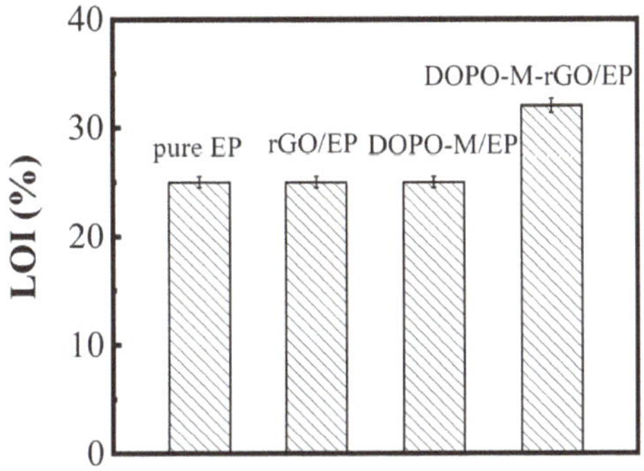

Figure 4. LOI values of the pure EP, rGO/EP composite, DOPO-M/EP composite, and DOPO-M-rGO/EP composite (the mass fractions of additive in the composites are all 1.5%).

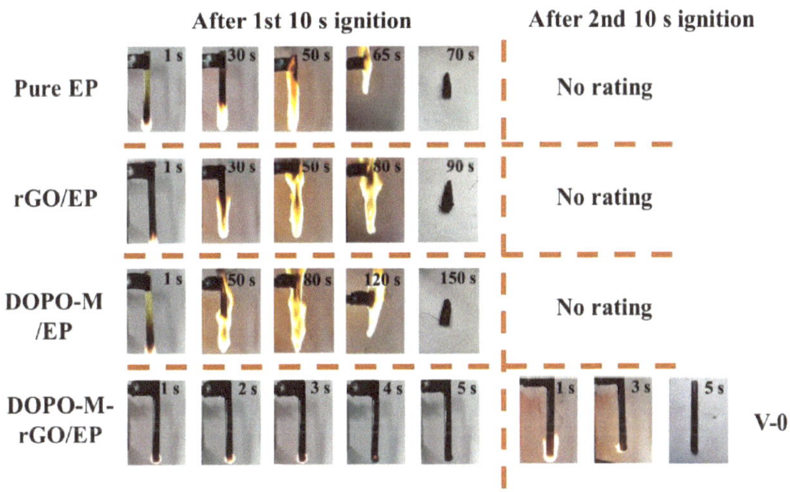

Figure 5. Digital photographs of the pure EP, rGO/EP composite, DOPO-M/EP composite, and DOPO-M-rGO/EP composite (the mass fractions of additive in the composites are all 1.5%) in the UL-94 test.

Table 1. Cone calorimeter data of the neat EP and the EP-based composites.

Sample	Pure EP	1.5% rGO/EP	1.5% DOPO-M/EP	1.5% DOPO-M-rGO/EP
pHRR (kW/m^2)	1593	991	839	719
Time to pHRR (s)	156	167	162	141
THR (MJ/m^2)	125	130	99	88
TSP (m^2/kg)	65	45	74	52

The HRR curve and THR curve of the DOPO-M-rGO/EP composite with 1.5% DOPO-M-rGO are shown in Figure 6a,b, respectively, which are compared with those of pure EP, rGO/EP composite, and DOPO-M/EP composite with the same additives content. It can be seen that the pure EP burns at 100 s, showing a single pHRR of 1593 kW/m^2 at 156 s, and the THR is up to 125 MJ/m^2. The pHRR for the composite containing 1.5 wt% rGO is reduced by 38% to 991 kW/m^2, while the THR shows no significant change compared with the pure EP. When 1.5 wt% DOPO-M is added into the EP, not only is the pHRR decreased to 839 kW/m^2, but also the THR is significantly reduced by 21% to 99 MJ/m^2 compared with that of pure EP. The incorporation of DOPO-M-rGO results in the further decreased pHRR and THR values of the EP-based composite (as shown in Table 1 and Figure 6). The 1.5 wt% DOPO-M-rGO addition decreases the pHRR value to 719 kW/m^2, and the THR value is decreased to 88 MJ/m^2. Therefore, as the synthesis product of GO and DOPO-M, DOPO-M-rGO simultaneously endows the EP-based composite with slower HRR and significantly reduced THR, presenting the best flame-retardant effect on EP.

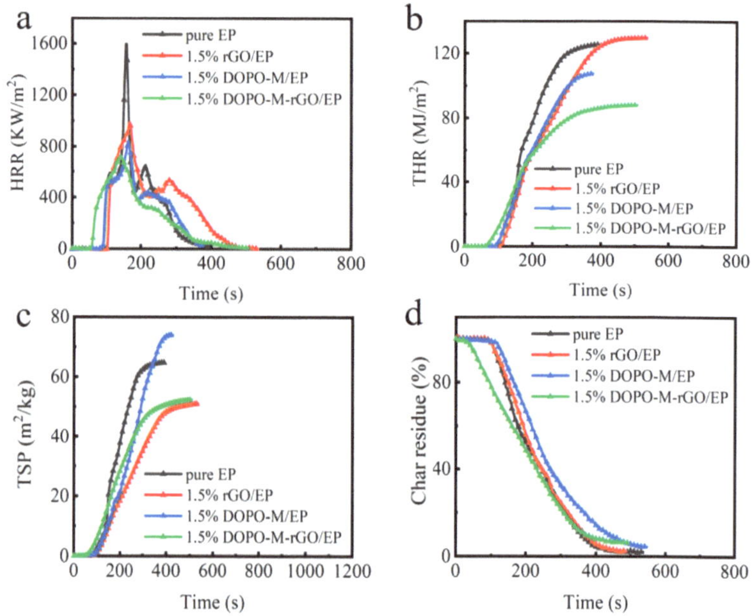

Figure 6. (**a**) Heat release rate; (**b**) total heat release; (**c**) total smoke production; (**d**) char residue curves of the pure EP, rGO/EP composite, DOPO-M/EP composite, and DOPO-M-rGO/EP composite (the mass fractions of additive in the composites are all 1.5%) under an external heat flux of 35 kW/m^2.

The smoke production of materials is regarded as one of the major factors leading to death, and the lower TSP denotes lower smoke risk and longer escaping time in fire disaster The TSP of the pure EP and the EP-based composites are shown in Figure 6c, and the detailed data are listed in Tables 1 and S5. It can be seen that the pure EP releases 65 m^2/kg of TSP during the combustion, and the composite with only DOPO-M as the additive exhibits an even higher TSP value of above 70 m^2/kg. In contrast, the addition of 1.5 wt% rGO or 1.5 wt% DOPO-M-rGO significantly decreases the TSP values of the EP-based composites to 50 m^2/kg and 52 m^2/kg, respectively, indicating that the rGO component is the key to the smoke suppression of DOPO-M-rGO, which should be attributed to its smoke absorbing effect during the combustion [54–57]. From the char residue values of the pure EP and the EP-based composites (see Figure 6d), it can be seen that the DOPO-M/EP composite and the DOPO-M-rGO/EP composite left more char residue after the combustion than the pure EP, which should be attributed to the char layer formation promotion effect of the DOPO component in DOPO-M.

Therefore, compared with rGO and DOPO-M, the addition of 1.5 wt% DOPO-M-rGO can provide the EP-based composite with the optimal fire-resistant performances: the V-0 grade in the UL-94 test, the highest LOI value of 32%, the lowest pHRR, THR, and TSP values, indicating the excellent flame-resistant effect and smoke-suppression effect of small quantities of DOPO-M-rGO on EP. Therefore, our synthesized DOPO-

M-rGO considerably progressed in improving the flame-resistance efficiency in the EP-based composite compared with other DOPO-functionalized rGO flame-retardant systems reported previously (see Table S3).

3.4. Mechanism for the Flame-Resistant Effect of DOPO-M-rGO

In order to study the mechanism of the flame resistance of DOPO-M-rGO on EP, the properties and structure of char residue after CCT of the pure EP and the EP-based composites were analyzed, and the digital photos and SEM images are shown in Figure 7. It can be seen from Figure 7a,e that the pure EP almost forms no char residue, leading to the weak barrier effect. The rGO/EP composite also forms a few char fragments after the combustion (see Figure 7b,f), indicating its inferior barrier effect similar to the pure EP. Thus, the rGO/EP composite presents no obvious difference in the THR value compared with the pure EP. However, the addition of rGO can achieve a smoke-suppression effect during combustion due to its smoke absorbing ability [44–47], and thus the rGO/EP composite presents an obvious decreased TSP value (as shown in Figure 7c). The DOPO-M/EP composite forms an integrated char layer after the combustion (see Figure 7c,g), which is mainly caused by the dehydration and carbonization effect of the P-O-C and N-Si bonds in DOPO-M [58,59]. The formed thermal stabilized char layer functions as an intact shield, which effectively suppresses the transfer of heat during combustion. In addition, the incombustible gas (including N_2 and NO_x) is generated from the disintegrated melamine component during combustion, which further suppresses the spreading of flame. Therefore, the DOPO-M/EP composite exhibits a significantly declined THR value compared with EP and the rGO/EP composite. Due to the synergistic effect of the therein rGO and DOPO-M components, the DOPO-M-rGO/EP composite also forms a relative integrated char (see Figures 7d,h and S6), and the heat release and smoke release are simultaneously decreased during combustion.

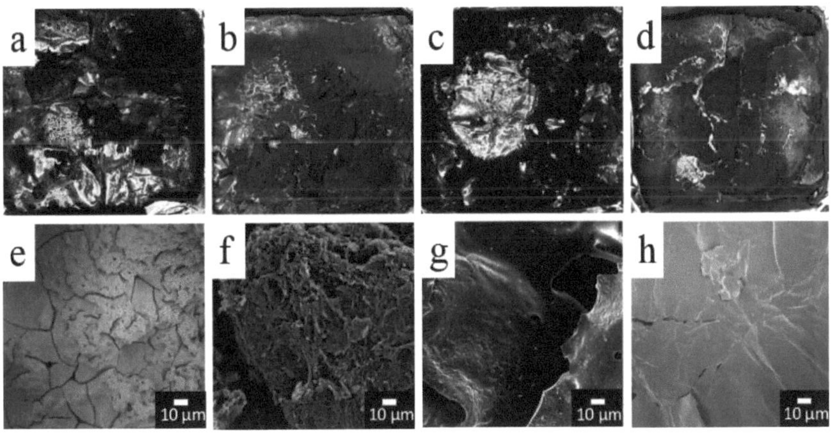

Figure 7. Digital images of char residues for (**a**) pure EP; (**b**) rGO/EP composite; (**c**) DOPO-M/EP composite; (**d**) DOPO-M-rGO/EP composite and SEM images of char residues for (**e**) pure EP; (**f**) rGO/EP composite; (**g**) DOPO-M/EP composite; (**h**) DOPO-M-rGO/EP composite after cone calorimeter test (the mass fractions of additive in the composites are all 1.5%).

In order to further clarify the flame-retardant mechanism of DOPO-M-rGO on EP, Raman spectroscopy of the char residues after CCT were conducted (see Figure 8). The graphitization degree of char residue can be reflected by the I_D and I_G ratio. The lower I_D/I_G value indicates the higher graphitization degree and the higher thermal stability of the formed char residue [60]. It can be seen from Figure 8a,b that the I_D/I_G value of the rGO/EP composite is 1.05, which is obviously lower than that of pure EP (1.33). This is mainly because rGO is a carbonaceous material; thus, showing an excellent graphitization effect.

For the DOPO-M/EP composites, the I_D/I_G value is 1.19 (see Figure 8c), indicating that DOPO-M enhances the graphitization degree and promotes the dense structure formation of the char layer, which is well inconsistent with the SEM images in Figure 7. Under the synergistic effect of the rGO component and the DOPO-M component in DOPO-M-rGO, the DOPO-M-rGO/EP composite exhibits the lowest I_D/I_G value of 1.01 (see Figure 8d).

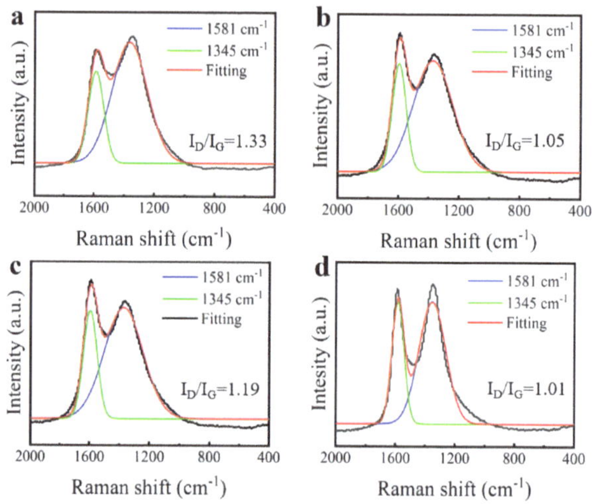

Figure 8. Raman spectra of char residues of (**a**) pure EP; (**b**) rGO/EP composite; (**c**) DOPO-M/EP composite; (**d**) DOPO-M-rGO/EP composite.

Furthermore, the chemical structure of char residues of the DOPO-M-rGO/EP composite was demonstrated by XPS (see Figure 9). As displayed in Figure 9a, C 1s of the char residue has peaks at 285.3 eV (C-O-C/C-O-P) and 284.4 eV (C-C). For the O 1s spectra (Figure 9b), two peaks are observed, which are attributed to the C=O/P=O (531.7 eV) or/and C-O-C (533.5 eV) groups. The P 2p peak in Figure 9c, which appears around 132.1 eV, is attributed to the P-O-C/PO_3 structure, which is derived from DOPO decomposition [11]. The peaks spectrum for N 1s (Figure 9d) are attributed to C-N (397.0 eV), C=N (397.8 eV), N-Si (398.5 eV), and oxidized N compounds (399.6 eV), respectively [11,61]. The result shows that P-O-C/PO_3 and N-Si formed through the decomposition of DOPO-M-rGO during the combustion process of DOPO-M-rGO/EP composite, simultaneously forming the char layer with high thermal stability, providing the DOPO-M-rGO/EP composite with efficient flame retardance [11].

Based on the above results, the flame-retardant mechanism for DOPO-M-rGO on EP is proposed, which is shown in Figure 10. The rGO component in DOPO-M-rGO functions as a skeleton, on which DOPO/melamine (DOPO-M) are grafted. During the combustion of the DOPO-M-rGO/EP composite, the incombustible gases, such as N_2 and NO_x, are firstly generated from the disintegrated melamine component in DOPO-M-rGO, which diluted the surrounded oxygen concentration; thus, significantly decreasing the pHRR value of the composite. With the further combustion of the composite, the decomposed DOPO-M component in DOPO-M-rGO enhances the graphitization degree and promotes the formation of a dense phosphorus-containing char layer, which presents a barrier effect and effectively suppresses the release of the heat. In addition, the rGO component in DOPO-M-rGO adsorbs the generated smoke during the combustion process to achieve the smoke-suppression effect. Therefore, rGO and DOPO-M components both play important roles in the high fire-resistant performances of the DOPO-M-rGO/EP composite. Due to its well dispersion in the EP matrix as the additive, DOPO-M-rGO in the composite presents high flame-resistance efficiency. In other words, only a small content (1.5 wt%) of DOPO-M-rGO can endow the EP-based composite with high flame-resistance performances.

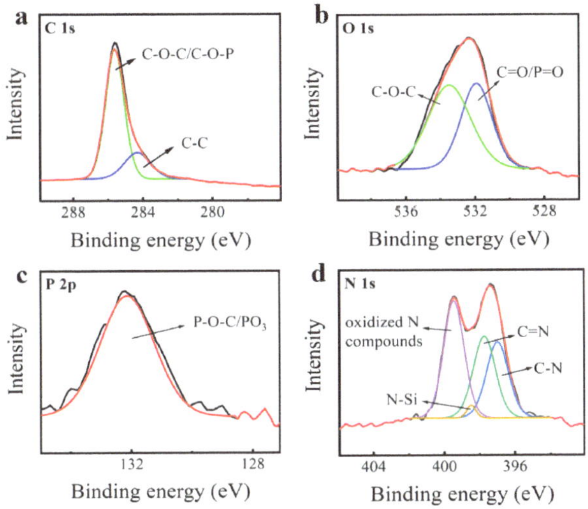

Figure 9. High resolution XPS spectra for (**a**) C 1s; (**b**) O 1s; (**c**) P 2p; and (**d**) N 1s of the char residue of the DOPO-M-rGO/EP composite.

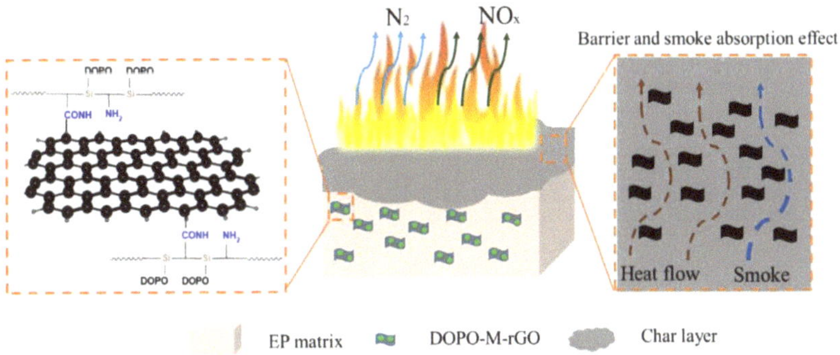

Figure 10. Schematic illustration of flame-retardant mechanism of DOPO-M-rGO in EP.

4. Conclusions

This study demonstrates a green and facile synthesis method for a novel flame retardant via a reduction of GO by the synthetic product of DOPO and melamine, which is thus regarded as a DOPO-functionalized rGO hybrid (DOPO-M-rGO). The EP-based composite with the addition of DOPO-M-rGO exhibited superior fire-resistant performance with extremely low loading of the flame retardant, which is attributed to the high dispersive and flame-resistance efficiency of DOPO-M-rGO. The DOPO-M-rGO/EP composite exhibits greatly reduced pHRR (decreased by 55%) and THR (decreased by 30%) values with only 1.5 wt% DOPO-M-rGO addition, and the LOI value is increased from 25% to 32%. In addition, the smoke production is also significantly decreased, and the TSP value is declined by 20%. The outstanding flame-retardant effect of DOPO-M-rGO in the EP-based composite is provided by the synergistic effect of the melamine, DOPO, and GO components, which promote the generation of the incombustible gases and the formation of a dense phosphorus-containing char layer during combustion. The char layer presents a barrier effect, effectively suppressing the heat release and absorbing the smoke.

Supplementary Materials: The following supporting information can be downloaded at: https://www.mdpi.com/article/10.3390/fire6010014/s1.

Author Contributions: Conceptualization, Z.Z.; data curation, Y.X., W.M. and J.G.; formal analysis J.G., F.X. and H.W.; funding acquisition, Z.Z., H.W. and R.W.; investigation, J.G.; methodology, Z.Z. and H.W.; resources, J.G.; software, J.G.; supervision, Z.Z. and H.W.; writing—original draft preparation, J.G.; writing—review and editing, H.W. All authors have read and agreed to the published version of the manuscript.

Funding: This work was supported by the Joint Research Fund for Overseas Chinese Hong Kong and Macao Young Scholars (51929301), the Young Elite Scientist Sponsorship Program by BAST (BYESS2022211), and the National Natural Science Foundation of China (51903006). The Beijing Scholar Program (RCQJ20303) and Postgraduate Education Quality Improving Program in BIFT (120301990132) should also be acknowledged.

Conflicts of Interest: None of the authors have any financial or scientific conflicts of interest with regard to the research described in this manuscript.

References

1. Saini, A.; Harner, T.; Chinnadhurai, S.; Schuster, J.K.; Yates, A.; Sweetman, A.; Aristizabal-Zuluaga, B.H.; Jiménez, B.; Manzano, C.A.; Gaga, E.O.; et al. GAPS-megacities: A new global platform for investigating persistent organic pollutants and chemicals of emerging concern in urban air. *Environ. Pollut.* **2020**, *267*, 115416. [CrossRef] [PubMed]
2. Qin, R.-X.; Tang, B.; Zhuang, X.; Lei, W.-X.; Wang, M.-H.; Zhang, L.-H.; Hu, K.-M. Organophosphate flame retardants and diesters in the urine of e-waste dismantling workers: Associations with indoor dust and implications for urinary biomonitoring. *Environ. Sci. Process. Impacts* **2020**, *23*, 357–366. [CrossRef] [PubMed]
3. Tang, B.; Christia, C.; Malarvannan, G.; Liu, Y.-E.; Luo, X.-J.; Covaci, A.; Mai, B.-X.; Poma, G. Legacy and emerging organophosphorus flame retardants and plasticizers in indoor microenvironments from Guangzhou, South China. *Environ. Int.* **2020**, *143*, 105972. [CrossRef] [PubMed]
4. Li, Q.Q.; Wang, T.; Zeng, Y.; Fan, Y.; Chen, S.J.; Mai, B.X. Brominated Flame *Retard.* (BFRs) PM2.5 associated with various source sectors in southern China. *Environ. Sci.-Proc. Imp.* **2021**, *23*, 179–187. [CrossRef]
5. Zhao, S.; Tian, L.; Zou, Z.; Liu, X.; Zhong, G.; Mo, Y.; Wang, Y.; Tian, Y.; Li, J.; Guo, H.; et al. Probing Legacy and Alternative Flame Retardants in the Air of Chinese Cities. *Environ. Sci. Technol.* **2021**, *55*, 9450–9459. [CrossRef]
6. Li, H.; Song, A.; Liu, H.; Li, Y.; Liu, M.; Sheng, G.; Peng, P.; Ying, G. Occurrence of Dechlorane series flame retardants in sediments from the Pearl River Delta, South China. *Environ. Pollut.* **2021**, *279*, 116902. [CrossRef]
7. Xie, J.; Sun, Y.; Cheng, Y.; Chen, Y.; Chen, L.; Xie, C.; Dai, S.; Luo, X.; Zhang, L.; Mai, B. Halogenated flame retardants in surface sediments from fourteen estuaries, South China. *Mar. Pollut. Bull.* **2021**, *164*, 112099. [CrossRef]
8. Li, Z.; He, C.; Thai, P.; Wang, X.; Bräunig, J.; Yu, Y.; Luo, X.; Mai, B.; Mueller, J.F. Organophosphate esters and their specific metabolites in chicken eggs from across Australia: Occurrence, profile, and distribution between yolk and albumin fractions. *Environ. Pollut.* **2020**, *262*, 114260. [CrossRef]
9. Pastore, R.; Albano, M.; Delfini, A.; Santoni, F.; Marchetti, M. Thermoplastic Polymeric Materials for Spacecraft Applications: Flame Retardant Properties and UV/AtOx Aging Analysis. *Appl. Sci.* **2021**, *11*, 949. [CrossRef]
10. Huang, P.-H.; Chang, S.-J.; Li, C.-C. Encapsulation of flame retardants for application in lithium-ion batteries. *J. Power Source* **2017**, *338*, 82–90. [CrossRef]
11. Li, S.; Chen, M.; Su, L.; Lin, X.; Liu, C. Highly efficient multielement flame retardant for multifunctional epoxy resin with satisfactory thermal, flame-retardant, and mechanical properties. *Polym. Adv. Technol.* **2019**, *31*, 146–159. [CrossRef]
12. Mu, X.; Zhou, X.; Wang, W.; Xiao, Y.; Liao, C.; Longfei, H.; Kan, Y.; Song, L. Design of compressible flame retardant grafted porous organic polymer based separator with high fire safety and good electrochemical properties. *Chem. Eng. J.* **2020**, *405*, 126946. [CrossRef]
13. Xiao, Y.; Ma, C.; Jin, Z.; Wang, J.; He, L.; Mu, X.; Song, L.; Hu, Y. Functional covalent organic framework for exceptional Fe2+, Co2+ and Ni2+ removal: An upcycling strategy to achieve water decontamination and reutilization as smoke suppressant and flame retardant simultaneously. *Chem. Eng. J.* **2020**, *421*, 127837. [CrossRef]
14. Geng, J.; Qin, J.; He, J. Preparation of Intercalated Organic Montmorillonite DOPO-MMT by Melting Method and Its Effect on Flame Retardancy to Epoxy Resin. *Polymers* **2021**, *13*, 3496. [CrossRef]
15. Wang, N.; Teng, H.; Li, L.; Zhang, J.; Kang, P. Synthesis of Phosphated K-Carrageenan and Its Application for Flame-Retardant Waterborne Epoxy. *Polymers* **2018**, *10*, 1268. [CrossRef]
16. Yuan, Y.; Yu, B.; Shi, Y.; Mao, L.; Xie, J.; Pan, H.; Liu, Y.; Wang, W. Insight into Hyper-Branched Aluminum Phosphonate in Combination with Multiple Phosphorus Synergies for Fire-Safe Epoxy Resin Composites. *Polymers* **2020**, *12*, 64. [CrossRef]
17. Zhang, Y.; Shi, C.; Qian, X.; Jing, J.; Jin, L. DOPO/Silicon/CNT Nanohybrid Flame Retardants: Toward Improving the Fire Safety of Epoxy Resins. *Polymers* **2022**, *14*, 565. [CrossRef]
18. Zhou, R.; Lin, L.; Zeng, B.; Yi, X.; Huang, C.; Du, K.; Liu, X.; Xu, Y.; Yuan, C.; Dai, L. Diblock Copolymers Containing Titanium-Hybridized Polyhedral Oligomeric Silsesquioxane Used as a Macromolecular Flame Retardant for Epoxy Resin. *Polymers* **2022**, *14*, 1708. [CrossRef]

19. Wu, F.; Bao, X.; Wang, J. One-Step Reduction of Graphene Oxide with Phosphorus/Silicon-Containing Compound and Its Flame Retardancy in Epoxy Resin. *Polymers* **2021**, *13*, 3985. [CrossRef]
20. Attia, N.F.; Elashery, S.E.; Zakria, A.M.; Eltaweil, A.S.; Oh, H. Recent advances in graphene sheets as new generation of flame retardant materials. *Mater. Sci. Eng. B* **2021**, *274*, 115460. [CrossRef]
21. Cai, W.; Wang, B.-B.; Wang, X.; Zhu, Y.-L.; Li, Z.-X.; Xu, Z.-M.; Song, L.; Hu, W.-Z.; Hu, Y. Recent Progress in Two-dimensional Nanomaterials Following Graphene for Improving Fire Safety of Polymer (Nano)composites. *Chin. J. Polym. Sci.* **2021**, *39*, 935–956. [CrossRef]
22. Haeri, Z.; Ramezanzadeh, B.; Ramezanzadeh, M. Recent progress on the metal-organic frameworks decorated graphene oxide (MOFs-GO) nano-building application for epoxy coating mechanical-thermal/flame-retardant and anti-corrosion features improvement. *Prog. Org. Coat.* **2021**, *163*, 106645. [CrossRef]
23. Madhad, H.V.; Vasava, D.V. Review on recent progress in synthesis of graphene–polyamide nanocomposites. *J. Thermoplast. Compos. Mater.* **2019**, *35*, 570–598. [CrossRef]
24. Szeluga, U.; Pusz, S.; Kumanek, B.; Olszowska, K.; Kobyliukh, A.; Trzebicka, B. Effect of graphene filler structure on electrical, thermal, mechanical, and fire retardant properties of epoxy-graphene nanocomposites—A review. *Crit. Rev. Solid State Mater. Sci.* **2020**, *46*, 152–187. [CrossRef]
25. Chen, C.; Xiao, G.; Zhong, F.; Dong, S.; Yang, Z.; Chen, C.; Wang, M.; Zou, R. Synergistic effect of carbon nanotubes bonded graphene oxide to enhance the flame retardant performance of waterborne intumescent epoxy coatings. *Prog. Org. Coat.* **2021**, *162*, 106598. [CrossRef]
26. Chen, Y.; Wu, H.; Duan, R.; Zhang, K.; Meng, W.; Li, Y.; Qu, H. Graphene doped Sn flame retardant prepared by ball milling and synergistic with hexaphenoxy cyclotriphosphazene for epoxy resin. *J. Mater. Res. Technol.* **2022**, *17*, 774–788. [CrossRef]
27. Dong, S.; Xiao, G.; Chen, C.; Chen, C.; Yang, Z.; Zhong, F.; Wang, M.; Zou, R. Zn-Al layered double metal hydroxide anchored reduced graphene oxide for enhancing the fire performance of composite coatings. *Colloids Surf. A Physicochem. Eng. Asp.* **2021**, *632*, 127736. [CrossRef]
28. Duan, R.; Wu, H.; Li, J.; Zhou, Z.; Meng, W.; Liu, L.; Qu, H.; Xu, J. Phosphor nitrile functionalized UiO-66-NH2/graphene hybrid flame retardants for fire safety of epoxy. *Colloids Surf. A Physicochem. Eng. Asp.* **2022**, *635*, 128093. [CrossRef]
29. Yang, P.; Wu, H.; Yang, F.; Yang, J.; Wang, R.; Zhu, Z. A Novel Self-Assembled Graphene-Based Flame Retardant: Synthesis and Flame Retardant Performance in PLA. *Polymers* **2021**, *13*, 4216. [CrossRef]
30. Fang, F.; Ran, S.; Fang, Z.; Song, P.; Wang, H. Improved flame resistance and thermo-mechanical properties of epoxy resin nanocomposites from functionalized graphene oxide via self-assembly in water. *Compos. Part B Eng.* **2019**, *165*, 406–416. [CrossRef]
31. Wang, W.; Wang, Z. In-situ preparation of layered zinc N, N′-piperazine (bismethylene phosphonate) functionalizing reduced graphene oxide for epoxy resin with reduced fire hazards and improved thermal/mechanical properties. *Compos. Part A Appl. Sci. Manuf.* **2021**, *149*, 106588. [CrossRef]
32. Zhu, M.; Liu, L.; Wang, Z. Iron-phosphorus-nitrogen functionalized reduced graphene oxide for epoxy resin with reduced fire hazards and improved impact toughness. *Compos. Part B Eng.* **2020**, *199*, 108283. [CrossRef]
33. Qu, L.; Sui, Y.; Zhang, C.; Li, P.; Dai, X.; Xu, B.; Fang, D. POSS-functionalized graphene oxide hybrids with improved dispersive and smoke-suppressive properties for epoxy flame-retardant application. *Eur. Polym. J.* **2019**, *122*, 109383. [CrossRef]
34. Xiao, Y.; Jin, Z.; He, L.; Ma, S.; Wang, C.; Mu, X.; Song, L. Synthesis of a novel graphene conjugated covalent organic framework nanohybrid for enhancing the flame retardancy and mechanical properties of epoxy resins through synergistic effect. *Compos. Part B Eng.* **2019**, *182*, 107616. [CrossRef]
35. Guo, W.; Yu, B.; Yuan, Y.; Song, L.; Hu, Y. In situ preparation of reduced graphene oxide/DOPO-based phosphonamidate hybrids towards high-performance epoxy nanocomposites. *Compos. Part B Eng.* **2017**, *123*, 154–164. [CrossRef]
36. Hu, C.; Yu, T.; Li, Y. Novel DOPO-Modified Graphene: Synthesis and Characterization. *J. Nanosci. Nanotechnol.* **2017**, *17*, 4894–4900. [CrossRef]
37. Yuan, G.; Yang, B.; Chen, Y.; Jia, Y. Synthesis of a novel multi-structure synergistic POSS-GO-DOPO ternary graft flame retardant and its application in polypropylene. *Compos. Part A Appl. Sci. Manuf.* **2018**, *117*, 345–356. [CrossRef]
38. Shi, X.; Peng, X.; Zhu, J.; Lin, G.; Kuang, T. Synthesis of DOPO-HQ-functionalized graphene oxide as a novel and efficient flame retardant and its application on polylactic acid: Thermal property, flame retardancy, and mechanical performance. *J. Colloid Interface Sci.* **2018**, *524*, 267–278. [CrossRef]
39. Wang, Y.; Qing, Y.; Sun, Y.; Zhu, M.; Dong, S. A study on preparation of modified Graphene Oxide and flame retardancy of polystyrene composite microspheres. *Des. Monomers Polym.* **2020**, *23*, 1–15. [CrossRef]
40. Zheng, P.; Wang, R.; Wang, D.; Peng, X.; Zhao, Y.; Liu, Q. A phosphorus-containing hyperbranched phthalocyanine flame retardant for epoxy resins. *Sci. Rep.* **2021**, *11*, 1–10. [CrossRef]
41. Xu, B.; Liu, Y.; Wei, S.; Zhao, S.; Qian, L.; Chen, Y.; Shan, H.; Zhang, Q. A Phosphorous-Based Bi-Functional Flame Retardant Based on Phosphaphenanthrene and Aluminum Hypophosphite for an Epoxy Thermoset. *Int. J. Mol. Sci.* **2022**, *23*, 11256. [CrossRef] [PubMed]
42. Liao, S.-H.; Liu, P.-L.; Hsiao, M.-C.; Teng, C.-C.; Wang, C.-A.; Ger, M.-D.; Chiang, C.-L. One-Step Reduction and Functionalization of Graphene Oxide with Phosphorus-Based Compound to Produce Flame-Retardant Epoxy Nanocomposite. *Ind. Eng. Chem. Res.* **2012**, *51*, 4573–4581. [CrossRef]

43. Luo, F.; Wu, K.; Guo, H.; Zhao, Q.; Lu, M. Simultaneous reduction and surface functionalization of graphene oxide for enhancing flame retardancy and thermal conductivity of mesogenic epoxy composites. *Polym. Int.* **2016**, *66*, 98–107. [CrossRef]
44. Zhi, M.; Liu, Q.; Chen, H.; Chen, X.; Feng, S.; He, Y. Thermal Stability and Flame Retardancy Properties of Epoxy Resin Modified with Functionalized Graphene Oxide Containing Phosphorus and Silicon Elements. *ACS Omega* **2019**, *4*, 10975–10984. [CrossRef] [PubMed]
45. Ji, P.; Cui, Y.; Liu, D.; Zhang, T.; Lv, J. The Bi-DOPO derivative functionalized graphene oxide: Preparation and its flame-retardation on epoxy resin. *Polym. Adv. Technol.* **2021**, *32*, 2843–2855. [CrossRef]
46. Feng, Y.; Li, X.; Zhao, X.; Ye, Y.; Zhou, X.; Liu, H.; Liu, C.; Xie, X. Synergetic Improvement in Thermal Conductivity and Flame Retardancy of Epoxy/Silver Nanowires Composites by Incorporating "Branch-Like" Flame-Retardant Functionalized Graphene. *ACS Appl. Mater. Interfaces* **2018**, *10*, 21628–21641. [CrossRef]
47. Qian, X.; Song, L.; Yu, B.; Wang, B.; Yuan, B.; Shi, Y.; Hu, Y.; Yuen, R.K.K. Novel organic–inorganic flame retardants containing exfoliated graphene: Preparation and their performance on the flame retardancy of epoxy resins. *J. Mater. Chem. A* **2013**, *1*, 6822–6830. [CrossRef]
48. Aljamal, A.; Szolnoki, B.; Marosi, G. Improving thermal and flame retardant properties of sorbitol-based bioepoxy systems by phosphorus-based flame retardants. *Fire Mater.* **2021**, *46*, 605–614. [CrossRef]
49. Hu, X.; Li, M.; Yang, J.; Liu, F.; Huang, H.; Pan, H.; Yang, H. In situ fabrication of melamine hydroxy ethylidene diphosphonate wrapped montmorillonite for reducing the fire hazards of epoxy resin. *Appl. Clay Sci.* **2021**, *201*, 105934. [CrossRef]
50. Qin, P.; Yi, D.; Hao, J.; Ye, X.; Gao, M.; Song, T. Fabrication of melamine trimetaphosphate 2D supermolecule and its superior performance on flame retardancy, mechanical and dielectric properties of epoxy resin. *Compos. Part B Eng.* **2021**, *225*, 109269. [CrossRef]
51. Wang, L.; Zhu, J.; Deng, H.; Lyu, R.; Wei, Y.; Lan, X. Improved Pyrolysis Performance of a Cyclic Phosphate-Melamine Intumescent Fire-Retardant Coating System Using Ceria as an Additive. *ChemistrySelect* **2022**, *7*, e202200062. [CrossRef]
52. Wang, W.; Liu, Y.; Wen, H.; Wang, Q. Synthesis of a hyperbranched polyamide charring agent and its flame-retarding and toughening behavior in epoxy resin. *Polym. Degrad. Stab.* **2021**, *184*, 109479. [CrossRef]
53. Hummers, W.S., Jr.; Offeman, R.E. Preparation of Graphitic Oxide. *J. Am. Chem. Soc.* **1958**, *80*, 1339. [CrossRef]
54. Jiang, S.-D.; Bai, Z.-M.; Tang, G.; Hu, Y.; Song, L. Synthesis of ZnS Decorated Graphene Sheets for Reducing Fire Hazards of Epoxy Composites. *Ind. Eng. Chem. Res.* **2014**, *53*, 6708–6717. [CrossRef]
55. Hu, J.; Zhang, F. Self-assembled fabrication and flame-retardant properties of reduced graphene oxide/waterborne polyurethane nanocomposites. *J. Therm. Anal.* **2014**, *118*, 1561–1568. [CrossRef]
56. Yu, S.; Wang, X.; Ai, Y.; Tan, X.; Hayat, T.; Hu, W.; Wang, X. Experimental and theoretical studies on competitive adsorption of aromatic compounds on reduced graphene oxides. *J. Mater. Chem. A* **2016**, *4*, 5654–5662. [CrossRef]
57. Sun, Y.; Yang, S.; Zhao, G.; Wang, Q.; Wang, X. Adsorption of Polycyclic Aromatic Hydrocarbons on Graphene Oxides and Reduced Graphene Oxides. *Chem. Asian J.* **2013**, *8*, 2755–2761. [CrossRef]
58. Varganici, C.-D.; Rosu, L.; Bifulco, A.; Rosu, D.; Mustata, F.; Gaan, S. Recent advances in flame retardant epoxy systems from reactive DOPO–based phosphorus additives. *Polym. Degrad. Stab.* **2022**, *202*, 110020. [CrossRef]
59. Wang, J.; Yu, X.; Dai, S.; Wang, X.; Pan, Z.; Zhou, H. Synergistic effect of chitosan derivative and DOPO for simultaneous improvement of flame retardancy and mechanical property of epoxy resin. *Cellulose* **2021**, *29*, 907–925. [CrossRef]
60. Hou, Y.; Hu, W.; Gui, Z.; Hu, Y. A novel Co(II)–based metal-organic framework with phosphorus-containing structure: Build for enhancing fire safety of epoxy. *Compos. Sci. Technol.* **2017**, *152*, 231–242. [CrossRef]
61. Sun, Q.; Wang, J.; Meng, X.; Zhang, J.; Yan, H. A novel high-efficient P/N/Si-containing APP-based flame retardant with a silane coupling agent in its molecular structure for epoxy resin. *Chin. J. Chem. Eng.* **2022**, *30*, e202200221. [CrossRef]

fire

MDPI

Article

Full-Scale Experiments of Water-Mist Systems for Control and Suppression of Sauna Fires

Paolo E. Santangelo [1,2,*], Luca Tarozzi [3] and Paolo Tartarini [4,5]

1 Dipartimento di Scienze e Metodi dell'Ingegneria, Università degli Studi di Modena e Reggio Emilia, 42122 Reggio Emilia, Italy
2 Centro Interdipartimentale per la Ricerca En&Tech, 42122 Reggio Emilia, Italy
3 Bettati Antincendio S.r.l., 42124 Reggio Emilia, Italy
4 Dipartimento di Ingegneria "Enzo Ferrari", Università degli Studi di Modena e Reggio Emilia, 41125 Modena, Italy
5 Centro Interdipartimentale per la Ricerca InterMech-MO.RE., 41125 Modena, Italy
* Correspondence: paoloemilio.santangelo@unimore.it; Tel.: +39-0522-52-2223

Abstract: Sauna is a common fixture in many facilities; a specific fire-protection system is typically designed and installed for this application, as short circuits or direct contact with incandescent materials may result in a fire. Water mist has been recently considered as a promising option for this purpose; so, assessing its control and suppression capability in a sauna configuration has become of paramount importance for designers and engineers. To this end, an unprecedented real-scale test rig was built and instrumented with thermocouples and a hot-plate thermometer towards the evaluation of water-mist performance against various fire scenarios and, ultimately, to provide guidelines to designers. Timber benches were employed as target materials, while the fire was initiated in a wood crib. Design parameters, such as initial room temperature, location of the ignition source, nozzle-to-wall distance, and air gap between benches and wall, were varied, also including natural ventilation in a dedicated experiment. The system proved successful in controlling and containing the fire: bench damage ratio—selected as a quantitative parameter to assess water-mist performance—was consistently lower than 5%. However, extinction was not always achieved, especially under the most challenging configuration in terms of ventilation, initial room temperature, and nozzle-to-wall distance.

Keywords: timber; ventilation; room temperature; air gap; damage ratio; thermometry

Citation: Santangelo, P.E.; Tarozzi, L.; Tartarini, P. Full-Scale Experiments of Water-Mist Systems for Control and Suppression of Sauna Fires. *Fire* **2022**, 5, 214. https://doi.org/10.3390/fire5060214

Academic Editors: Song Lu, Changcheng Liu, Guohui Li and Paweł Wolny

Received: 14 November 2022
Accepted: 7 December 2022
Published: 10 December 2022

1. Introduction

Saunas have been used to provide relax and health benefits for a long time, with sweat houses being operated by ancient civilizations even 3000 years ago [1]. Over the last 50 years, they have become a typical amenity in the hospitality (e.g., hotels) and fitness (e.g., gyms, spas) industries; their market is expected to experience a steady growth at a CAGR (compound annual growth rate) of about 7%, at least until 2026 [1]. However, the involved structural components and equipment make the likelihood of fire incidents in saunas relatively high, and the learning of sauna fires through media or technical publications is currently not uncommon [2,3]. The source of ignition consists of the presence of electrical heaters or incandescent stones; a substantial amount of combustibles is also often included within sauna scenarios, which mostly comprise timber (e.g., benches and walls), fabric (e.g., towels and linen), and chemicals (e.g., detergents). A fire may start quite easily as a result of short circuits or direct contact between hot materials and combustibles, frequently caused by human errors. The potentially high ambient temperature—even exceeding 90 °C—tends to contribute to its spread over the whole compartment. Passive fire-protection systems are often employed, since their effectiveness in avoiding or limiting fatalities appears established; nevertheless, fire events may cause remarkable loss or damage of property, together with lengthy business interruption, if not properly contained and possibly suppressed by an action of some kind.

So, the inclusion of active fire-protection systems in the design of saunas has become ever increasing, which is also promoted by specific requirements imposed through laws and standards. Quite an expected outcome, the challenge for designers and engineers has turned into an attractive opportunity within the fire-protection system market, as almost every fitness center and most resorts or recreational facilities are currently endowed with saunas. In this frame, performance-based design applied to the various elements of the built environment in terms of fire protection seems to foster the use of innovative technologies able to control the fire both thermally and spatially until emergency responders arrive on the scene. Yet, saunas are also designed to reduce the costs associated with the amount and storage of the selected fire-extinguishing agent [4]. Among them, the popularity of water mist is increasing for a variety of stationary applications [5]: its control and suppression capability is the subject of a substantial body of the literature that comprises scenarios relating to high-hazard storages [6], car parks [7], timber buildings [8], and pipelines containing natural gas [9–11]. Mitigation of explosions was investigated as an additional action by water mist in highly hazardous storage facilities [12]. Its success can be mainly attributed to a discharge of generally less water than that by sprinkler systems, since fragmentation into tinier droplets yields a faster evaporative action [13,14]. Dedicated studies were also focused on water-mist action in suppressing compartment fires, mainly emphasizing the need for controlled discharge [15] and systems that govern activation on a detection-based feedback [16].

Unfortunately, the research specifically focused on sauna fires is still relatively limited and mostly consists of case studies, even though it is seminal [2,3]: for instance, Howe and Lloyd showed that sprinklers are mainly installed in saunas located within a sprinkler-protected site, whereas water mist is often chosen if no sprinkler system is already present or can be integrated [2]. So, this work was primarily inspired by the available experimental research on water-mist fire suppression. It includes several studies where the discharge is released against canonical fire scenarios, both solid—(e.g., wood cribs, lithium-ion batteries) [17,18] and liquid-fueled (e.g., pool fires) [19,20], together with works on real-scale—actually large-scale and highly hazardous—scenarios [6,7,9–11]. These latter works present compartments where the characteristic length—usually the room height in compartment-fire problems [21]—is of the same order of magnitude as that of saunas. Since a spray discharge is involved, the acquired knowledge of water-mist spray characterization in terms of both atomization and dispersion [22–24], and the achievements about spray/flame interaction [25], also served as a foundation for the present work. As an additional source of guidance to design a test rig, several standards are currently available for water-mist systems [26,27], which provided technical guidelines in devising the experimental setup assembled for the present work. On the other hand, no standardized protocol to evaluate water-mist performance for sauna fire scenarios is currently available; so, this work is also aimed at proposing an approach that may become a standard reference to the purpose, being that it is inspired by the Annex B of standard NFPA 750 [27]. Overall, the latter was followed, as it reports a strategy to challenge water-mist systems against generic fire scenarios and assess their performance.

An experimental setup was developed and built as fully representative of a sauna environment from both the structural (i.e., involved materials, dimensions, and arrangement of the included components) and the thermo-hygrometric (i.e., temperature and vents) standpoint. In consistency with the aim of reproducing the actual action by water mist within that scenario, the discharge system was also designed and scaled accordingly. Several physical quantities were measured or evaluated; various parameters were considered and varied throughout the test series: location of the ignition source, presence of drywall boards, bench-to-wall and nozzle-to-wall distance, ventilation, and initial room temperature. An evaluation of fire spread and intensity under the tested configurations was quantitatively performed by introducing a damage ratio that ultimately allowed assessing the impact of the varied parameters. The present work may serve as a reference for future assessments of control and suppression capability in a scenario that is currently not covered by standardized procedures. Moreover, the parametric study may also provide guidance to fire-protection system designers, especially if water-mist technology is selected for this application.

2. Materials and Methods

The main objective of the conducted experiments consisted of assessing quantitatively the ability of a water-mist system to successfully control and possibly suppress a sauna fire, with extinction being considered unexpected by this technology [5,13], yet welcome. So, the main sources of ignition were identified as the first step, then a test scenario was developed and realized to reproduce the most relevant configurations in terms of geometry and involved combustibles. A testing procedure was also devised as founded on the recommendations provided by recognized standards [26] and aimed at combining the following tasks:

- evaluation of fire hazards within the tested environment;
- evaluation of the compartment conditions;
- expression of performance objectives in terms of thermal/spatial control of the fire and suppression;
- emphasizing the worst-case scenario through an *a posteriori* analysis of the outcomes from the test series.

It is worth noting that this kind of real-scale experiment is relatively expensive, especially in terms of material consumption and waste production; therefore, a limited number of tests—lower than ten—was carried out. Nevertheless, some statistical analysis was performed to assess repeatability, mostly referred to the initial free-burn condition (i.e., unhindered evolution of the fire).

2.1. Scenario and Geometry

The test chamber is presented in the technical sketches of Figure 1, with some photos also being shown in Figures 2 and 3. The dimensions involved in Figure 1 are reported in Table 1, together with the chamber height. The employed benches and internal walls were made up of timber beams; drywall boards were attached to the back vertical surface of the benches. Plywood boards were used to upholster the ceiling and walls. The facility was assembled and placed within a corrugated-iron container, the door of which—kept closed in all the conducted experiments, except for those featuring natural ventilation—was of 0.7 × 1.9 m size. The developed setup was designed to include the structural components and the items of typical saunas, as described in studies dedicated to fire protection of those compartments [2,3].

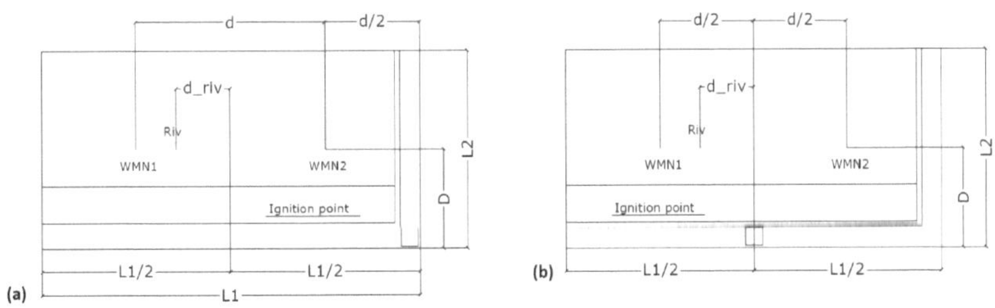

Figure 1. Plan view of the test chamber: (**a**) ignition source at the corner behind benches (configuration C1); (**b**) ignition source at the center of the wall behind the benches (configuration C2); *WMN*: water-mist nozzle location, *Riv*: heat-detector location.

Figure 2. Test chamber: benches and drywall/plywood boards.

Figure 3. Test chamber: nozzles and heat detector.

Table 1. Geometry of the test chamber: design parameters (dimensions are in m).

Height	*L1*	*L2*	*d*	*d_riv*	*D*
2.4	5.9	2.3	3.6	1.0	*L2*/2 in all tests but the last one; *d*/2 in the last one

Since the electrical heater was considered one of the main sources of fire hazard in terms of probability and potential extent of the fire among those reviewed in Section 1, an ignition source was devised to resemble it and reproduce the fire arisen from electrical heaters typical of actual saunas. To the purpose of also complying with available standards, a wood crib was selected as the ignition source [17,28], following the recommendations by Annex G of a recognized standard for water-mist tests [29]. The basic foundation of this design choice consists of a solid-fueled fire being more spatially spread than a liquid-fueled one, thus making the former more similar to an electric-heater fire. So, a 4-layer crib (overall size of 300 × 300 × 150 mm) was employed; each stick featured a square base of 38 mm side (thickness), and its length was 300 mm. The peak heat release rate of wood-crib fires is yielded by a well established formulation [28]:

$$HRR = MLR \cdot \Delta h_c = C\left(\frac{S}{H}\right)\left(\frac{m_i}{t}\right) \cdot \Delta h_c, \quad (1)$$

where *HRR* is heat release rate, *MLR* is mass-loss rate, Δh_c is wood heat of combustion (=12 MJ kg^{-1} [28]), *C* is a constant (=7.44 × 10^{-4} [28]), *S* is the clear spacing between sticks in the same layer (=50 mm), H is the crib height, m_i is the initial crib mass (=2.9 kg on average, for the cribs used in this experiment), and t is the stick thickness. The average heat release rate resulted from calculation as 230 kW. The crib was placed onto a metal container, hosting 0.47 L of water—employed to balance the potential unevenness of the base—and 0.24 L of heptane.

The latter served as the accelerant and was manually ignited to start the fire; Figure 4 shows a photo of the whole crib/container assembly. As shown in the sketch of Figure 1, the test chamber included two configurations (C1 and C2), embodying two fire scenarios, each of which presents different locations of the ignition source (i.e., the wood crib). Notably, the wood-crib assembly was located behind the benches and on the floor in both configurations, with it being placed at the corner in configuration C1, while being placed at the center of the wider drywall board in configuration C2.

Figure 4. Wood crib/container assembly, including the container for water and accelerant (heptane).

Those scenarios were aimed at reproducing typical locations of the electric heater within saunas, thus allowing one to assess the impact of the position of the ignition source on fire spread. Testing a sauna fire within a container may imply boundary and initial conditions different from those of an actual fire occurring in a real sauna. The thermo-hygrometric operating conditions of most saunas consist of 80 ± 5 °C ambient temperature and 10 ± 2% relative humidity [30], as opposed to steam baths that feature lower temperature and higher humidity. Initial room temperature and heat transfer may bear the main discrepancy with respect to real scenarios. As an assessment of the potential impact of this parameter on fire development, initial room temperature was governed in one of the performed tests and raised above 80 °C by an industrial heater. However, when not imposed, its variations and variations of the global heat-transfer coefficient were assumed negligible with respect to temperature rise due to a fire event [31]: in fact, even significant changes in the heat-transfer coefficient between a compartment and its outer environment, under natural convection, may affect room temperature by less than 10 °C, whereas typical flame temperature is in the order of 1000 °C. Moreover, no major impact by variations of relative humidity was also expected; relative humidity within the test chamber was not recorded, but tests were carried in summer mornings, which made relative humidity within the container arguably low. On the other hand, the use of a hydrocarbon (i.e., heptane) as the accelerant for the wood-crib fire yields water (steam) as a product of combustion reaction, but the fuel concentration was sufficiently low (i.e., about 8 mL heptane per 1 m^3 ambient air) to neglect the related raise of humidity. As a conservative measure, timber benches were left to dry out before installation, thus making them more prone to involvement with the fire, were flames to ever reach them. Notably, their moisture content may be assumed as lower than 5%, following a similar approach to the drying of timber components [17].

The discharge system consisted of a high-pressure delivery unit, stainless steel piping water-mist nozzles, and the detection set; a sketch including all the items is presented in Figure 5. Notably, the delivery unit included four cylinders, each of which is of 80 L capacity: three of them hosted water, whereas the other one contained nitrogen to pressurize the system. The nitrogen's initial pressure was 150 bar, which allowed the initial water pressure at the nozzle to be greater than 100 bar. As the core of the discharge system two water-mist nozzles manufactured by Bettati Antincendio S.r.l. (code NWMO14) were inserted in the enclosure at the ceiling height (i.e., 2.4 m, as reported in Table 1).

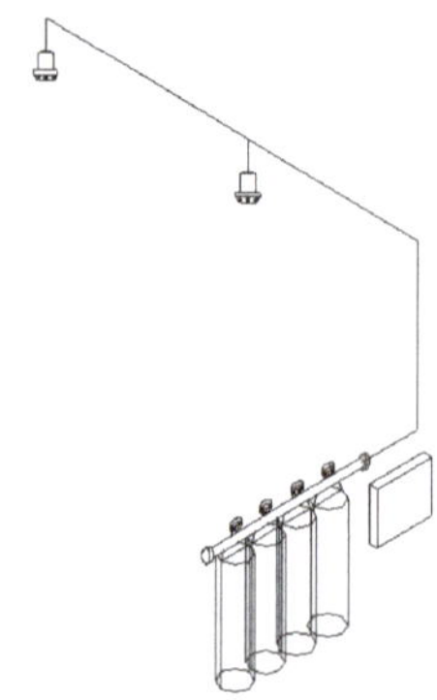

Figure 5. Discharge system, including delivery unit, piping, nozzles, and detection set.

The devised arrangement within the chamber—presented in Figure 1—made the area coverage of each nozzle 3.6 $\times$ 3.6 m. This type of nozzle features seven pressure-swirl injectors (one central and six peripheral, as shown in the technical sketch of Figure 6) and overall K-factor of 1.4 L min^{-1} $bar^{-0.5}$. Therefore, a 10 min minimum duration of the discharge could be achieved within the enclosure under the reported operating conditions The released spray was characterized in previous works [23,24]; the characteristic droplet size (Sauter Mean Diameter) at 1 m distance from the nozzle outlet could be estimated, as in the order of 50 μm.

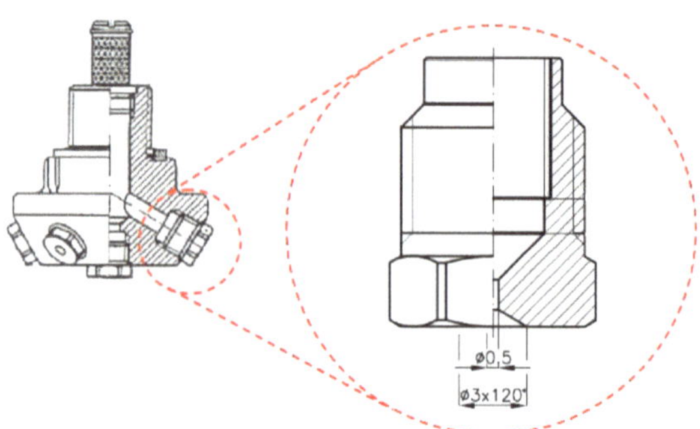

Figure 6. Technical sketch of the employed nozzle type (NWMO14 by Bettati Antincendio s.r.l.), with a detail of the injector.

The detection set consisted of a heat detector (by Kidde-Fenwal Inc., threshold temperature of 165 °C, in accordance with recent regulations for residential and shipboard applications [32]), installed at the ceiling height within the test chamber, as shown in Figure 1. In actual saunas, the discharge is also activated upon temperature feedback, and two detectors would be typically employed, each one at *d_riv* distance from the closer nozzle. However, only one detector was inserted in the developed experimental setup to anticipate the worst-case scenario, both for configuration C1 (Figure 1a) and for configuration C2 (Figure 1b). Overall, it is worth noting that the designed discharge system was merely inspired by the Annex B of a generic standard on water mist [26], since no guidance is available for this specific application.

2.2. Instruments and Experimental Procedure

As a typical measurement in fire experiments, temperature was recorded at various locations and within specific devices by seven thermocouples (type K, wire diameter of 0.5 mm, 1 Hz acquisition frequency, and accuracy complying with standard IEC 584-2). Notably, the following list identifies the probes and reports their respective positions in the test chamber, as well as their scope in terms of measured parameter:

- *T_gas*—three probes used to record gas temperature 76 mm below the ceiling at the vertical symmetry axis of the ignition source (wood crib); gas temperature 76 mm below the ceiling at the heat-detector location (i.e., *Riv* in Figure 1); gas temperature 76 mm below the ceiling at the virtual heat-detector location (not inserted, as noted in Section 2.1), symmetric to *Riv* with respect to the symmetry axis in the sketches of Figure 1, respectively;
- *T_crib*—probe used to record gas temperature at the center of the wood-crib top surface;
- *T_sts*—probe used to record surface temperature of the timber bench bottom surface, at the vertical symmetry axis of the ignition source;
- *T_clg*—probe used to record surface temperature at the ceiling and at the symmetry axis of the ignition source;
- *T_rad*—probe associated with hot-plate thermometer and located at 500 mm height from the floor.

All the three *T_gas* thermocouples were conveniently shielded to limit and virtually prevent wetting by water droplets throughout the discharge. The hot-plate thermometer was developed as a device to quantitatively evaluate heat flux in configurations where radiation prevails over other heat-transfer modes (e.g., furnaces) [33]; its applicability to fire scenarios was also successfully validated [34], and hot-plate thermometry has been employed in a variety of full-scale tests, since radiation is largely predominant as flames are involved [7]. This instrument features a steel plate exposed to radiation and an insulating slab on its back side, with a thermocouple (*T_rad*, in this experiment) welded onto it. The hot-plate thermometer was placed in front of the presumed fire location, between the wood crib and the involved bench, at 500 mm height from the floor; a photo of its location in tests carried out against configuration C1 is presented in Figure 7.

Figure 7. Hot-plate thermometer located in the corner (configuration C1).

Pressure measurements were taken within the nitrogen tank and at the nozzle outlet. Moreover, a load cell was employed to measure wood-crib (±0.5 g accuracy) and bench mass (±0.5 kg accuracy) before and after tests to estimate the burning ratio. In consistency with the procedure adopted prior to starting each experiment (Section 2.1), the materials were left out to dry out after each test before weighing in order to rule out the effect of water absorbed during the discharge.

A procedure was developed to organize the experiments through a test matrix based upon varying the selected governing parameters towards an identified worst-case scenario. In a similar manner to the design of the discharge system (Section 2.1), this approach was founded on the guidelines from Annex B of a well established standard on water mist, which recommends developing a path that leads to the worst case, also setting some pass/fail criteria for the tested system [26]. A limited number of parameters was varied throughout the experiments so as to emphasize the most significant factor among the other numerous factors in a real scenario. Notably, the following set was identified:

- Location of the ignition source—varied between configuration C1 and C2, as reported in Section 2.1 and Figure 1;
- Initial room temperature (T_i)—almost equal to outside ambient temperature in most tests (20–30 °C), above 80 °C in two of them (operating an industrial heater, as mentioned in Section 2.1);
- Discharge activation time (τ_{act})—governed by the heat-detector threshold or equal to a fixed, pre-assigned value (180 s);
- Ventilation—closed or open door (natural ventilation);
- Distance between the nozzles and the wall behind benches (D)—1.15 m (=$L2/2$) or 1.8 m (=$d2/2$), as shown in Figure 1 and reported in Table 1;
- Presence of drywall boards attached to the back of the benches;
- Distance between benches and the wall behind (δ)—benches against the wall (δ = 0), or δ = 0.250 m.

The discharge was manually activated 5 s after the heat-detector alarm turned on, which was deemed a conservative offset to estimate the delay in actual systems; however, delayed manual discharge was also performed to challenge the system against a longer free burn.

As previously reported, nozzle-to-wall distance was generally kept constant; however, tests with a larger distance were also conducted, which also implied droplets potentially bouncing or evaporating against the opposite wall (i.e., the one closer to the nozzle). Drywall boards were attached to the back of the benches in the tests, where a gap between benches and wall (δ) was applied. Table 2 shows the test matrix with all variables included. This approach leads one to identify a worst-case scenario (i.e., Test no. 7 in Table 2), as recommended by the followed guidance [26].

Table 2. Test matrix with experimental conditions.

Test no.	Ignition Source	T_i (°C)	τ_{act} (s)	Ventilation	D (m)	Drywall Boards	δ (mm)
1	C1	20–30	alarm + 5	NO	1.15	NO	0
2	C2	20–30	alarm + 5	NO	1.15	NO	0
3	C2	20–30	180	NO	1.15	NO	0
4	C2	20–30	alarm + 5	NO	1.15	NO	0
5	C2	20–30	alarm + 5	NO	1.15	YES	250
6	C2	>80	alarm + 5	NO	1.15	NO	0
7	C2	>80	alarm + 5	YES	1.80	NO	0

2.3. *Approach to Data Analysis*

The analysis of both the most immediate outcomes of each test and of the quantitative dataset was carried out through an approach including a set of actions and the implementation of relationships and formulations. Notably, the developed methodology consisted of:

- Opening the test chamber at the end of each test and checking if any flames, smoldering fires, or charring materials were still present after the discharge, thus allowing, first, observations about thermal control and extinction, as defined by a recognized standard [27];
- Distinguishing successful from unsuccessful suppression primarily by the presence of a sharp decay in temperature trend at the locations where gas or surface temperature was measured, which also allows emphasizing instances of fire regrowth; as an additional approach, the same observation was applied to the heat-flux trend;
- Identifying the time at which temperature readings by thermocouple *T_crib* fall below wood-ignition temperature (~220 °C [35]) that is the physical threshold to determine suppression on timber surfaces [17], thus marking the absence of flames at the ignition source (i.e., the most hindered location and a surely deep-seated fire); it is acknowledged that the open literature actually presents a set of values for the various categories and types of timber, so the selected value may be considered as conventionally representative;
- Carrying out a close comparative examination of the photos taken before and after each test to qualitatively estimate the amount of materials involved with fire (i.e., based on the extent of the visibly burnt surface), as suggested to assess damage ratio of timber parts [2,3];
- Evaluating, quantitatively, wood-crib and bench damage ratio as $(m_i - m_f)/m_i$, where m is mass, while indexes i and f refer to initial (pre-test) and final (post-test) load-cell reading, respectively [7,17,36];
- Calculating incident radiant heat flux q relating to the hot-plate thermometer through the relationship proposed in the literature [34]:

$$q = \frac{\varepsilon_{PT}\sigma T_{PT}^4 + (h_{PT} + K_{cond})(T_{PT} - T_\infty) + \rho_{st}c_{st}s(\Delta T_{PT}/\Delta t)}{\varepsilon_{PT}}, \quad (2)$$

where ε_{PT} is emissivity of the plate thermometer (=0.95, *de facto* a blackbody surface), σ is the Stefan-Boltzmann constant (=5.67 × 10^{-8} W m^{-2} °C^{-4}), T_{PT} is temperature of the plate thermometer recorded by thermocouple *T_rad*, h_{PT} is convective heat-transfer coefficient (=10 W m^{-2} °C^{-1} [34]), K_{cond} is conduction correction factor (=22 W m^{-2} °C^{-1} [34]), T_∞ is room temperature (i.e., thermocouple reading prior to starting the fire), ρ_{st} is steel density (=8100 kg m^{-3}), c_{st} is steel specific heat capacity (=400 J kg^{-1} °C^{-1}), s is thickness of the steel plate included in the assembly (= 0.7 mm), and t is time.

It is worth noting that a quantitative expression of successful (or unsuccessful) suppression would rigorously stem from an evaluation of the heat release rate (HRR) trend throughout free burn and discharge, in accordance with the definition provided by available standards [27]. However, this measurement is often hardly feasible in a real-scale scenario, since calorimetric methods usually apply to configurations with smaller length scale. So, temperature and heat-flux measurements were employed, as they allow inferring HRR trends, even though HRR—an integral quantity—cannot be explicitly extrapolated out of local quantities, such as the readings from thermocouples and hot-plate thermometer.

3. Results

As an example, the dataset of gas-temperature and ceiling-surface temperature readings, together with pressure signature, is shown in Figure 8, where the profiles are presented for Test no. 1 as a function of time. Notably, temperature T is actually reported as temperature difference $\Delta T = T - T_{amb}$, where T_{amb} is ambient temperature (i.e., thermocouple reading prior to starting the fire). This strategy is aimed at offsetting discrepancies—even though relatively small—in ambient temperature between tests conducted on different days; it was also applied to Tests no. 6 and 7, where initial room temperature was imposed as greater than 80 °C (Table 2). The origin of the time coordinate is set at the start of the fire (i.e., manual ignition of the heptane pool fire below the wood crib); the actual discharge initiated as pressure (i.e., relative pressure) at the nozzle suddenly rose up from

0 to nitrogen pressure, which implied that pressurized water reached the nozzle exit. As noted in Section 2.3, the time at which gas temperature at the center of the crib fell below wood-ignition temperature is also marked in the plot as the conventional achievement of wood-crib fire extinction. It is worth noting that gas temperature at the ceiling raised more rapidly than surface temperature, but it also decreased down to smaller values as the discharge was activated. That can be explained by the higher thermal inertia of solid materials, even when involved with fire, together with the ability of water mist to perform convective flame cooling as soon as the two flows interact [17]. The fluctuations exhibited by gas temperature at the ceiling, even when thermal control was reached (in the 150–250 s timespan), suggest that high-pressure water mist emphasizes mixing of hot and cooler gas streams as a result of turbulence, a finding also reported in previous works [6,7].

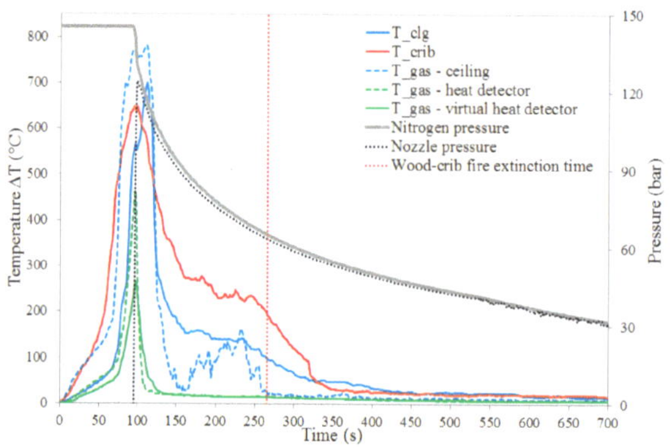

Figure 8. Temperature and pressure trends for Test no. 1.

Gas temperature at the wood crib showed the slower decay under a milder slope with respect to the other temperature profiles, which supports the use of that reading as a mark to assess the effective action of the discharge (Section 2.3).

As the most relevant parameters to quantify thermal stress and degradation imposed to the involved materials, surface-temperature trends are shown both at the bottom of the timber benches and at the ceiling for Tests no. 2–4 (Figure 9) and Tests no. 5–7 (Figure 10) As no significant difference occurred between temperature profiles from Test no. 1 and Tests no. 2 and 4, the data from the former are not displayed in Figure 5. The consistency between those three tests also allowed the assumption that the location of the ignition source was not critical in determining the outcome of water-mist action, thus focusing on configuration of C2 as generally representative of a sauna fire scenario.

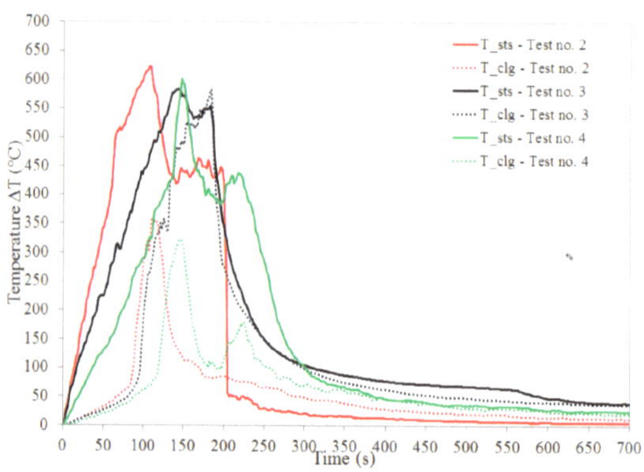

Figure 9. Timber-bench and ceiling surface temperature trends for Tests no. 2–4.

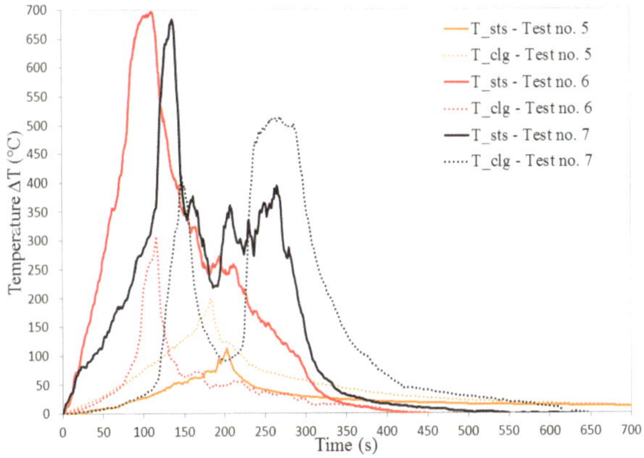

Figure 10. Timber-bench and ceiling surface temperature trends for Tests no. 5–7.

It is worth noting that maxima at the bench and maxima at the ceiling are consistent between Tests no. 2–4, with the exception of ceiling surface temperature in Test no. 3. As expected, temperature at the ceiling was lower than that at the bench, since this latter was involved with fire earlier, with its exposure being longer before discharge activation. However, ceiling temperature in Test no. 3 (i.e., that with longer free burn imposed prior to starting the discharge) reached almost the same peak as bench temperature, which suggests that timely activation is instrumental in containing the involvement of the combustibles with fire. The prolonged free-burn period due to a fixed discharge activation well beyond heat-detector alarm made the fire spread further than when an alarm-governed discharge was operated. The comparison between Test no. 2 and Test no. 3 in terms of extent of burnt surface is shown in Figure 11, which emphasizes much wider involvement as activation was delayed.

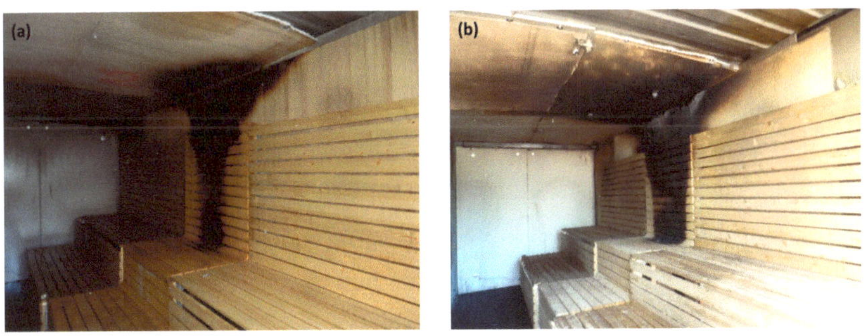

Figure 11. Bench and plywood boards at the ceiling and against the walls at the end of: (**a**) Test no. 2; (**b**) Test no. 3.

Surface temperature in Tests no. 6 and 7 exhibited a behavior somewhat similar to that of Tests no. 2 and 4 (Figure 10), with bench temperature trend being overall higher than that of ceiling. However, the peaks were bigger by about 100 °C, which suggests that the path towards worst-case scenario paved in Table 2 was well representative. Notably, initial room temperature in the order of 80 °C implied an environment generally more prone to involvement with the fire. Most importantly, the fire evolution throughout Test no. 7 was of particular interest, since it was identified as the worst-case scenario by higher initial room temperature, larger nozzle-to-wall distance, and ongoing natural ventilation. As already mentioned, higher initial room temperature implied higher maxima, but did not affect water-mist performance remarkably: as shown by the trends related to Test no. 6 (Figure 10), temperature decay was mildly slower than in Tests no. 1–5, but the system effectively suppressed the fire, arguably by spray penetration and room saturation. However, the presence of natural ventilation and a larger distance between nozzles and ignition source clearly made it more difficult to suppress the fire in Test no. 7: the timber bench kept burning above wood-ignition temperature for about 200 s before its suppression and an instance of fire regrowth—actually reignition, since ceiling temperature fell below wood-ignition temperature prior to raising back up in the timespan between 200 and 300 s—clearly occurred at the ceiling (Figure 10). So, Test no. 7 can be regarded as the worst-case scenario, with ventilation and area coverage of the nozzles embodying the challenge: the former drives more oxygen that feeds combustion reaction, while the latter deserves appropriate design and selection to meet the requirements from the specific scenario. This aspect becomes apparent by comparing the extent of burned surface at the end of Test no. 7 (Figure 12) with that at the end of Tests no. 2 and 3 (Figure 11). On the other hand, the comparison between ceiling and bench temperature for Test no. 5 appears poorly significant, since the respective maxima were far lower than wood-ignition temperature. So, a gap between benches and behind the wall did emphasize fire spread onto the solid surfaces, even though it certainly made oxygen available in that region. In fact, the flames exhibited a rather vertical development, so water droplets were able to suppress them by convective flame cooling [17] once the discharge was activated.

Figure 12. Bench and plywood boards at the ceiling and against the walls at the end of Test no. 7.

The heat flux incident onto the hot-plate thermometer is shown in Figure 13 as a function of time for all the performed experiments; in that plot, the time reference (0) corresponds to discharge activation, thus highlighting fire history before (free burn) and after that (water-mist discharge). Even though heat-flux profiles present a certain degree of noise, their trend firmly supports the observations proposed out of the analysis of gas and surface temperature. All the profiles exhibit an initial increase during the free-burn phase, followed by a rather steep decrease upon discharge activation; a clear instance of regrowth appears for the heat-flux trend of Test no. 7 prior to reaching final decay, which is related to reignition at the ceiling combined with the still-burning timber bench.

Therefore, hot-plate thermometry proved overall quite effective in capturing both the free burn and the suppression phase, whether achieved or not.

As the most significant observation, all temperature trends featured a dramatic decay upon discharge, which is typical of high-pressure (i.e., greater than 35 bar [27]) water-mist systems when successful in suppressing a fire: the spray momentum is instrumental in allowing droplets to reach the hot surfaces and realize surface cooling, also emphasized by surface-temperature trends (Figures 9 and 10) while performing flame cooling, as shown in the trends of Test no. 1 (Figure 8) [6,7,17]. Overall, only a small portion of the drywall boards at the ceiling was involved by the fire and resulted as mildly blackened at the end (Figure 11a).

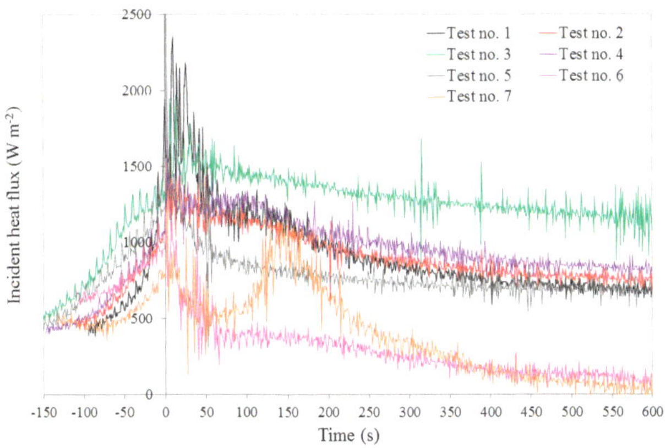

Figure 13. Incident heat-flux trend for all the conducted tests.

However, in Tests no. 3 and 7, flames engulfed the ceiling considerably, and gasification was comparatively more intense (Figures 11b and 12), which hints at the prolonged free burn (Test no. 3) and at the most challenging conditions among those tested (Test no. 7).

Table 3 presents a summary of all the experimental outcomes, with all the time values referring to ignition of the heptane pool fire as the origin. The reported data support the ability of the designed system to control and, overall, suppress the fire over the various tested scenarios. The results from Tests no. 1 and no. 2 do not significantly differ, which allows consideration of the location of the ignition source as almost irrelevant in achieving successful suppression. The presence of smoldering materials at the end of some experiments (Tests no. 4 and 7) strengthens the concept of thermal control and suppression as the results from water-mist discharge, rather than extinction; that is particularly apparent as the worst-case scenario is concerned. Overall, the timber-bench damage ratio was always kept below 5%, with the maximum (3%) occurring in Test no. 3, where discharge activation was delayed on purpose. As expected, the wood-crib fire extinction required a longer time in the tests, where initial room temperature was made consistent with that of actual saunas (about 80 °C); interestingly, that did not translate into wood-crib and bench damage ratio higher than in the other tests, which emphasizes the performance by the developed system. The heat detector issued its alarm at a time reasonably consistent with the whole test series. However, the time was somewhat longer in Test no. 5: the smaller involvement of the benches with fire as a result of the air gap between benches and wall arguably yielded a slower fire evolution, also shown by temperature trends in Figure 10; therefore, the threshold triggering the alarm signal was reached later. In general, the location of the ignition source, presence of drywall boards behind benches, and presence of a gap between benches and the wall behind were not determining factors; higher initial room temperature and ventilation led to an impact on some aspects (i.e., surface temperature, extent of involved materials, and regrowth/reignition phenomena); discharge activation time proved relatively significant with regard to gasified (burnt) mass, which suggests that early activation is instrumental in containing the loss of property, even when high-pressure water mist is operated.

Table 3. Summary of the experimental outcomes.

	Test No. 1	Test No. 2	Test No. 3	Test No. 4	Test No. 5	Test No. 6	Test No. 7
Heat-detector activation time (s)	91	100	107	143	179	107	129
Discharge activation time (s)	96	105	182	148	184	112	134
Smoldering materials at the end	NO	NO	NO	YES	NO	NO	YES
Overall suppression	YES	YES	YES	YES	YES	YES	YES
Wood-crib fire extinction time (s)	267	284	226	273	256	311	327
Initial wood-crib mass (g)	2813.2	2745.5	2750.5	2849.5	2907.5	3230.0	3175.5
Wood-crib damage ratio	12%	14%	11%	7%	12%	5%	12%
Initial bench mass (kg)	43.5	43.5	43.5	43.5	43.0	43.0	41.5
Bench damage ratio	1%	2%	3%	1%	0%	1%	1%

As recalled in Section 2, the cost associated with full-scale fire experiments is not negligible, which often makes repeating tests hardly sustainable. Moreover, the inherent statistical nature of turbulent-flame evolution shall be taken into account when assessing repeatability in such experiments. Nevertheless, a qualitative comparison between the free-burn phase of Tests no. 2–4—expressed by the temperature profiles in Figure 9—represents a certain degree of repeatability: even though the initial slope of surface temperature at the benches is more variable than that of surface temperature at the ceiling, overall consistency is displayed. Moreover, the already mentioned consistency between maxima of surface temperature at the benches also suggests that the behavior of unhindered fire is somewhat repeatable in the proposed experimental setup. As for the suppression phase, the performed tests (Table 2) overall feature different conditions, which does not allow proposing any conclusive observations on repeatability in that regard. Nevertheless, Tests no. 2 and 4—actually conducted under the same experimental conditions—exhibit very similar surface-temperature trends (Figure 9) and similar results in terms of damage ratio and wood-crib extinction time (Table 3). Those outcomes give evidence that even suppression may be identified as a repeatable mechanism, at least to a certain extent.

4. Conclusions

A series of experiments was devised and conducted to assess the ability of water-mist systems to control and suppress fires within a sauna scenario; the approach and the variety of considered parameters appear unprecedented in the open literature, especially as the study focuses on such a particular application. The compartment was designed to reproduce a sauna where the failure of an electrical heater causes a fire; the design of the employed fire-protection system—combining detection and discharge—and the test procedure were developed, as inspired by recognized standards [26,27,29], even though no specific guidance is currently available for this scenario. Several measurements were taken to assess, quantitatively, the performance of the discharge system. Notably, gas and surface temperature were evaluated at various locations of interest, both at the ceiling and at the benches, together with employing a hot-plate thermometer to quantify heat flux emitted through the fire evolution. Those quantities were deemed representative of HRR, thus allowing one to distinguish unsuccessful from successful suppression. A wood-crib fire accelerated by a heptane pool fire was used as the ignition source, and its extinction was also evaluated as a mark of achieved suppression, since its location was arguably the hardest to reach by released water. Damage ratio was measured for bench and wood crib; qualitative evaluation of fire spread by comparing pre- and post-fire photos was performed. A number of design parameters were varied through the test series towards an identified worst-case scenario: location of the ignition source, air gap between benches and wall, initial room temperature, ventilation, nozzle-to-wall distance, and activation of the discharge.

Overall, the designed water-mist system was capable of controlling and suppressing the fire in all tests, with evidence of sharp temperature decay upon release of the spray at all sampling locations. The wood-crib fire was extinguished in all the performed experiments The presence of smoldering materials at the end of some tests corroborated the reference to suppression, rather than extinction, as the main scope of water-mist technology. In general, the ceiling was mildly involved with the fire, even though a significant extent of its surface resulted as blackened in the case of prolonged exposure to free burn (i.e., delayed discharge activation), and this related to the worst-case scenario (i.e., initial room temperature above 80 °C, natural ventilation ongoing, and larger distance between nozzles and benches). The location of the ignition source, as well as the presence of drywall boards behind timber benches, together with a gap between benches and the wall behind the sauna, did not prove critical in determining the degree of success in suppression. A discharge governed by the feedback from a heat detector was effective in containing the timber-bench damage ratio; even though it never exceeded 3% over the whole series, a delayed activation of the discharge yielded higher damage, which emphasizes the need for early activation, even when a high-pressure water mist discharge is operated. Initial room temperature—aligned to that of actual saunas in some experiments—does not appear a determining factor, even though surface-temperature profiles demonstrate that the higher room temperature, the more prone the materials are to be involved with fire. The worst-case scenario showed that natural ventilation and a larger distance between nozzles and timber benches did not prevent the system from effectively suppressing the fire, but may have yielded regrowth and reignition phenomena, as they were observed at the ceiling surface while the benches were still burning. Hot-plate thermometry also proved able to capture those instances. Even though the timber-bench damage ratio did not vary with respect to the other tested conditions, the test run under the worst-case scenario suggests that water-mist systems in saunas may be combined with self-closing doors, thus limiting oxygen influx due to natural ventilation.

This work proved water-mist technology capable of controlling and suppressing a fire occurring in a sauna scenario, and it also provided designers with guidelines and—arguably an even more significant contribution—with a framework that allows relating the discharge system with both structural and ambient parameters within the compartment. As expected even for water mist, the effectiveness of an active fire-protection system is emphasized by early detection. To this end, inserting heat detectors is strongly recommended. However, the use of vision-based fire detection—currently a subject of many studies [37,38] and a very promising approach—may also be strategic, especially when combining feedback from surveillance systems with image-processing algorithms that allow identification of the presence of smoke and flames, together with assessing the fire size.

Author Contributions: Conceptualization, P.E.S., L.T. and P.T.; methodology, P.E.S. and L.T.; formal analysis, P.E.S. and L.T.; investigation, P.E.S. and L.T.; resources, L.T.; writing—original draft preparation, P.E.S.; writing—review and editing, P.E.S.; supervision, P.E.S. and P.T.; project administration, L.T.; funding acquisition, L.T. All authors have read and agreed to the published version of the manuscript.

Funding: This research was funded by Bettati Antincendio S.r.l., Italy.

Institutional Review Board Statement: Not applicable.

Informed Consent Statement: Not applicable.

Data Availability Statement: Not applicable.

Acknowledgments: The authors wish to thank M. Bettati and F. Dignatici for their helpful suggestions throughout the experimental tests.

Conflicts of Interest: The authors declare no conflict of interest.

References

1. *Sauna and Spa Market—Growth, Trends, COVID-19 Impact, and Forecasts (2021–2026)*; Mordor Intelligence: Hyderabad, India, 2021.
2. Howe, G.; Lloyd, S. Application of water mist to saunas. *Int. Fire Prof.* **2014**, *9*, 15–18.
3. Howe, G.; Palle, A. Fixed fire protection of saunas. In Proceedings of the 14th International Water Mist Conference, Istanbul, Turkey, 22–23 October 2014.

4. Wang, Y.C.; Marsden, J.; Kelly, M. Challenges of fire fighting in fire engineered built environment. *Procedia Eng.* **2011**, *11*, 583–592. [CrossRef]
5. Liu, Z.; Kim, A.K. A review of water mist fire suppression technology: Part II—Application studies. *J. Fire Prot. Eng.* **2001**, *11*, 16–42. [CrossRef]
6. Santangelo, P.E.; Tartarini, P. Full-scale experiments of fire suppression in high-hazard storages: A temperature-based analysis of water-mist systems. *Appl. Therm. Eng.* **2012**, *45*, 99–107. [CrossRef]
7. Santangelo, P.E.; Tarozzi, L.; Tartarini, P. Full-scale experiments of fire control and suppression in enclosed car parks: A comparison between sprinkler and water-mist systems. *Fire Technol.* **2016**, *52*, 1369–1407. [CrossRef]
8. Elsagan, N.; Ko, Y. A parametric study of numerical modelling of water mist systems in protection of wood frame buildings. In *Wood & Fire Safety*; Osvaldova, L.M., Markert, F., Zelinka, S.L., Eds.; Springer: Cham, Switzerland, 2020; pp. 166–172.
9. Liu, Y.P.; Wang, X.S.; Zhu, P.; Li, G.C.; Ni, X.M.; Zhang, J. Experimental study on gas jet suppressed by water mist: A clean control technique in natural gas leakage incidents. *J. Clean. Prod.* **2019**, *223*, 163–175. [CrossRef]
10. Li, G.; Pan, C.; Liu, Y.; Zhu, X.; Ni, X.; Zhao, X.; Chen, G.; Wang, X. Evaluation of the effect of water mist on propane/air mixture deflagration: Large-scale test. *Process Saf. Environ. Prot.* **2021**, *147*, 1101–1109. [CrossRef]
11. Liu, Y.; Shen, J.; Ma, J.; Li, G.; Zhao, Z.; Ni, X.; Wang, X. Laser-based measurement and numerical simulation of methane-air jet flame suppression with water mist. *Process Saf. Environ. Prot.* **2021**, *148*, 1033–1047. [CrossRef]
12. Li, G.; Wang, X.; Xu, H.; Liu, Y.; Zhang, H. Experimental study on explosion characteristics of ethanol gasoline-air mixture and its mitigation using heptafluoropropane. *J. Hazard. Mater.* **2019**, *378*, 120711. [CrossRef]
13. Grant, G.; Brenton, J.; Drysdale, D. Fire suppression by water sprays. *Prog. Energy Combust. Sci.* **2000**, *26*, 79–130. [CrossRef]
14. Mahmud, H.M.I.; Thorpe, G.; Moinuddin, K.A.M. The behaviour of water-mists in hot air induced by a room fire: Effect of the initial size of droplets. *Fire* **2022**, *5*, 116. [CrossRef]
15. Hopkin, C.; Spearpoint, M.; Muhammad, Y.; Makant, W. Estimating the suppression performance of an electronically controlled residential water mist system from BS 8458:2015 fire test data. *Fire* **2022**, *5*, 144. [CrossRef]
16. Kuznetsov, G.; Volkov, R.; Sviridenko, A.; Zhdanova, A. Compartment fire behavior at the stages of detection, containment and suppression using water mist. *Fire* **2022**, *5*, 155. [CrossRef]
17. Santangelo, P.E.; Jacobs, B.C.; Ren, N.; Sheffel, J.A.; Corn, M.L.; Marshall, A.W. Suppression effectiveness of water-mist sprays on accelerated wood-crib fires. *Fire Saf. J.* **2014**, *70*, 98–111. [CrossRef]
18. Liu, T.; Liu, Y.; Wang, X.; Kong, X.; Li, G. Cooling control of thermally—induced thermal runaway in 18,650 lithium ion battery with water mist. *Energy Convers. Manag.* **2019**, *199*, 111969. [CrossRef]
19. Jenft, A.; Collin, A.; Boulet, P.; Pianet, G.; Breton, A.; Muller, A. Experimental and numerical study of pool fire suppression using water mist. *Fire Saf. J.* **2014**, *67*, 1–12. [CrossRef]
20. Cordeiro, I.M.D.C.; Liu, H.; Yuen, A.C.Y.; Chen, T.B.Y.; Li, A.; Wang, C.; Cao, R.; Yeoh, G.H. On the Large Eddy Simulation modelling of water suppression systems droplet impact and coverage area. *Fire* **2022**, *5*, 165. [CrossRef]
21. Quintiere, J.G. Scaling applications in fire research. *Fire Saf. J.* **1989**, *15*, 3–29. [CrossRef]
22. Paulsen Husted, B.; Petersson, P.; Lund, I.; Holmstedt, G. Comparison of PIV and PDA droplet velocity measurement techniques on two high-pressure water mist nozzles. *Fire Saf. J.* **2009**, *44*, 1030–1045. [CrossRef]
23. Santangelo, P.E. Characterization of high-pressure water-mist sprays: Experimental analysis of droplet size and dispersion. *Exp. Therm. Fluid Sci.* **2010**, *34*, 1353–1366. [CrossRef]
24. Santangelo, P.E. Experiments and modeling of discharge characteristics in water-mist sprays generated by pressure-swirl atomizers. *J. Therm. Sci.* **2012**, *21*, 539–548. [CrossRef]
25. Santangelo, P.E.; Tartarini, P.; Pulvirenti, B.; Valdiserri, P.; Marshall, A.W. Fire suppression by water-mist sprays: Experimental and numerical analysis. In Proceedings of the 14th International Heat Transfer Conference (IHTC-14), Washington, DC, USA, 8–13 August 2010.
26. *CEN/TS 14972:2011*; Fixed Firefighting Systems—Watermist Systems—Design and Installation. European Committee for Standardization: Brussels, Belgium, 2011.
27. *NFPA 750*; Standard on Water Mist Fire Protection Systems. National Fire Protection Association: Quincy, MA, USA, 2015.
28. Xu, Q.; Griffin, G.J.; Jiang, Y.; Bicknell, A.D.; Bradbury, G.P.; White, N. Calibration burning of wood crib under ISO9705 hood. *J. Therm. Anal. Calorim.* **2008**, *91*, 355–358. [CrossRef]
29. *Class Number 5560*; Approval Standard for Water Mist Systems. FM Global: Norwood, MA, USA, 2014.
30. Nore, K.; Kraniotis, D.; Brückner, C. The principles of sauna physics. *Energy Procedia* **2015**, *78*, 1907–1912. [CrossRef]
31. Quintiere, J.G. *Principles of Fire Behavior*; CRC Press: Boca Raton, FL, USA, 2016.
32. *Document Number 2016–15229*; Harmonization of Standards for Fire Protection, Detection, and Extinguishing Equipment. USA Coast Guard, Department of Homeland Security: Washington, DC, USA, 2016.
33. Wickström, U. The plate thermometer—A simple instrument for reaching harmonized fire resistance tests. *Fire Technol.* **1994**, *30*, 195–208. [CrossRef]
34. Ingason, H.; Wickström, U. Measuring incident radiant heat flux using the plate thermometer. *Fire Saf. J.* **2007**, *42*, 161–166. [CrossRef]
35. Babrauskas, V. Ignition of wood: A review of the state of the art. In Proceedings of the 9th International Fire Science and Engineering Conference (Interflam), Edinburgh, UK, 17–19 September 2001.

36. Cannio, M.; Righi, S.; Santangelo, P.E.; Romagnoli, M.; Pedicini, R.; Carbone, A.; Gatto, I. Smart catalyst deposition by 3D printing for Polymer Electrolyte Membrane Fuel Cell manufacturing. *Renew. Energy* **2021**, *163*, 414–422. [CrossRef]
37. Bu, F.; Gharajeh, M.S. Intelligent and vision-based fire detection systems: A survey. *Image Vis. Comput.* **2019**, *91*, 103803. [CrossRef]
38. Geetha, S.; Abhishek, C.S.; Akshayanat, C.S. Machine vision based fire detection techniques: A survey. *Fire Technol.* **2021**, *57*, 591–623. [CrossRef]

fire

MDPI

Article

The Behaviour of Water-Mists in Hot Air Induced by a Room Fire: Effect of the Initial Size of Droplets

H. M. Iqbal Mahmud [1,2,*], Graham Thorpe [2] and Khalid A. M. Moinuddin [2]

[1] Department of Civil Engineering, Khulna University of Engineering & Technology, Khulna 9203, Bangladesh
[2] Institute for Sustainable Industries and Liveable Cities, Victoria University, P.O. Box 14428, Melbourne, VIC 8001, Australia
* Correspondence: iqbal.mahmud@ce.kuet.ac.bd

Abstract: This paper presents work on investigating the effect of the initial size of water mist droplets on the evaporation and removal of heat from the fire-induced hot gas layer while travelling through the air in a compartment. The histories of the temperature, diameter and position of droplets with different initial diameters (varied from 100 μm to 1000 μm) are determined considering surrounding air temperatures of 75 °C and 150 °C and a room height of 3.0 m. A water droplet evaporation model (WDEM) developed in a previous study (Fire and Materials 2016; 40:190–205) is employed to navigate this work. The study reveals that tiny droplets (for example, 100 μm) have disappeared within a very short time due to evaporation and travelled a very small distance from the spray nozzle because of their tiny size. In contrast, droplets with a larger diameter (for example, 1000 μm) reached the floor with much less evaporation. In the case of this study, the relative tiny droplets (≤200 μm) have absorbed the highest amount of energy from their surroundings due to their complete evaporation, whereas the larger droplets have extracted less energy due to their smaller area/volume ratios, and their traverse times are shorter. One of the key findings of this study is that the smaller droplets of spray effectively cool the environment due to their rapid evaporation and extraction of heat from the surroundings, and the larger droplets are effective in traversing the hot air or smoke layer and reaching the floor of the compartment in a fire environment. The findings of this study might help in understanding the behaviour of water-mist droplets with different initial diameters in designing a water-mist nozzle.

Keywords: water-mist; droplet size; evaporation; equilibrium temperature; fire; hot air

Citation: Mahmud, H.M.I.; Thorpe, G.; Moinuddin, K.A.M. The Behaviour of Water-Mists in Hot Air Induced by a Room Fire: Effect of the Initial Size of Droplets. *Fire* **2022**, *5*, 116. https://doi.org/10.3390/fire5040116

Academic Editors: Song Lu, Changcheng Liu, Guohui Li and Paweł Wolny

Received: 5 June 2022
Accepted: 10 August 2022
Published: 15 August 2022

1. Introduction

Water in the form of a mist has been shown to be an effective fire extinguishing agent. The water-mist fire suppression system (WMFSS) uses smaller size of water droplets compared to the conventional water sprinkler droplets and suppresses the fire by the displacement of oxygen by means of evaporation of water droplets [1,2]. The National Fire Protection Association (NFPA) defined water mist as a water spray in which 99% of the water is in droplets whose diameter (D_{v99}) is less than 1000 μm [3]. During the past several years, water mist technology has been developed and regarded as a promising substitute since it can extinguish fires quickly with little water and, at the same time, without damaging the environment [4,5]. In the analysis of the suppression mechanism of fire by water droplets, it is imperative to understand the evaporation behaviour of a single droplet travelling through a fire-induced smoke layer and the phenomena connected with the evaporation process.

The literature is replete with studies on the heating and evaporation of water from single droplets suspended in high-temperature environments [6–10]. Strizhak et al. [8] conducted a series of experiments to investigate the heating and evaporation of suspended water droplets with diameters varying from 1.8 to 3 mm in a hot air flow with temperatures up to 800 °C. Volkov and Strizhak [7] performed experiments on water droplets suspended in the air with the size ranging from 1 to 2 mm at high air temperature varying from 100 to 800 °C, and measured the evaporation rate and profile of droplet size, and recorded the time of their existence. Sobac et al. [9] developed a quasi-steady model to analyse the evaporation process of a suspended water droplet in air. Thielens et al. [10] developed a two-zone model for heating an evaporating water droplet suspended and exposed to hot airflow. They used the model to predict the lifetime and saturation temperatures of droplets.

However, it is worthy to note that in these studies, the droplets were suspended in a stationary position; hence, the effect of momentum transfer of the droplets was not of primary importance. A defining feature of water-mist nozzles is that they produce fine mists consisting of tiny droplets with diameters of less than 1000 μm. The fine mists exhibit fog-like behaviour that renders their fire suppression mechanism quite different from conventional water sprays that comprise larger droplets [11]. Studies of the interaction of traditional sprinkler spray with hot air or smoke layers [12–19] have focused mainly on the convective heat transfer phenomena between the large water droplets and the layer of hot air. It was found that the evaporation of larger droplets discharged from conventional sprinklers had not been affected significantly by the fire plume [12–17]. It has also been confirmed from the previous study that large size of water droplets produced by sprinklers can remove only between 11% and 26% of the heat produced by fire [19–21].

Water-mist droplets can cool the surrounding air and attenuate thermal radiation, and the water vapour produced by evaporation reduces the fuel vapour/air ratio by displacing oxygen [22–27]. Evidence supports the idea that water-mist spray can remove 100% of the heat produced by a fire either by extinguishing it or allowing enough evaporation to keep the compartment's temperature at or below the boiling point of water [21,28]. However, not all droplets evaporate before striking burning surfaces, and this provides a direct method of suppressing fires [23,24,29,30]. In particular, the initial diameter of the droplets and the surrounding environment are the main factors determining whether a droplet can travel through the hot smoke layer or evaporate entirely before reaching the floor. However, it is imperative to understand the evaporative behaviour of a single water mist droplet travelling through the hot air in a fire-induced room environment based on heat and mass transfer fundamentals in the droplet.

Therefore, the objective of this paper is to investigate the evaporation behaviour of a single droplet while travelling through a fire-induced smoke layer in a room. The profiles of the temperature, diameter and position of the droplet are studied with different initial drop sizes under two different room environments. In addition, the suspension time in the air and the evaporation rate of the droplets are explored. A previously developed water droplet evaporation model (WDEM) by the same authors [11] is used to facilitate this investigation.

Several important parameters should be considered while analysing the behaviour of a single droplet travelling through the hot smoke layer. The tiny droplets of water that comprise fine mists have a higher surface area/volume ratio. This results in their rapid evaporation, and their movement is highly responsive to their local environment [11]. As a result, water droplets emanating from a nozzle, the diameter and velocity change continuously due to evaporation, and this affects the drag coefficient. Furthermore, the Reynolds number (Re) affects the heat and mass transfer coefficient between water and air. The relative humidity (RH) of the surrounding air significantly affects the evaporation of droplets. Lower humidity contributes to a higher evaporation rate of the water droplet. In a fire scenario, when sprinkler systems are activated at a temperature of 60 °C, say, the relative humidity (RH) of the surrounding air is very low, typically 5%. It is worth noting that some of these effects were not considered in the many previous studies, for example, the studies in the reference [31–34]. However, the above-mentioned phenomena should be considered in the movement of water droplets in hot air induced by a room fire.

The novelty of the model used in this study is that the following parameters are considered: (i) the contribution of radiation emanating from the flame and the surrounding boundary walls to the rate of evaporation of water droplets, (ii) the change of Re and drag coefficient of the droplets with time and position, (iii) the effect of high mass transfer rate due to the high evaporation of droplets resulting from a high air temperature and low relative humidity, (iv) the change of droplet momentum, due to the change of droplet velocity and diameter and (v) the variable thermo-physical properties of water and air. The law of conservation of mass, momentum and energy was used to evaluate the heat and mass transfer phenomena in an air-droplet system in connection with the effect of change of momentum of an evaporating droplet.

2. Description of the Model

The evaporation processes experienced by a moving droplet encompass simultaneous heat, mass, and momentum transfer between the particle and surrounding air. Momentum transfer affects the motion of the particle, mass transfer results in changes in the size of the particle, and heat transfer determines the temperature of the particle. Interestingly, these mechanisms are inextricably related [31]. The model developed by these authors accounts for the effect of a mass transfer rate due to high temperature and low humidity. The accuracy of the predictions of temperature, velocity, diameter, evaporation rate and other parameters related to the behaviour of droplets is enhanced by considering the changes in the diffusivity of the air-water system, density and latent heat of vaporisation of water with the change of temperature.

In this model, the shape of droplet is assumed to be spherical as this would not give any significant error in the computation [35]. The droplet is considered as a 'lumped mass' as the Biot number is very low [36]; as a result, the temperature distribution in the droplet was assumed to be uniform during the evaporation. This considerably simplified the analysis of the overall computational process, since it avoided the need for a conjugate heat-conduction analysis for the internal transient temperature-distribution inside the droplet [34]. The air velocity in the hot air layer is supposed to be insignificant as the droplet is supposed to travel through hot air layer, which is not nearby the fire source. The heat transfer from the ambient air to the droplet is due to forced convection.

In this study, the collision and coalescence of droplets are considered to be negligible. This assumption is reasonable for a low volume of water spray that results in dilute droplet loading in the air [37–39]. In fact, the water mist spray nozzle is designed to use a small amount of water. As a result, the mass concentration of water in the air in terms of droplet number per cubic meter of air is comparatively low. This causes a higher value of the ratio of droplet–droplet separation distance to droplet diameter. Under this condition, the probability of droplet coalescence is negligible [37]. This phenomenon was also addressed by Ananth and Mowrey [38], and Sikanen et al. [40].

The smoke layer or hot air is assumed to be in quasi steady-state, and this attribute to a stable smoke layer which is formed finally when the ceiling jet reaches the boundary wall and rebounded several times [32]. This assumption is more appropriate for the nozzles and smoke layers, which are located away from the fire source or burning object [11]. This is also supported by the experimental observations in articles of [18,41].

The details of the mathematical models of mass and heat transfer and momentum of a droplet are summarized in the following sub-sections and are also available in details in the previously published paper [11].

According to the theory of mass transfer, mass flux per unit area from the interfacial surface of a water droplet is proportional to the mass concentration difference across the boundary layer of the droplet [42]. Therefore, the rate of change of droplet diameter with time can be determined from the following equation:

$$\frac{dD}{dt} = 2h_m \frac{(\rho_s - \rho_\infty)}{\rho_w} \quad (1)$$

In the above equation, h_m is the mass transfer coefficient, ρ_w is the density of water, and ρ_s and ρ_∞ are the mass concentration of water vapour on the droplet surface and in the air, respectively. The mass concentration of water particle at the surface of the droplet depends on the partial pressure of vapour at the droplet surface. In thermodynamic equilibrium state, the partial pressure of the vapour at the droplet surface depends on the surface temperature [32]. Under this condition, evaporation keeps the droplet surface in a saturated condition until the droplet is totally vaporized due to heat and mass transfer [43]. The vapour concentration at the surface is the saturated mass fraction of air at temperature of droplet. As the mass concentration of water particle depends on vapour pressure of water, this can be found from the ideal gas equation of state [11].

The mass transfer coefficient, h_m, can be calculated by using the correlation for Sherwood number, Sh, where Sh is $h_m D/D_{AB}$ [44] and D_{AB} is the mass diffusivity coefficient in a binary system of A and B; here, binary system is air and water. The correlation of Sherwood number can be expressed as [45],

$$Sh = 2.0 + 0.216\left(Re^{\frac{1}{2}}\,Sc^{\frac{1}{3}}\right)^2,\ For\ \left(Re^{\frac{1}{2}}\,Sc^{\frac{1}{3}}\right) < 1.4 \tag{2}$$

$$Sh = 1.56 + 0.616\left(Re^{\frac{1}{2}}\,Sc^{\frac{1}{3}}\right),\ For\ \left(Re^{\frac{1}{2}}\,Sc^{\frac{1}{3}}\right) \geq 1.4 \tag{3}$$

In the above equations, the Schmidt number, Sc, is a dimensionless number defined as the ratio of momentum diffusivity (viscosity) and mass diffusivity, and is used to characterize fluid flows in which there are simultaneous momentum and mass diffusion convection processes.

When a droplet is exposed to a higher temperature, it receives heat from the surrounding and temperature increases to a threshold limit, at a given pressure, and this is known as steady state or equilibrium temperature. At this temperature, the water droplet changes its phase from liquid to vapour, keeping the droplet surface in steady state condition of temperature until the droplet is entirely vaporised [46]. Under this condition, the heat of vaporisation is supplied to the droplet surface from surrounding air, flame and hot objects. Therefore, according to the law of conservation of energy, the net convective and radiative heat transfer to the droplet surface is equal to the heat accumulated in the droplet plus heat leaving the droplet due to evaporation of water particles from the surface of the droplet. Considering the rate of change of temperature and mass, the transient equation of the conservation of heat can be expressed as:

$$\begin{aligned} c_{pw} m \tfrac{dT}{dt} = {} & h_c A(T_\infty - T) \\ & + [\sigma\varepsilon F A\left\{\left(T_f + 273.15\right)^4 - (T + 273.15)^4\right\} \\ & + \sigma\varepsilon A(1 - F)\left\{(T_{bw} + 273.15)^4 - (T + 273.15)^4\right\}] - \tfrac{dm}{dt} L \end{aligned} \tag{4}$$

Here, c_{pw} is the specific heat capacity of water, h_c is the convective heat transfer coefficient, and L is the latent heat of vaporisation of water. In the above equation, the first part on the right side of the equation is due to convective heat transfer from air to droplet, the second and third parts are due to radiative heat transfer from fire flame and boundary wall to the droplet, respectively, and the fourth part contributes to the evaporation of water droplets.

The convective heat transfer coefficient, h_c, can be calculated by using the correlation for the Nusselt number, Nu, where Nu is $h_c D/k_a$ [44]. The correlation for the Nusselt number can be expressed as [45],

$$Nu = 2.0 + 0.216\left(Re^{\frac{1}{2}}\,Pr^{\frac{1}{3}}\right)^2,\ For\ \left(Re^{\frac{1}{2}}\,Pr^{\frac{1}{3}}\right) < 1.4 \tag{5}$$

$$Nu = 1.56 + 0.616\left(Re^{\frac{1}{2}}\,Pr^{\frac{1}{3}}\right),\ For\ \left(Re^{\frac{1}{2}}\,Pr^{\frac{1}{3}}\right) \geq 1.4 \tag{6}$$

These relationships were found to be in good agreement with the numerical and experimental results [47]. In those above equations, *Pr* is Prandtl number which is the ratio of viscous diffusion rate (v) to thermal diffusion rate (α) of air, i.e., $Pr = \frac{v}{\alpha} = \frac{c_p\mu}{k}$ and *Re* is Reynolds number, which is the ratio of inertia force to viscous force, i.e., $Re = \frac{\rho v D}{\mu}$.

The velocity of the droplet can be obtained by solving the equation of conservation of momentum. When a body is falling from a height, body force (or weight) works in the downward direction and resistance of air drag and buoyancy force work in the upward direction. The equation of momentum for a water droplet, traveling in air, can be obtained by Newton's second law of motion. Therefore, the momentum equation for a water droplet with mass m, diameter D and relative velocity v, can be written as:

$$\frac{d(mv)}{dt} = mg - \frac{1}{2}\rho_a v^2 C_d A_{proj} - \frac{1}{6}\rho_a \pi D^3 g \tag{7}$$

In the above equation, ρ_a is the density of air, C_d is the coefficient of drag, g is the acceleration due to gravity, A_{proj} is the projected area of the droplet and v is the velocity of the droplet. The term in the left side of the equation is the rate of change of momentum of the droplet; The first term on the right side is the force acting on the droplet due to gravity, the second term is the drag, sometimes called air resistance or fluid resistance, refers to forces which act on the droplet in the opposite direction of the movement, and the third term is the buoyancy force work in the upward direction. In case of the movement of water droplet, mass, m, and velocity of droplet, v, both of them are changing with time. Therefore, simplifying the above equation, the rate of change of velocity of the droplet can be expressed as:

$$\frac{dv}{dt} = g\frac{(\rho_w - \rho_a)}{\rho_w} - \frac{3}{4}\frac{C_d \rho_a v^2}{\rho_w D} - \frac{3v}{D}\frac{dD}{dt} \tag{8}$$

It is to be noted that C_d for a droplet depends on *Re*, which is based on the air-droplet relative velocity. Brown and Lawler [48] proposed a correlation between drag coefficient and *Re*, and compared it with 178 experimental data points. The proposed correlation was found to be quite satisfactory in relation to the experimental data in the range of $0 \le Re < 2 \times 10^5$. The correlation by Brown and Lawler [48] is used here for the calculation of C_d. It is:

$$C_d = \frac{24}{Re}\left(1 + 0.15\, Re^{0.681}\right) + \frac{0.407}{1 + \frac{8710}{Re}} \tag{9}$$

The velocity of a droplet is given by:

$$\frac{dy}{dt} = v \tag{10}$$

in which y is the vertical distance from where the mist is dropping down. Overall, Equations (1), (4), (8) and (10) are solved simultaneously to obtain D, T, v and y, respectively, of the droplet with respect to time. A Lagrangian approach is used to track the movement of water droplets, as this is well suited to tracking the dispersed particle flow.

The use of coefficients, h_m and h_c are limited to the case of low-mass-transfer-rate. In case of high temperature and low humidity, the evaporation rate is high and this invokes the high mass transfer rate and affects the heat transfer rate, as well. Therefore, corrections are also applied to calculate the mass and heat transfer coefficient to account for the effect of high mass and heat mass rate in the evaporation of droplets. The density of humid air, saturation vapour pressure, latent heat of vaporisation of water and other thermos-physical properties of air, such as viscosity, thermal conductivity and specific heat capacity of air were accounted in the model. The details of the correction factors, thermos-physical properties or vapour and air, and computational procedures of the model are presented in reference [11].

3. Validation of the Model

The developed model (WDEM) was validated and verified against experimental data by Gunn and Kinzer [49] and theoretical data of adiabatic saturation temperature, respectively. The results of this model were also compared with that of a computational fluid dynamics (CFD)-based fire model, fire dynamics simulator (FDS) [50]. The prediction of the developed model agreed well with the calculated values by FDS. The model (WDEM) was also compared with two other models by Li and Chow [32] and Barrow and Pope [34]. The details of the validation and verification of the model are available in the previous version of the paper as in reference [11].

In this section, the proposed model is further validated using the data of CARAIDAS experiments [51]. The details of the experimental set-up and results are also presented in the publication by Plumecocq et al. [52]. The conditions of the experiments used in the CARAIDAS are mentioned in Table 1.

Table 1. Test conditions of three EVAP tests of the CARAIDAS program [52] used as benchmarks in this work.

Test Condition	Air Conditions			Spray Droplets at Injection (z = 0 m)		
	P (Pa)	T (°C)	RH (%)	T (°C)	D (µm)	v (m/s)
EVAP4	10^5	47	12	25	387	1.44
EVPA8	10^5	106	<1	25	414	1.55
EVPA11	10^5	147	<1	28	423	1.59

These three experiments were performed at atmospheric pressure and low relative humidity which are usually corresponding to gas conditions that could be met in the fire room prior the activation of a spray system [52]. These conditions of the experiments are incorporated in the evaporation model (WDEM) to evaluate the droplet diameters. The predictions of the model are compared with the experimental data and the results are presented in Table 2.

Table 2. Comparison between the evaporation model predictions and CARAIDAS test results [52] for the evolution of droplets diameter.

Test Condition	D (µm) at z = 2.51 m			D (µm) at z = 4.39 m		
	Test	WDEM	Diff	Test	WDEM	Diff
EVAP4	375	375	0%	363	366	1%
EVPA8	393	382	−3%	363	356	−2%
EVPA11	393	377	−4%	342	335	−2%

The results show that the predicted data is very close to the experimental measurements. The difference between predicted data and experimental results is not more than 4%, which validating the model on the CARAIDAS experiments. In the following section, the developed model (WDEM) is used to investigate the parameters relating to the behaviour of a droplet travelling through two different hot air environments with variable initial droplet sizes.

4. Results and Discussion

Evaluation of the effect of initial sizes of droplets is conducted in this study at high air temperature in the vicinity of fire in a room. The initial air condition of the room, i.e., temperature and relative humidity, is taken to be 20 °C and 50%, respectively, and the initial temperature of the water drops is taken to be 20 °C. However, smoke will be produced in the room due to a fire, and the room temperature will be raised. Here two temperature conditions are considered for the analysis.

(i) Condition 1: the smoke layer temperature is raised to 75 °C, at which temperature the water-mist nozzle is activated. This activation temperature is typical of commercial water mist and sprinkler systems [53]; Due to the high temperature, the relative humidity of the air falls to 3%.
(ii) Condition 2: the smoke layer temperature is raised to 150 °C. The temperature of a smoke layer produced by fire usually varies between 100 and 180 °C and can reach up to 200 °C [54–57] based on the distance from the fire source. Therefore, an average value of 150 °C is considered for the analysis. When the air temperature is beyond 100 °C, there is no longer any limit to the amount of water vapour that can be stored in the air [32].

The potential travel path of the droplets is taken to be the height of a room in a residential or commercial building, i.e., 3.0 m. The elapsed time begins when the water droplets emanate from a ceiling-mounted nozzle and terminate when they strike the floor. The airflow is assumed to be in a quiescent state, as the droplets are considered to be located away from the fire. In the experimental study, it can also be observed that the hot air layer is almost steady when the nozzle is located away from the fire source [18,41]. A schematic of the computational domain of droplet movement is presented in Figure 1.

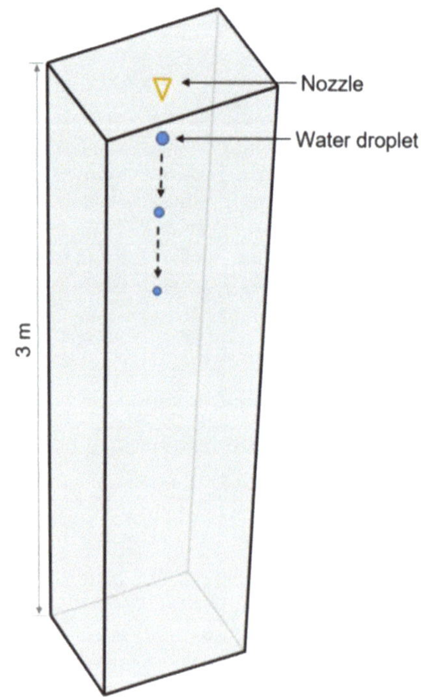

Figure 1. Schematic of the computational domain for the movement of droplet.

A sensitivity analysis has been conducted before performing the analysis using the computational model. The governing differential equations are discretised and solved explicitly using a forward differencing technique temporally. The initial conditions of a droplet, i.e., initial diameter (D), temperature (T), velocity (v) and position (y), as well as the relevant thermo-physical boundary conditions of water and air, are specified. The discretisation of the equation depends on the change of parameters with time steps (Δt).

Therefore, the histories of D, T, v, and y should be independent of the time step, Δt. In this study, it has been found that if the ratio of droplet diameter to time step is less than or equal to 0.01, the solution is independent of the time step. A time-step independence test for a droplet with a diameter of 200 µm was performed as an example. The temperature

(T) of a droplet with three different time steps of 0.1, 0.01 and 0.001 s are computed and compared. Numerical instabilities are observed when the time step is 0.1 s. However, for time steps of 0.01 and 0.001 s, the temperature and velocity measurements are similar and consistent. As a result, a calculating time step of 0.01 s can be used. A graphical presentation of the results is shown in Figure 2.

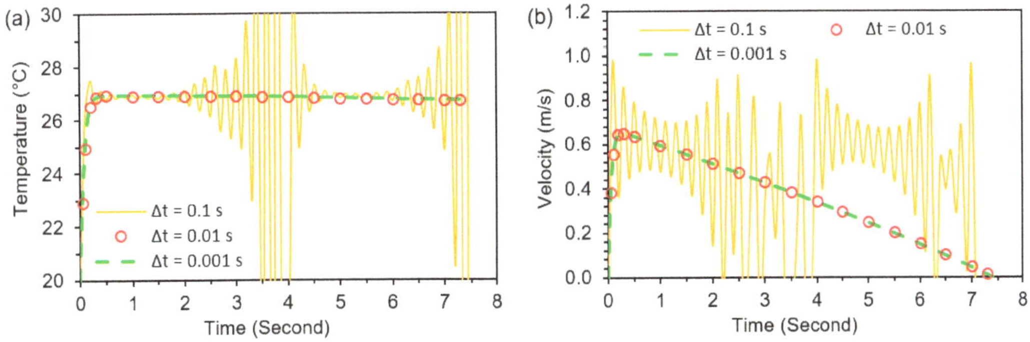

Figure 2. Sensitivity analysis of the computational model; (**a**) temperature (T), and (**b**) velocity (v).

Once the time step is selected, the parametric study is conducted using the model and the results are presented in the following sub-sections. The temperature histories of droplets for two different temperature conditions are calculated with initial diameters of 100, 200, 300, 400, 500, 750 and 1000 µm, and the results are illustrated in Figure 3. It is found that the temperature of the droplets has increased until they have reached an equilibrium temperature at which the heat gained by convection for the air is equal to that of the heat loss by the evaporation of water. The temperature at this condition is known as equilibrium temperature or steady-state temperature.

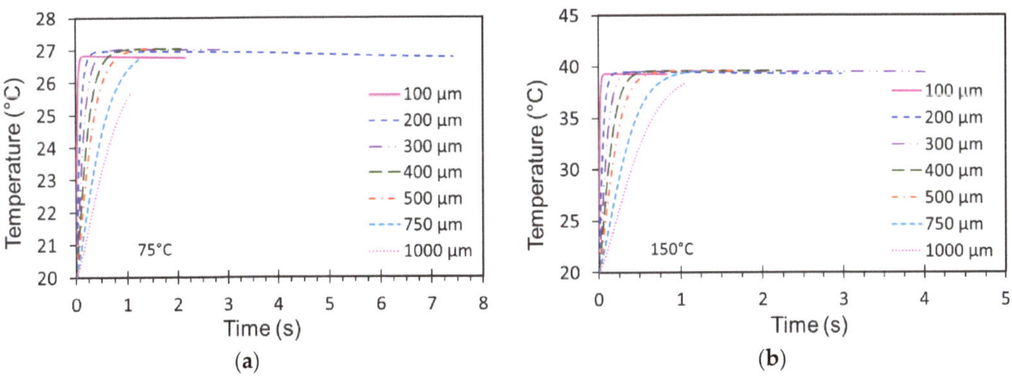

Figure 3. The effect of the initial size of diameter on the temperature trajectories of droplets; (**a**) condition 1 (75 °C); and (**b**) condition 2 (150 °C).

From the analysis, it has also been found that the smaller the initial diameter of the droplets, the less time it has taken to reach the equilibrium thermal condition. Concomitantly, as the size of the droplets has increased, the time to reach the equilibrium thermal condition has become higher. However, the equilibrium temperature has only a weak dependence on the size of the droplets, and it depends principally on the temperature and relative humidity of the air. For the condition of 75 °C air temperature, 20 °C water temperature and 3% relative humidity, the equilibrium temperature of droplets is about 27 °C, and for the second condition of 150 °C air temperature, 20 °C water temperature and 0.5% relative humidity, the equilibrium temperature is about 39 °C. It is important to note that the smaller size of droplets have reached to the equilibrium temperature earlier compared to the bigger droplets. Furthermore, the droplets have reached the thermal equilibrium state earlier at condition 2.

The diameter and position history of the droplet with different initial sizes are analysed for condition 1, and the results are presented in Figure 4. Two factors influence the longevity of the droplets in the air of a room, namely the rate at which they evaporate, and the time it takes to reach the floor. For example, it can be seen from Figure 4 that droplet with an initial diameter of 100 μm has been evaporated completely within 2.16 s, but fall a mere 0.25 m from the nozzle. On the other hand, the droplet with an initial diameter of 200 μm has had a relatively longer existence of 7.42 s in the air; and because of this, they have travelled 2.6 m distance from the nozzle. However, the droplets with an initial diameter of 300 μm or more scarcely change their diameters before impacting the floor. Furthermore, the higher the diameter above 300 μm, the shorter the suspension time in the air.

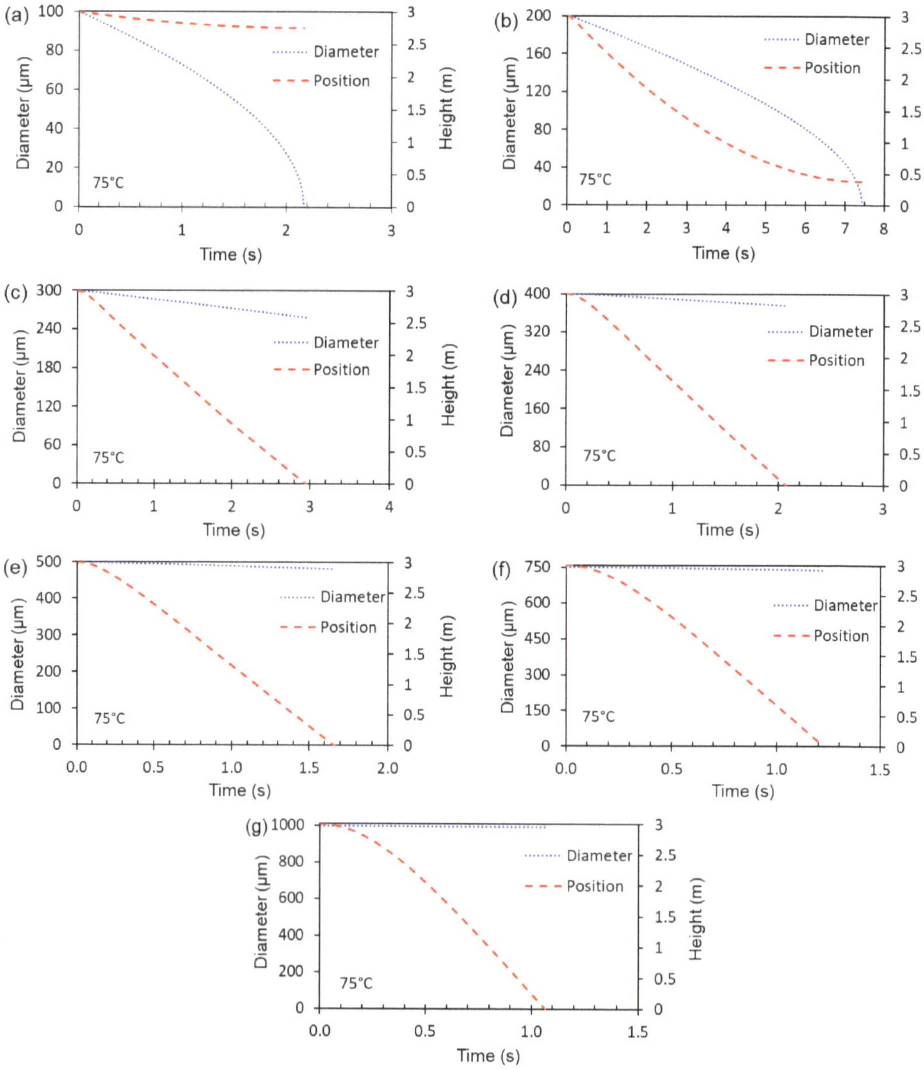

Figure 4. Diameter and position history of droplets with different initial diameters at condition 1 (75 °C); (**a**) 100 μm; (**b**) 200 μm; (**c**) 300 μm; (**d**) 400 μm; (**e**) 500 μm; (**f**) 750 μm; and (**g**) 1000 μm.

This is because the smaller droplet has higher value of surface area to volume ratio compared to the larger droplet and as a result, it enhances the rate of evaporation of the droplets by extracting heat from the hot gases and smoke layer [13,58]. The finer the droplets the greater the surface area to volume ratio of water. For example, the surface area to volume ratio of a 300 μm droplet is 20, whereas it is 30 and 60 for 200 μm and 100 μm diameter of droplet, respectively. Furthermore, the smaller droplet reaches to equilibrium temperature earlier compared to the larger droplet (as seen in Figure 3). As a result, the evaporation of water particles from the droplet surface also starts earlier in the smaller droplet. Furthermore, as the smaller droplet has lower momentum, it is suspended in the air for a longer time, if it has not completely disappeared due to evaporation. This also leads to a higher suspension time for the smaller droplet and results in a decrease in the diameter of the droplet due to evaporation.

Similar phenomena are also observed for condition 2, as shown in Figure 5. The droplets of diameter 100 and 200 μm are evaporated completely before they have reached the floor. In particular, the 100 μm droplet disappears within 1 s after leaving the nozzle. Thus, the droplet has travelled only about 0.1 m in the room. In contrast, the 200 μm droplet is sustained in the air for about 3 s and travelled about 1 m. Compared to condition 1, these two sizes of droplets have taken a lower time to evaporate as intuitively expected. They have also travelled a shorter distance due to faster evaporation at higher temperatures. The evaporation of 300 μm droplet is also significant. The diameter of this droplet has been reduced by half due to evaporation. The suspension time of this droplet is about 4 s which is the highest lifetime compared to the other size of droplets. However, the droplets with an initial diameter of 400 μm or more scarcely change their diameters before impacting the floor in both conditions. Furthermore, the travel time of the droplets with the initial diameter of 400 to 1000 μm is almost the same for both conditions.

It can be seen from Figures 4 and 5 that the higher the diameter above 300 μm, the shorter the suspension time in the air. Therefore, from the point of view of fire extinguishment, it appears that the smaller size of droplets (in this case, 100 and 200 μm) may be effective in cooling the air within an enclosure but they are less likely to penetrate a layer of hot air and extinguish conflagrating surfaces on the floor of an enclosure. On the other hand, the larger the diameter, the higher the possibility of penetrating a layer of hot air and reaching the floor, however, the larger droplet does not absorb heat from the surrounding. Actually, water mists (containing tiny droplets) and water sprays (containing large droplets) have quite different *modi operandi*. The fine particles with diameters that constitute water mists not only attenuate thermal radiation [59] but also reduce their environment's temperature due to their high surface area per unit volume and their high rate of vaporisation. In contrast, the larger droplets that comprise water sprays are likely to penetrate fires and cool surfaces of the undergoing combustion.

The velocity and corresponding vertical positions of droplets with different initial sizes in a room for condition 1 are illustrated in Figure 6. It is observed that the droplets with initial diameters of 100 and 200 μm initially accelerate in the gravitational field but decelerate quite markedly as their diameters decrease due to evaporation. These two smaller diameters of droplets reach their thermal equilibrium states after 0.081 and 0.284 s, respectively, and they continue to evaporate until they disappear. As the diameter of 100 and 200 μm droplets reduces due to evaporation, their velocity also reduces with time. However, the change in velocity of droplets with diameters greater than 400 μm is insignificant because there is little change in their diameters, as found in Figure 4.

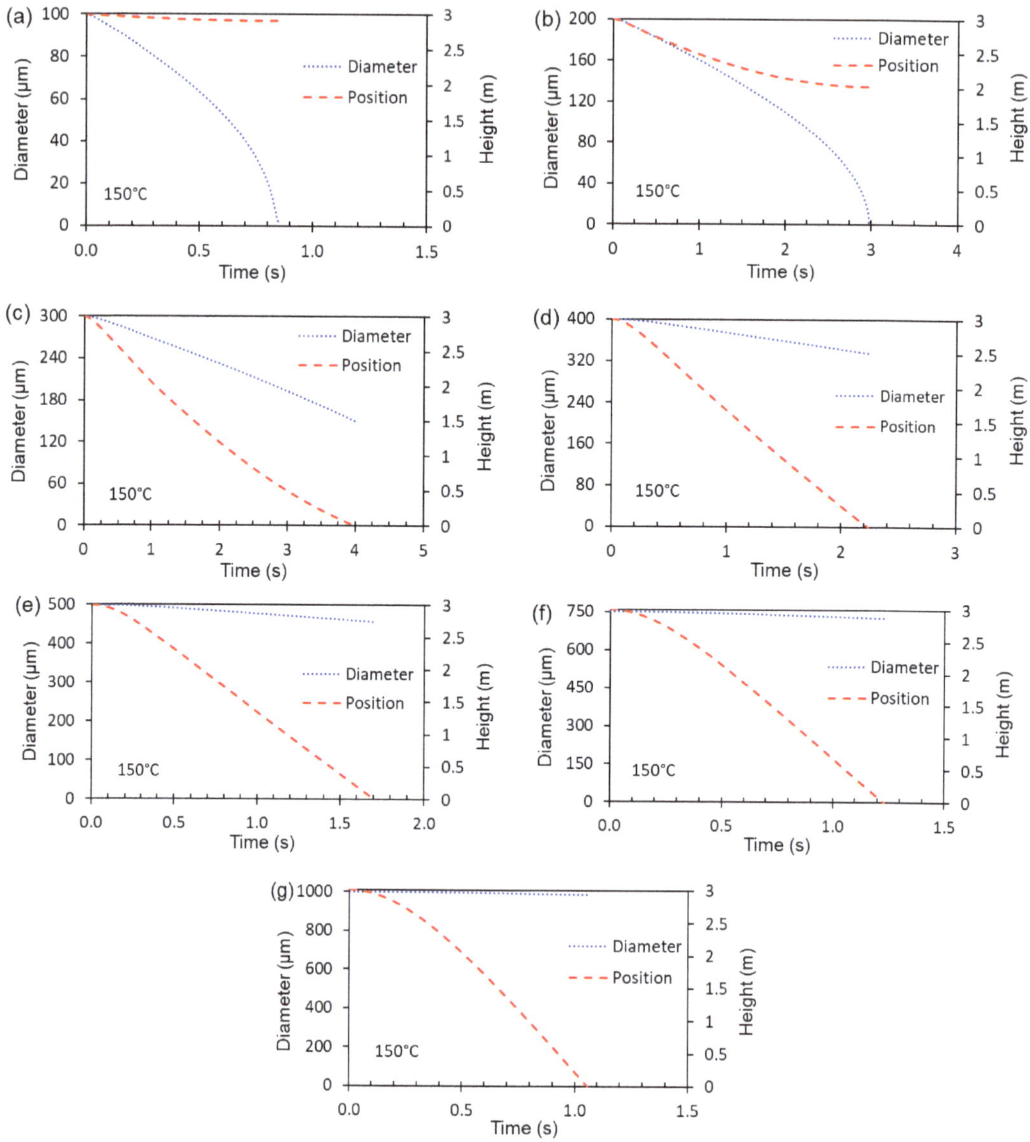

Figure 5. Diameter and position history of droplets with different initial diameters at condition 2 (150 °C); (**a**) 100 μm; (**b**) 200 μm; (**c**) 300 μm; (**d**) 400 μm; (**e**) 500 μm; (**f**) 750 μm; and (**g**) 1000 μm.

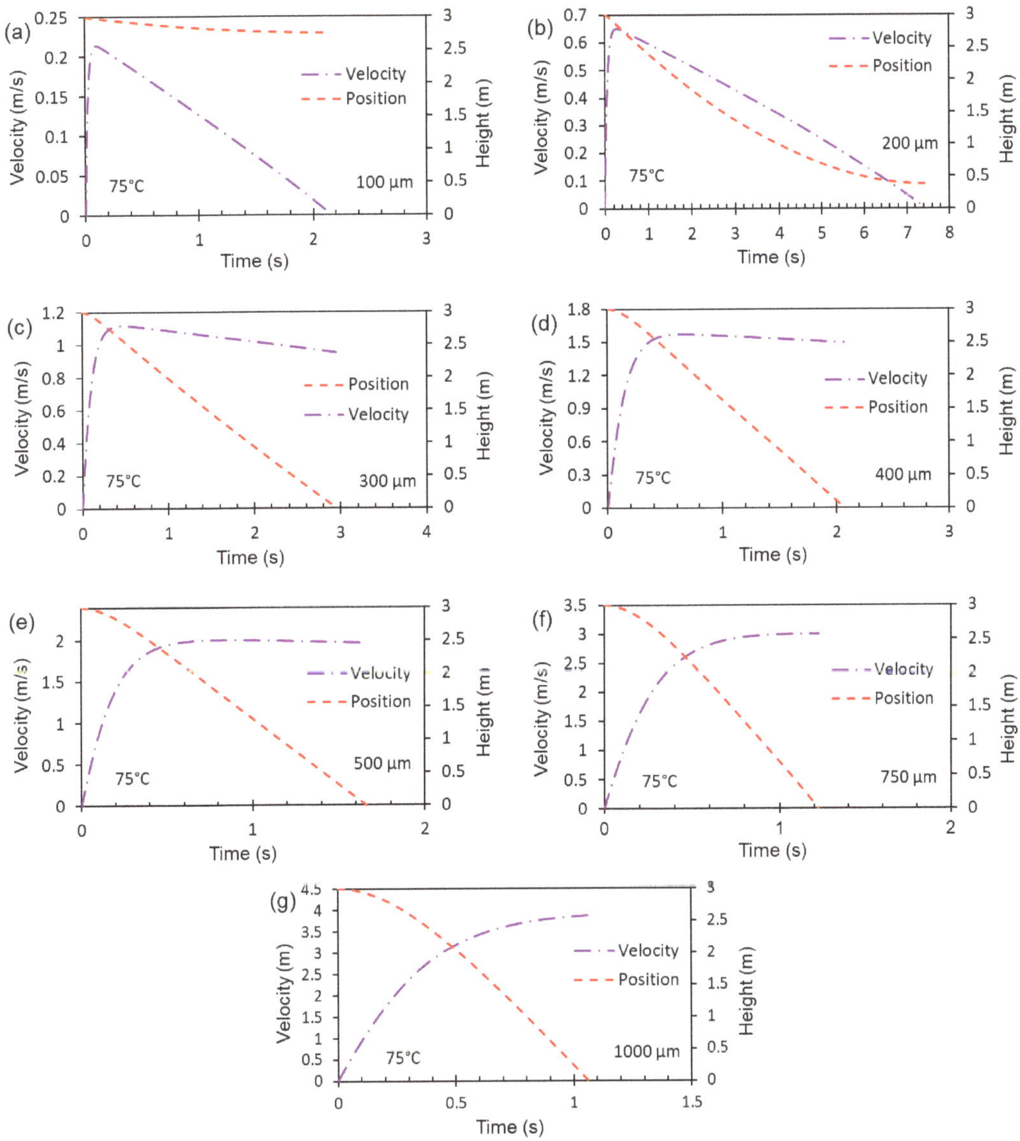

Figure 6. Velocity and position history of droplets with different initial drop sizes for condition 1 (75 °C); (**a**) 100 μm; (**b**) 200 μm; (**c**) 300 μm; (**d**) 400 μm; (**e**) 500 μm; (**f**) 750 μm; and (**g**) 1000 μm.

The velocity and position histories of droplets with different initial sizes for temperature condition 2 are demonstrated in Figure 7. The velocity of 100 μm droplets decreases quickly due to the reduction in the diameter of droplets caused by the rapid evaporation at high temperature. The velocity of the droplet of 200 μm is also reduced to zero, and the droplet is sustained in the air for 3 s. However, the droplet of 400 μm and larger have higher speeds due to their larger size. Similarly to condition 1, the change in velocity of droplets with diameters greater than 400 μm is negligible because there is little change in their diameters.

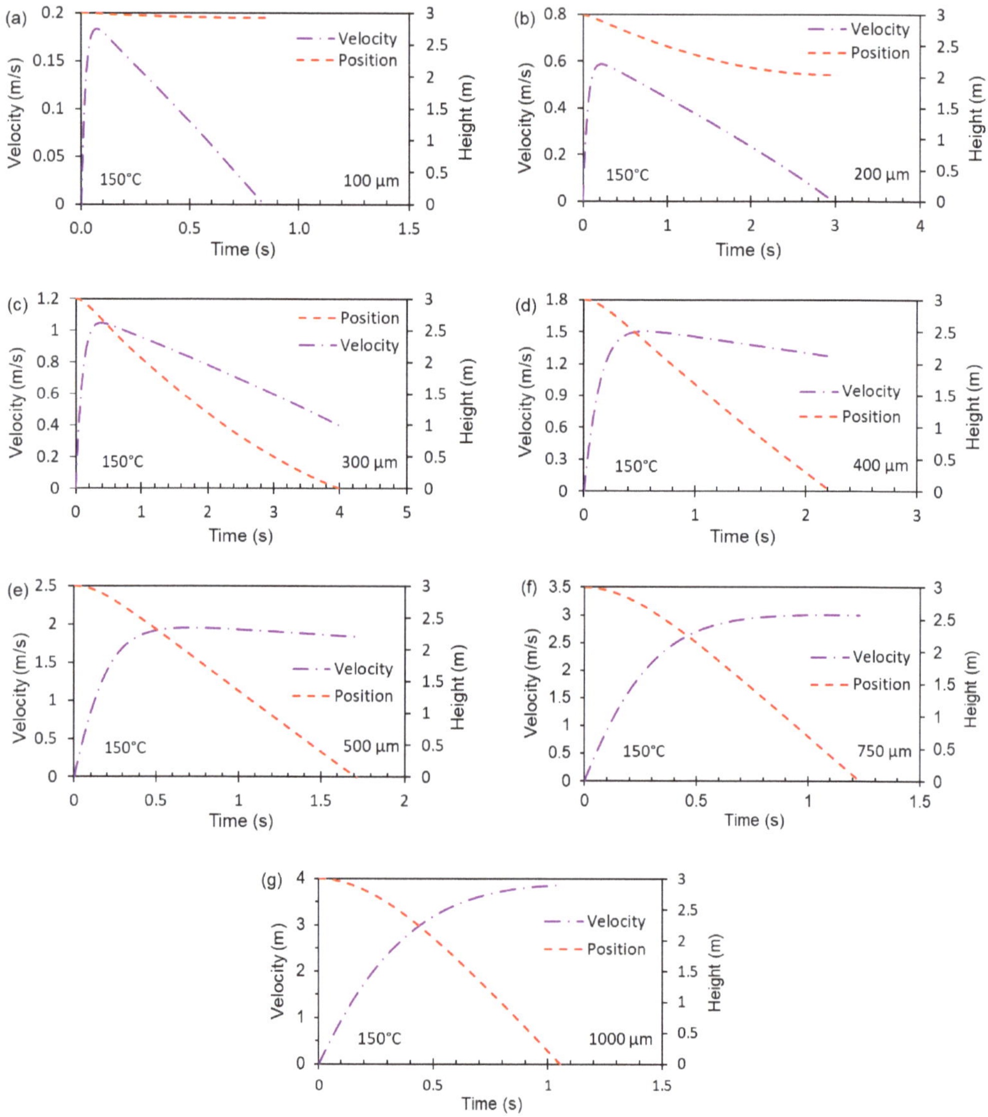

Figure 7. Velocity and position history of droplets with different initial drop sizes for condition 2 (150 °C); (**a**) 100 µm; (**b**) 200 µm; (**c**) 300 µm; (**d**) 400 µm; (**e**) 500 µm; (**f**) 750 µm; and (**g**) 1000 µm.

Interestingly, the analysis shows that, in condition 1, the 200 µm droplet has the highest lifetime compared to the other size of droplets; whereas in condition 2, the 300 µm droplet has the highest life time of the droplets. This is because, due to higher temperature in condition 2, the 200 µm droplet has taken less time to evaporate compared to condition 1. However, the 300 µm droplet does not evaporate significantly in condition 1, whereas it does in condition 2 as the air temperature is higher. As a result, due to the reduction in diameter of the 300 µm droplet in condition 2, the momentum of the droplet also reduces. This results in longer suspension time of the 300 µm droplet in the air in condition 2. Therefore, if the boundary condition of environment (smoke temperature, relative humidity, room height) changes, the suspension time of droplets also will be different.

The net heat flux rates in the droplets for condition 1 and 2 are presented in Figure 8. This heat flux is the resultant of the convective and radiative heat transfer to the droplet and the heat leaving the droplet due to evaporation of water particle from the droplet. The initial peak of the heat flux rate is due to the rise in temperature by the convection and radiation without significant evaporation of water particle from the droplet. When the droplet has attained the equilibrium temperature, the resultant is lowered due to leaving of heat by the evaporation of water particle from the droplet. The results show that the larger the droplet sizes, the higher the initial peak and also the higher the air temperature, the greater the net heat flux rate.

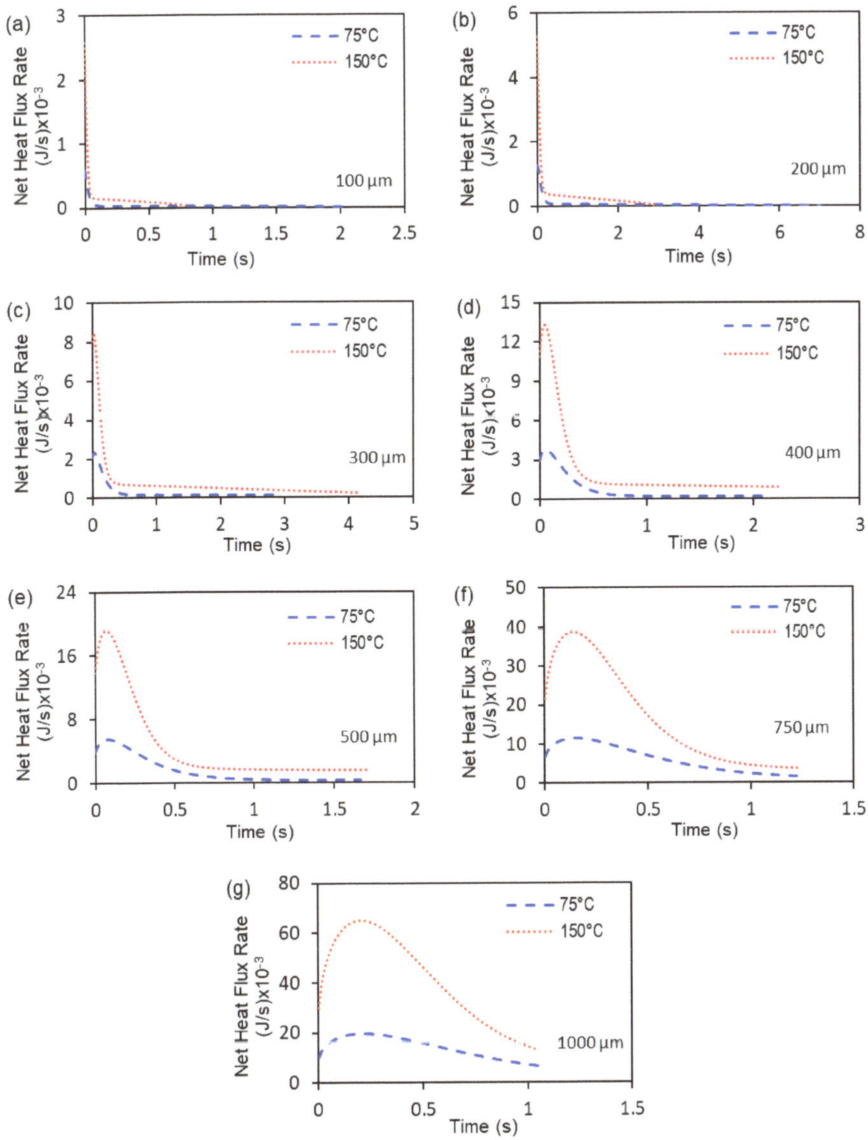

Figure 8. The net heat flux rate in the droplets for condition 1 and 2; (**a**) 100 µm; (**b**) 200 µm; (**c**) 300 µm; (**d**) 400 µm; (**e**) 500 µm; (**f**) 750 µm; and (**g**) 1000 µm.

The heat and mass transfer coefficients of the droplets for condition 1 and 2 are presented in Figure 9. The smaller size of the droplet has exhibited a higher heat and mass transfer rate. This is because smaller size of the droplet has reached equilibrium temperature earlier and started to evaporate significantly compared to the larger size of droplet; eventually, this causes higher heat and mass transfer rate for the smaller size of droplets. Specifically, in case of the 100 and 200 µm droplets a sudden rise in the heat and mass transfer coefficient is observed. This happens when the droplets have become very tiny due to evaporation; as a result, this has caused rapid evaporation to the tiny droplets.

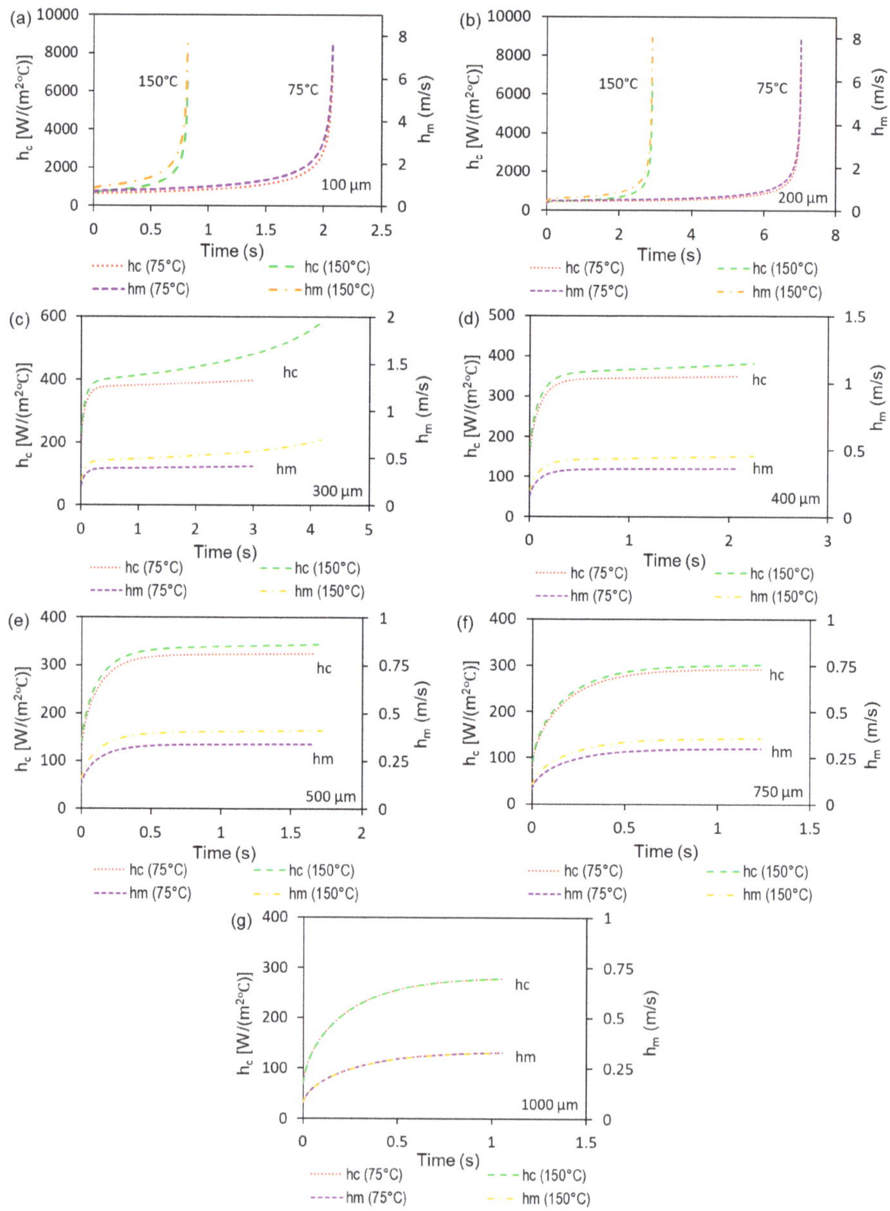

Figure 9. The heat and mass transfer coefficient of the droplets for condition 1 and 2; **(a)** 100 µm; **(b)** 200 µm; **(c)** 300 µm; **(d)** 400 µm; **(e)** 500 µm; **(f)** 750 µm; and **(g)** 1000 µm.

These phenomena are also reflected in Figure 10, where the percent mass loss rates of the water droplets are presented for conditions 1 and 2. The highest percent mass loss rate was for the smallest droplet (100 μm), and the lowest percent mass-loss rate was for the largest droplet (1000 μm). The figure shows that the droplet with initial diameters of 100 μm and 200 μm remain suspended for 2.16 and 7.42 s, respectively, for condition 1, and 0.85 and 2.98 s, respectively, for condition 2. At these times, they have disappeared because of their completely evaporation. The data presented in Figure 10 demonstrate that, for condition 1, the percentage mass loss of droplets initially with diameters of 300, 400, 500 and 750 μm are 36%, 18%, 11% and 4%, respectively, and these losses are 88%, 42%, 24% and 9%, respectively, for condition 2. As a result, the droplets can reach the floor of the enclosure. The evaporation of the 750 μm and 1000 μm diameter droplets results in a negligible decrease in their diameters because of their relatively small surface-to-volume ratios and reaching the floor in the shortest time.

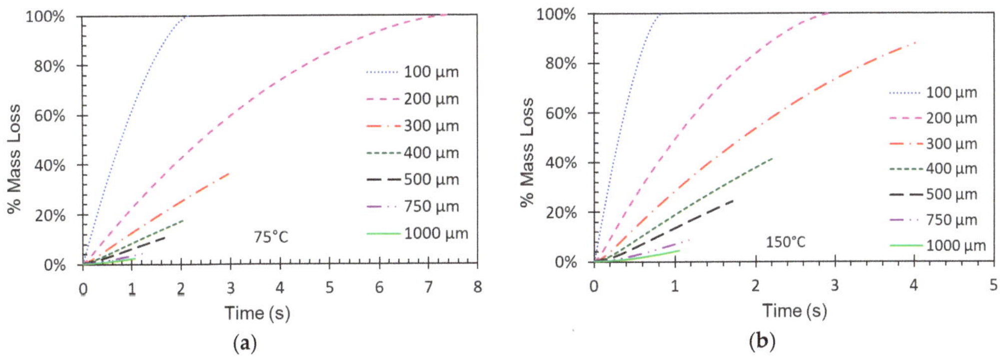

Figure 10. The mass loss of droplets with different initial diameters; (**a**) condition 1 (75 °C); (**b**) condition 2 (150 °C).

Water mists and water sprays both extract heat from the hot environment by convective heat transfer. The temperature of droplets is maintained lower resulting from the evaporation, which requires latent heat. The required heat for evaporation of individual droplets with the different initial diameters and one kilogram of droplets consisting of different initial sizes for conditions 1 and 2 is presented in Figure 11. The figure demonstrates that the amount of heat absorbed by a tiny individual droplet is much lower than that absorbed by an individual larger droplet in both cases. For example, in condition 1, a droplet with an initial diameter of 100 μm absorbs 1.35×10^{-3} J of heat, whereas a droplet with 1000 μm absorbs 40×10^{-3} J of heat. However, a given mass of water consists of droplets with 100 μm size containing 1000 times more droplets than the same amount of water consisting of droplets with 1000 μm size. As a result, 1 kg of droplets with 100 μm extracts about 2600 kJ of energy from its surrounding environment, whereas the same mass of water containing 1000 μm droplets extracts only about 76 kJ. This indicates one of the key attributes of fine sprays in suppressing fires: 1 kg of water consisting of 200 μm droplets also absorbs the same amount of energy as that of 100 μm droplets. This is due to the reason that both sizes of the droplets are evaporated entirely in the air.

The droplets in condition 2 absorbed a larger amount of heat compared to that of condition 1, due to the higher air temperature of condition 2. The difference of absorbed heat by the single droplets and 1 kg of droplets is not much (about 3%) for 100, and 200 μm droplets as both have evaporated in both conditions. However, this difference is significant (about 130%) for 300 μm and higher diameter because the equilibrium temperature at condition 2 is higher than that of condition 1. As a result, the droplets have absorbed a higher amount of heat. However, the heat absorbed by the individual droplets and 1 kg of water consisting of droplets has followed a similar trend.

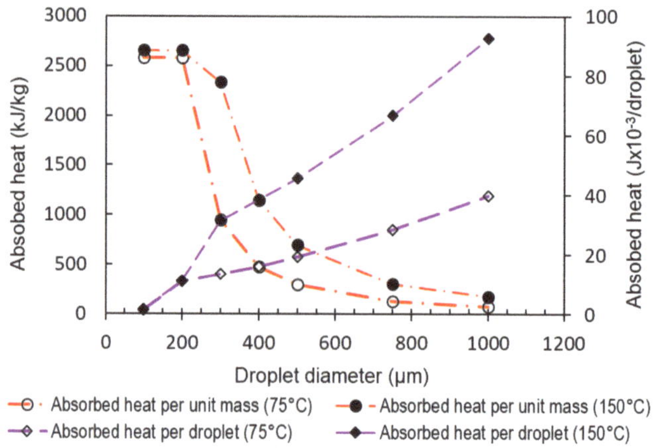

Figure 11. Absorbed heat by the droplets with different initial diameters for condition 1 and 2.

5. Conclusions

In this study, investigations have been performed on the behaviour of individual water-mist droplets travelling through a hot air layer induced by a room fire. A previously developed and validated quasi-physical model has been used to perform the work. The initial size of droplets has been varied, and the temperature profiles, locations, velocity trajectories, evaporation rates and the absorption of heat by the falling droplets have been predicted. The finding of the study can be summarised as below:

(a) The equilibrium thermal condition of droplets is independent of the initial size and temperature of droplets. Instead, it depends principally on the temperature and relative humidity of ambient air.

(b) The smaller size of droplets (in this study, 100 and 200 μm) can evaporate entirely before reaching the floor. This is because the low terminal velocity of droplets and rapid evaporation due to tiny in size have facilitated them to disappear completely before reaching the floor. On the other hand, the larger size of droplets (≥300 μm) has reached the floor within a shorter time and with a small amount of evaporation.

(c) Notably, smaller droplets are more effective in absorbing heat energy from the hot air per unit mass of water. This is because the smaller droplet has a significant amount of evaporation and takes a longer time to reach the floor or may evaporate entirely before reaching the floor.

(d) Therefore, in the case of fire extinguishment, the smaller size of droplets (in this case, 100 and 200 μm) may be effective in cooling the air within an enclosure; but they are less likely to penetrate a layer of hot air and extinguish conflagrating surfaces on the floor of an enclosure. On the other hand, droplets with a larger diameter are more effective in impingement of a smoke layer, but the larger droplet does not contribute significantly in absorbing heat from the surrounding.

The WDEM model has been shown to be useful for analysing the effects of initial diameters of droplets when they are exposed to fire-induced hot environments. Furthermore, the generality of the heat and mass transfer model enables it to be used to develop parametric models for a wide range of environments and droplet sizes. The results show that the smaller droplets of spray effectively cool the environment due to their rapid evaporation, and the larger droplets are effective in impinging the layer hot air or smoke in a fire environment. The findings of this study can be helpful in evaluating and designing the effective sizes of droplets or a group of droplets for a particular fire situation.

This study will be further extended to consider the effect of the interaction of droplets, injection velocity, distribution of droplet sizes, the effect of different initial velocity of air and droplets, and also the effect of humidification of air due to evaporation of droplets.

Author Contributions: H.M.I.M. developed the model. H.M.I.M., G.T. and K.A.M.M. contributed in conceptualisation, development of the background theory and formulations of the model. The text of the paper was formed by H.M.I.M.; K.A.M.M. and G.T. edited the paper and supervised the research direction. All authors have read and agreed to the published version of the manuscript.

Funding: The authors wish to acknowledge the technical and financial assistance provided by Defence Science and Technology Organisation (DSTO), Australia.

Institutional Review Board Statement: Not applicable.

Informed Consent Statement: Not applicable.

Data Availability Statement: Not applicable.

Conflicts of Interest: The authors declare no conflict of interest.

Nomenclature

A	surface area, m^2
Bi	biot number
C_d	drag coefficient
c_{pa}	specific heat capacity of air, J/(kg·°C)
c_{pw}	specific heat capacity of water, J/(kg·°C)
D	diameter, m
F	view factor
g	acceleration due to gravity, m/s^{-2}
h_c	convective heat transfer coefficient, W/(m^2·°C)
h_m	mass transfer coefficient, m/s
k	conductivity, J/m·°C
L	latent heat of vaporisation of water, J/kg
m	mass, kg
Nu	Nusselt number
p	vapour pressure, Pa
Pr	Prandtl number
P	pressure, Pa
R	universal gas constant, J/(K·mol)
Re	Reynolds number
RH	relative humidity, %
Sc	Schmidt number
Sh	Sherwood number
T	temperature, °C
t	time, s
V	volume, m^3
v	air-droplet relative velocity, m/s
y	vertical distance from the mist nozzle, m
z	height, m
Greek symbols	
α	thermal diffusivity, m^2/s
μ	dynamic viscosity, Pa·s
v	kinematic viscosity, m^2/s
ρ	density, kg/m^3
ε	emissivity factor
D	mass diffusivity coefficient
σ	Stefan–Boltzmann constant, W/(m^2·K^4)

Subscripts

a	air
AB	binary system of A and B
bw	boundary wall
c	convective heat
d	drag
e	evaporation
f	flame
g	gravitation/gas
proj	projected area
s	droplet surface
w	water
∞	refers to the far-field value

References

1. Yang, P.; Liu, T.; Qin, X. Experimental and numerical study on water mist suppression system on room fire. *Build. Environ.* **2010**, *45*, 2309–2316. [CrossRef]
2. Kim, S.C.; Ryou, H.S. An experimental and numerical study on fire suppression using a water-mist in an enclosure. *Build. Environ.* **2003**, *38*, 1309–13016. [CrossRef]
3. *NFPA 750*; Standard on Water Mist Fire Protection Systems. National Fire Protection Association: Quincy, MA, USA, 2010.
4. Liu, J.; Liao, G.; Li, P.; Fan, W.; Lu, Q. Progress in research and application of water-mist fire suppression technology. *Chin. Sci. Bull.* **2003**, *48*, 718–725. [CrossRef]
5. Mahmud, H.M.I.; Moinuddin, K.A.M.; Thorpe, G.R. Experimental and numerical study of high-pressure water-mist nozzle sprays. *Fire Saf. J.* **2016**, *81*, 109–117. [CrossRef]
6. Beji, T.; Thielens, M.; Merci, B. Assessment of heating and evaporation modelling based on single suspended water droplet experiments. *Fire Saf. J.* **2019**, *106*, 124–135. [CrossRef]
7. Volkov, R.S.; Strizhak, P.A. Planer laser-induced fluorescence diagnostics of water droplets heating and evaporation at high-temperature. *Appl. Therm. Eng.* **2017**, *127*, 141–156. [CrossRef]
8. Strizhak, P.; Volkov, R.; Castanet, G.; Lemoine, F.; Rybdylova, O.; Sazhin, S. Heating and evaporation of suspended water droplets: Experimental studies and modelling. *Int. J. Heat Mass Tran.* **2018**, *127*, 92–106. [CrossRef]
9. Sobac, B.; Talbot, P.; Haut, B.; Rednikov, A.; Colinet, P. A comprehensive analysis of the evaporation of a liquid spherical drop. *J. Colloid Interface Sci.* **2015**, *438*, 306–317. [CrossRef]
10. Thielens, M.; Merci, B.; Beji, T. Development of a novel two-zone model for the heating of an evaporating liquid droplet. *Fire Saf. J.* **2021**, *120*, 103019. [CrossRef]
11. Mahmud, H.M.I.; Moinuddin, K.A.M.; Thorpe, G.R. Study of water-mist behaviour in hot air induced by a room fire: Model development, validation and verification. *Fire Mater.* **2016**, *40*, 190–205. [CrossRef]
12. Chow, W.K.; Yao, B. Numerical modeling for interaction of a water spray with smoke layer. *Numer. Heat Transf.* **2001**, *39*, 267–283. [CrossRef]
13. Cooper, L.Y. The interaction of an isolated sprinkler spray and a two-layer compartment fire environment. *Int. J. Heat Mass Transf.* **1995**, *38*, 679–690. [CrossRef]
14. Chow, W.K.; Tong, A.C. Experimental studies on sprinkler water spray-smoke layer interaction. *J. Appl. Fire Sci.* **1995**, *4*, 171–184. [CrossRef]
15. Chow, W.K.; Cheung, Y.L. Simulation of Sprinkler-Hot Layer Interaction Using a Field Model. *Fire Mater.* **1994**, *18*, 359–379. [CrossRef]
16. Morgan, H.P. Heat transfer from a buoyant smoke layer beneath a ceiling to a sprinkler spray. 1—A tentative theory. *Fire Mater.* **1979**, *3*, 27–33. [CrossRef]
17. Morgan, H.P.; Baines, K. Heat Transfer from a Buoyant Smoke Layer Beneath a Ceiling to a Sprinkler Spray. 2—An Experiment. *Fire Mater.* **1979**, *3*, 34–38. [CrossRef]
18. John, V.V. A full-scale experiment study of water mist spray convection. In Proceedings of the 2nd International Conference on Fire Research and Engineering, Gaithersburg, MD, USA, 3–8 August 1997; pp. 282–289.
19. Chow, W.K. On the evaporation effect of a sprinkler water spray. *Fire Technol.* **1989**, *25*, 364–373. [CrossRef]
20. Li, S.C.; Yang, D.; Huo, L.H.; Li, Y.Z.; Wang, H.B. Studies of cooling effects of sprinkler spray on smoke layer. In *Fire Safety Science, Proceedings of the Ninth International Symposium, International Association of Fire Safety Science*; University of Karlsruhe: Karlsruhe, Germany, 2008; pp. 861–872.
21. Floyd, J.; McDermott, R. Development and evaluation of two new droplet evaporation schemes for fire dynamics simulations. *Fire Saf. J.* **2017**, *91*, 643–652. [CrossRef]
22. Liu, Z.; Kim, A.K. A review of water mist fire suppression systems-fundamental studies. *J. Fire Prot. Eng.* **2000**, *10*, 32–50. [CrossRef]
23. Grant, G.J.; Brenton, J.; Drysdale, D. Fire suppression by water sprays. *Prog. Energy Combust. Sci.* **2000**, *26*, 79–130. [CrossRef]

24. Liu, Z.; Kim, A.K. A review of water mist fire suppression technology: Part II- Application studies. *J. Fire Prot. Eng.* **2001**, *11*, 16–42. [CrossRef]
25. Rashbash, D.J.; Rogowski, Z.W.; Stark, G.W.V. Mechanism of extinction of liquid fires with water sprays. *Combust. Flame* **1960**, *4*, 223–234. [CrossRef]
26. Shu, Y.L.; Jeng, W.J.; Chiu, C.W.; Chen, C.H. Assessment of fire protection performance of water mist applied in exhaust ducts for semiconductor fabrication process. *Fire Mater.* **2005**, *29*, 295–302. [CrossRef]
27. Yao, B.; Chow, W.K. Extinguishment of a PMMA fire by water spray with high droplet speeds. *Int. J. Therm. Sci.* **2005**, *44*, 410–419. [CrossRef]
28. Back, G.G.; Beyler, C.L.; DiNenno, P.J.; Hansen, R.; Zalosh, R. *Full-Scale Testing of Water Mist Fire Suppression Systems in Machinery Spaces*; CG-D-26-98; United States Coast Guard: Washington, DC, USA, 1998.
29. Mawhinney, J.R.; Dlugogorski, B.Z.; Kim, A.K. A closer look at the fire extinguishing properties of water mist. Fire Safety Science. In Proceedings of the 4th International Symposium on Fire Safety Science, Ottawa, ON, Canada, 13–17 July 1994; pp. 47–60.
30. Wighus, R. Engineering relations for water mist fire suppression systems. In Proceedings of the Halon Alternatives Technical Working Conference, Albuquerque, NM, USA, 30 April–1 May 1995; p. 397.
31. Novozhilov, V. Flashover control under fire suppression conditions. *Fire Saf. J.* **2001**, *36*, 641–660. [CrossRef]
32. Li, Y.F.; Chow, W.K. Study of water droplet behaviour in hot air layer in fire extinguishment. *Fire Technol.* **2008**, *44*, 351–381. [CrossRef]
33. Vaari, J. A transient one-zone computer model for total flooding water mist fire suppression in ventilated enclosures. *Fire Saf. J.* **2002**, *37*, 229–257. [CrossRef]
34. Barrow, H.; Pope, C.W. Droplet evaporation with reference to the effectiveness of water-mist cooling. *Appl. Energy* **2007**, *84*, 404–412. [CrossRef]
35. Katsios, X.K.; Krikkis, R.N. Effect of surface tension and evaporation on phase change of fuel droplets. *Heat Transf. Eng.* **2001**, *22*, 33–40. [CrossRef]
36. Holman, J.P. *Heat Transfer*, 9th ed.; McGraw-Hill Book Company: New York, NY, USA, 2002; p. 133.
37. Chelliah, H.K. Flame inhibition/suppression by water mist: Droplet size/surface area, flame structure, and flow residence time effects. *Proc. Combust. Inst.* **2007**, *31*, 2711–2719. [CrossRef]
38. Ananth, R.; Mowrey, R.C. Ultra-Fine Water Mist Extinction Dynamics of a Co Flow Diffusion Flame. *Combust. Sci. Technol.* **2008**, *180*, 1659–1692. [CrossRef]
39. Hua, J.S.; Kumar, K.; Khoo, B.C.; Xue, H. A numerical study of the interaction of water spray with a fire plume. *Fire Saf. J.* **2002**, *37*, 631–657. [CrossRef]
40. Sikanen, T.; Vaari, J.; Hostikka, S.; Paajanen, A. Modeling and Simulation of High Pressure Water Mist Systems. *Fire Technol.* **2014**, *50*, 483–504. [CrossRef]
41. Bullen, M.L. The effect of a sprinkler on the stability of a smoke layer beneath a ceiling. *Fire Technol.* **1977**, *13*, 21–34. [CrossRef]
42. *ASHRAE Handbook: Fundamentals*; American Society of Heating, Refrigeration and Air Conditioning Engineers. Inc.: New York, NY, USA, 1985; p. 5.5.
43. William, C.H. *Aerosol Technology*; John Wiley & Sons Inc.: New York, NY, USA, 1999.
44. Bird, R.B.; Stewart, W.E.; Lightfoot, E.N. *Transport Phenomena*; John Willy and Sons, Inc.: New York, NY, USA, 1960; pp. 59, 505.
45. Beard, K.V.; Pruppacher, H.R. A wind tunnel investigation of the rate of evaporation of small water drops falling at terminal velocity in air. *J. Atmos. Sci.* **1971**, *28*, 1455–1464. [CrossRef]
46. Cengel, Y.A.; Turner, R.H. *Fundamental of Thermal Fluid Sciences*, 2nd ed.; McGraw-Hill Book Company: New York, NY, USA, 2005; pp. 1088–1089.
47. Smolik, J.; Dzumbova, L.; Schwarz, J.; Kulmala, M. Evaporation of ventilated water droplet: Connection between heat and mass transfer. *Aerosol Sci.* **2001**, *32*, 739–748. [CrossRef]
48. Brown, P.P.; Lawler, D.F. Sphere drag and settling velocity revisited. *J. Environ. Eng. (ASCE)* **2003**, *129*, 222–231. [CrossRef]
49. Gunn, R.; Kinzer, G.D. The terminal velocity of fall for water droplets in stagnant air. *J. Meteorol.* **1949**, *6*, 243–248. [CrossRef]
50. McGrattan, K.; McDermott, R.; Vanella, M.; Hostikka, S.; Floyd, J. *Fire Dynamics Simulator (Version 6), Technical Reference Guide, Volume 1: Mathematical Model*, 6th ed.; NIST Special Publication 1018-1, Revision: FDS6.7.4-0-gbfaa110; National Institute of Standards and Technology (NIST), U.S. Department of Commerce: Gaithersburg, MD, USA, 2021.
51. Vendel, J. Aerosols collection by spray water drops: Experimental results and model. In Proceedings of the Cooperative Severe Accident Research Program Meeting, Bethesda, MD, USA, May 1998.
52. Plumecocq, W.; Audouin, L.; Joret, J.P.; Pretrel, H. Numerical method for determining water droplets size distributions of spray nozzles using a two-zone model. *Nucl. Eng. Des.* **2017**, *324*, 67–77. [CrossRef]
53. *NFPA 13*; Automatic Sprinkler Systems Handbook. National Fire Protection Association: Quincy, MA, USA, 2007; p. 119.
54. Li, D.; Zhu, G.; Zhu, H.; Yu, Z.; Gao, Y.; Jiang, X. Flame spread and smoke temperature of full-scale fire test of car fire. *Case Stud. Therm. Eng.* **2017**, *10*, 315–324. [CrossRef]
55. Pretrel, H.; Varrall, K.; Vauquelin, O.; Audouin, L. Smoke induced flow in two rooms mechanically ventilated and linked with a horizontal vent type opening. In Proceedings of the 11th International Symposium of Fire Safety Science, University of Canterbury, Christchurch, New Zealand, 10–14 February 2014; pp. 486–498. [CrossRef]

56. Jeong, J.Y.; Ryou, H.S. A Study on Smoke Movement in Room Fires with Various Pool Fire Location. *KSME Int. J.* **2002**, *1*
1485–1496. [CrossRef]
57. Zhou, J.; Mao, J.; Huang, Y.; Xing, Z. Studies on Smoke Temperature Distribution in a Building Corridor Based on Reduced-scal
Experiments. *J. Asian Archit. Build. Eng.* **2017**, *16*, 341–348. [CrossRef]
58. Lefebvre, A.H.; McDonell, V.G. *Atomization and Sprays*; CRC Press: Boca Raton, FL, USA, 2017; p. 247. [CrossRef]
59. Zhou, Y.; Bu, R.; Zhang, X.; Fan, C.; Gong, J. Performance evaluation of water mist fire suppression: A clean and sustainabl
fire-fighting technique in mechanically-ventilated place. *J. Clean Prod.* **2019**, *209*, 1319–1331. [CrossRef]

fire

Article

Simulation Study on Aircraft Fire Extinguishing Pipeline with Different Filling Conditions and Pipeline Characteristics

Rulin Liu [1,2], Changyu Yuan [1], Weitong Ma [1], Shaonan Liu [1], Song Lu [1,*], Heping Zhang [1] and Jun Gong [2]

1 State Key Laboratory of Fire Science, University of Science and Technology of China, Hefei 230026, China; liurl001@avic.com (R.L.); yuanchangyu@mail.ustc.edu.cn (C.Y.); wtma@mail.ustc.edu.cn (W.M.); lsnan@mail.ustc.edu.cn (S.L.); zhanghp@ustc.edu.cn (H.Z.)
2 Chengdu Aircraft Design Institute, Chengdu 610091, China; gongj003@avic.com
* Correspondence: lusong@ustc.edu.cn

Abstract: The filling conditions and pipeline characteristics of the aircraft fire extinguishing system determine the pressure of the fire extinguishing cylinder outlet, the discharge quality of the extinguishing agent, and the flow distribution during the discharge process. The simulation model of the fire extinguishing system pipeline of an aircraft was established by Amesim. The influence of filling conditions and pipeline characteristics was studied. It was found that the mass curves of the fire extinguishing agent were similar under filling pressures of 4, 5, and 6 MPa with a filling amount of 5.55 kg. The lower the initial temperature is, the pressure at the outlet of the cylinder decreases, but the emptying time is similar to 1.22 s. The lower the roughness is, the faster the discharge is. Under the ideal smooth pipe (ε = 0 mm), the emptying time of the fire extinguishing cylinder is 0.72 s. When the diameter of the short branch pipe is 10 mm, and the diameter of the long branch pipe is 14 mm, the discharge quality of the two pipes is close. The larger the diameter of the main pipe, the higher the discharge rate. The research results have a certain guiding significance for the pipeline design of certain aircraft.

Keywords: aircraft fire extinguishing; one-dimensional simulation; Amesim; filling conditions; piping characteristics

Citation: Liu, R.; Yuan, C.; Ma, W.; Liu, S.; Lu, S.; Zhang, H.; Gong, J. Simulation Study on Aircraft Fire Extinguishing Pipeline with Different Filling Conditions and Pipeline Characteristics. *Fire* **2022**, *5*, 86. https://doi.org/10.3390/fire5040086

Academic Editor: Alistair M. S. Smith

Received: 11 May 2022
Accepted: 22 June 2022
Published: 23 June 2022

1. Introduction

In the event of a fire accident in an aircraft, if the fire is not extinguished in time, the fire will develop rapidly and even cause an explosion and cause a serious crash [1]. Reliable fire suppression systems play an important role in aircraft safety. Halon 1301 is the alias of the agent Bromotrifluoromethane (CBrF3) [2], which is easy to store, has low toxicity, and its comprehensive performance is unmatched by other agents. Halon 1310 was banned by the international community by the Montreal Protocol but remained widely used as a fire extinguishing agent globally. This dichotomy exists because several countries, including the United States, continue to use Halon 1310 for specialty fire extinguishing by recycling existing stocks [3]. Despite its environmental damage, the guideline "Halon replacement in the aviation industry" [4] developed by the European Commission and EASA in 2019 states that the mainstream extinguishing agent for aircraft fire suppression systems is still Halon1301. The status quo will not change considerably until the next generation of Halon replacement agents is clarified. Nearly 40 years have passed since the U.S. proposed research on the next generation of fire suppressant programs [5], and Halon replacements are still unclear [6]. Therefore, the current research on the flow characteristics of Halon 1301 in the aircraft fire suppression system still has its value.

The objective of this research is to propose a method to study the discharge of extinguishing agents in aircraft fire suppression systems and to optimize the system design based on this method to reduce the actual use of Halon 1301 and to provide benchmarks standard for potential halon substitutes. In addition, by modifying some of the parameters, the method can be broadly transferred to other halon substitutes, for example, HFC-125.

Different structural designs of the Halon 1301 fire extinguishing system in the engine nacelle, such as pipe diameter, nozzle size, and filling conditions of the agent, will affect the flow of the agent in the pipe and then affect the flow and distribution of the agent in the fire extinguishing system [7,8]. The mass flow of the agent at the nozzle will greatly affect the evaporation, flow, and diffusion of the agent in the engine nacelle and affect the concentration distribution of the agent and the fire extinguishing effect [9]. In the pipeline design process of an aircraft fire extinguishing system, the traditional-formula-based method [10] for calculating the pipeline flow of fire extinguishing systems can no longer meet the needs of engineering applications.

It is the most direct research method to carry out the discharge experiment of agents in the built test pipe network or the real aircraft fire extinguishing system pipe network to carry out research on the discharge rules of the fire extinguishing system pipeline. Many scholars have carried out extensive research work through experimental methods. Williamson found that when Halon 1301 flowed in the pipeline below 2.48 MPa, the pressure decreased nonlinearly, the boiling of the agent slowed down the pressure drop, and the release rate increased with the increase in the cylinder volume [11]. Elliot, et al. [12] built an experimental pipe network to discharge Halon 1301 under nitrogen pre-pressure and provided pressure decay data as a function of time so that the discharge flow rate could be estimated. The tests of Yang, et al. [13] and Pitts, et al. [14] at NIST conducted multiple discharge experiments with different haloalkane agents, obtaining a large amount of data under various pipeline configurations and operating conditions.

Since the discharge of Halon 1301 can cause damage to the environment, compared with experimental studies, simulation technology is more efficient and appropriate in the development and commissioning of fire suppression systems. Scholars also carry out a lot of research. Yang, et al. [15] developed a method for predicting the pressure inside a cylinder and formed a computer code called PROFISSY. Elliott, et al. [12] developed a computer program called HFLOW to predict the discharge of Halon 1301 from the discharge vessel through the piping system. Tuzla, et al. [16] reported a one-dimensional two-fluid two-phase flow model based on an agent pipeline flow calculation program, and the program allows the user to choose any one of the five liquids: water, Halon 1301, carbon dioxide, HFC-227ea, or HFC-125. Lee [17] reported a one-dimensional two-phase flow software Hflowx for the calculation of the flow of agents in pipelines, and Hflowx is an extension of Hflow. However, the computing software reported above is difficult to obtain publicly, and some scholars have begun to develop computing models using publicly available commercial software. Kim [18] and used Fluent to study the flow process of the agent in fire extinguishing cylinder and fire extinguishing pipeline. Min Hua and Jia-ming JIN [7,8] of Nanjing University of Science and Technology used the one-dimensional simulation software Amesim to study the effect of nozzle type and size of nozzle pressure and flow rate as well as the effect of filling state on the outlet pressure and discharge time of cylinder.

It can be seen from the above research reports that the application of simulation technology provides great convenience for the calculation of the pipeline of the aircraft fire extinguishing system. However, to meet the requirements of the compact structure of modern aircraft, the layout of the aircraft fire extinguishing system is often complicated, the pipeline length is large, and there are often branch pipes to protect different target nacelles [19]. In the process of 3D CFD research, the establishment of the simulation model is complicated [20,21], and the calculation period of the simulation results is relatively long. One-dimensional simulation based on multiphase flow theory is an important means for the structural design of future aircraft fire extinguishing systems. It has the advantages of simple modeling and fast simulation speed.

At present, most of the research on one-dimensional simulation focuses on the experimental pipeline networks, and there are few public reports on the research on the discharge of the actual pipeline of the aircraft fire extinguishing system; the influence of filling conditions and pipeline characteristics on the pressure of the fire extinguishing cylinder outlet, the discharge quality of the agent, and the flow distribution during the discharge process have not been comprehensively studied [15,18]. To optimize the system design and reduce the amount of Halon 1301 carried in the actual aircraft fire extinguishing system. In this paper, one-dimensional simulation software Amesim is used to model a certain type of aircraft fire extinguishing system based on multiphase flow theory, cubic equation of state, and heat transfer model based on Nusselt number [22]. The influence of filling pressure, initial temperature, absolute roughness of pipeline, branch pipe diameter, and main pipe diameter on the discharge quality of agent, the pressure of cylinder outlet, and flow distribution of agent at each nozzle were analyzed, respectively. Furthermore, the results of the simulation study can be used as an input function for the subsequent simulation of the fire extinguishing agent concentration distribution in the nacelle, which contributes to reducing the consumption of Halon 1301 for the actual airworthiness experiments.

2. Simulation Model

2.1. Fire Extinguishing System Parameters

In this paper, a certain type of aircraft fire extinguishing system is used as the simulation object. The agent is Halon 1301, the volume of the cylinder is 7.0 L, the agent filling is 5.55 ± 0.2 kg, and the filling pressure is 4.2 MPa at room temperature. The pipeline of the fire extinguishing system can be classified into the main pipe and the branch pipe. The end of each branch pipe is connected with multiple nozzles. The pipeline structure is shown in Figure 1.

Figure 1. Schematic diagram of the pipeline structure of the fire extinguishing system.

2.2. Theoretical Model

The main flow form of Halon 1301 in the pipeline is unsteady gas–liquid two-phase flow. The state of Halon 1301 in the pipeline is defined by the Peng–Robinson equation [23] applicable to gas–liquid two-phase fluid:

$$P = \frac{RT}{v_c - b} - a\frac{\alpha}{v_c^2 + 2v_c b - b^2} \tag{1}$$

$$v_c = \frac{1}{\rho} + c \tag{2}$$

$$\alpha = 1 + m\left(1 - \sqrt{\frac{T}{T_c}}\right)^2 \tag{3}$$

where P is the absolute pressure, the unit is barA, T is the fluid temperature, and the unit is K; R is the ideal gas constant, its value is 8.314 J/mol K; α is temperature-dependent gravitational parameter; T_c is the fluid critical temperature, the unit is K; v_c, a, b, c, and m are constants depending on Halon 1301.

Based on the above assumptions, without considering the gravitational pressure drop of Halon 1301 in the pipeline and ignoring the acceleration pressure drop, the flow pressure drop of Halon 1301 in the pipeline $\left(\frac{dp}{dz}\right)_F$ can be expressed as the sum of the frictional pressure drop $\left(\frac{dP}{dz}\right)_{Freg}$ and the local pressure drop $\left(\frac{dP}{dz}\right)_{F_{sing}}$:

$$\left(\frac{dP}{dz}\right)_F = \left(\frac{dP}{dz}\right)_{F_{reg}} + \left(\frac{dP}{dz}\right)_{F_{sing}} \tag{4}$$

When there is single-phase flow in the pipeline, the frictional pressure drop can be expressed as:

$$-\left(\frac{dP}{dz}\right)_{F_{reg}} = \frac{fG^2\overline{v}}{2D} \tag{5}$$

where Z is the length; G is the mass flow velocity; $\overline{v}$ is the average velocity of the pipe and f is the friction coefficient.

According to Churchill's formula [24] for laminar and turbulent flow, the friction coefficient f can be expressed as:

$$f = 8 \cdot \left[\left(\frac{8}{Re}\right)^{12} \frac{1}{\left(\left[2.457 \cdot ln\left(\frac{7}{Re}\right)^{0.9} + 0.27\left(\frac{\varepsilon}{D}\right)\right]^{16} + \left[\frac{37530}{Re}\right]^{16}\right)^{\frac{3}{2}}}\right]^{\frac{1}{12}} \tag{6}$$

where Re is the Reynolds number; D is the actual diameter of the pipeline; ε is the absolute roughness of the pipeline.

When the agent flows in the pipeline, as the pressure decreases, Halon 1301 gradually boils, and then the flow pattern in the pipeline changes from single-phase to two-phase flow. At this time, the pressure drop $\left(\frac{dP}{dz}\right)_{F_{reg}}$ along the path can be calculated using the phase separation model [25], the principle of which is a special interpolation between single-phase liquid and single-phase gas:

$$\left(\frac{dP}{dz}\right)_{F_{reg}} = D^{(1-x)^{\frac{1}{3}} + Bx^3} \tag{7}$$

$$D = A + 2(B - A)x \tag{8}$$

$$A = \frac{f_{L0}G^2 v_L}{2D_h} \tag{9}$$

$$B = \frac{f_{v0}G^2 v_v}{2D_h} \tag{10}$$

where x is the gas mass fraction; f_{L0} is the liquid phase friction coefficient; f_{v0} is the gas phase friction coefficient; v_L is the liquid flow velocity; v_v is the gas flow velocity; D_h is the hydraulic diameter.

The two-phase mass flow rate at the nozzle at the end of the pipe m can be expressed as

$$\mathrm{m} = \frac{1}{\sqrt{k}} S\varphi \sqrt{\frac{2\Delta P\rho}{k_{dp}}} \tag{11}$$

where k is the local pressure drop coefficient, S is the cross-sectional flow area, φ is the flow coefficient, and k_{dp} is the coefficient of frictional resistance, and ρ is the density of the incoming flow, and ΔP is the pressure difference between the two ends of the restriction, then the flow velocity vel can be expressed as

$$vel = \frac{\mathrm{m}}{S\varphi\rho} \tag{12}$$

2.3. One-Dimensional Simulation Model Establishment Parameters

This chapter uses the Amesim software of Siemens to simulate the flow of Halon 1301 in a certain type of aircraft fire extinguishing system. The agent Halon 1301 is described in the model based on the Peng–Robinson equation of state. The flow characteristics of the agent are calculated based on the two-phase flow library, and the heat transfer coefficient is defined based on the Nusselt number.

In the modeling process, the same assumptions are made as in the literature [7]: 1. Nitrogen in the cylinder is incompatible with the agent; 2. Only the agent Halon 1301 flows in the pipeline, and the flow of nitrogen and air in the pipeline is not considered; 3. There is no relative slip between the two phases of the agent to maintain thermodynamic equilibrium; 4. The influence of gravity on the jet flow is not considered; 5. Only consider the heat exchange between the agent and the pipeline; ignore the heat exchange between the pipeline and the outside.

This software product is organized as follows: 1. Simplify the actual physical model: Simplify the geometric parameters of the actual physical model; 2. Build the component line: Model the component using the software's built-in component library; 3. Assign the component model: Select a suitable sub-model for the component; 4. Set the component parameters: Set the boundary conditions and geometric conditions; 5. Set simulation conditions (time step, tolerance).

The built simulation model is shown in Figure 2. The cylinder consists of a nitrogen piston, an agent piston, and a mass block representing the gas–liquid interface. The lengths of the two branch pipes are different. For the convenience of description, the two types of branch pipes are named as long branch pipes and short branch pipes according to their lengths. The agent starts discharging 1 s after the simulation starts. The straight pipes, elbows, solenoid valves, and nozzles in the pipeline are all set to contain flow resistance elements. In addition to its own phase change, Halon 1301 also exchanges heat with the pipeline to cause temperature changes. The initial value of the pipeline temperature is 20 °C. The downstream nozzle is directly connected to an infinite space; that is, the downstream boundary conditions of the flow model are constant pressure and temperature values of 1.013 barA and 20 °C, respectively. The multi-nozzle structure at the end of the branch pipe is simplified into a single expanding pipe with the same flow area.

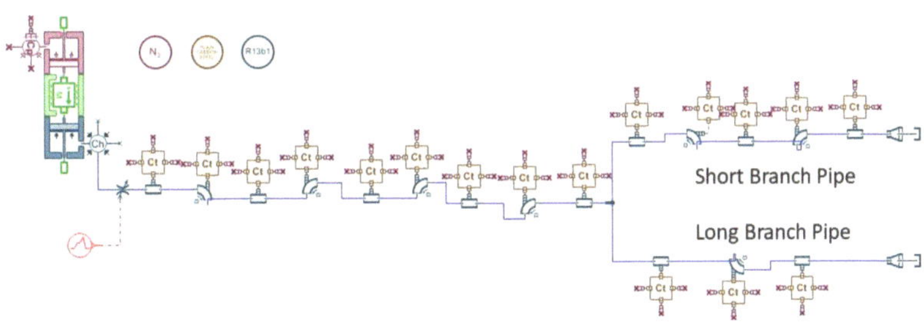

Figure 2. Amesim simulation model diagram.

It is worth noting that the model has the potential to be extended to other Halon substitute discharge applications, for example, HFC-125, when changing some of the above physical parameters such as critical temperature (Equation (3)), constants partially related to the agents (Equation (3)) and flow coefficient (Equation (11)).

2.4. Limitations of Model

The nitrogen in the cylinder is out of phase with the extinguishing agent. This assumption ignores the effect of the dissolved amount of nitrogen in the initial state and the possible pressure fluctuations in the tube caused by nitrogen precipitation during the discharge process, which then has an effect on the calculation of the flow rate. According to the study of Amatriain, A. [26], the characteristic time of nitrogen precipitation is about 10–9 s, which has a small effect compared to the discharge time.

Only the extinguishing agent Halon1301 flows in the pipeline without considering the flow of nitrogen and air in the pipeline. When the height of the mass block describing the gas–liquid interface in the one-dimensional model is reduced to a minimum, the cylinder empties all Halon1301 enters the pipe, the nitrogen in the air gap in the model does not continue to escape, and the pressure at the outlet of the cylinder plummets from nitrogen pressure to Halon1301 vapor pressure. This assumption affects the change in pressure of the cylinder after the extinguishing agent has been discharged completely.

No relative slip between the two phases of the extinguishing agent maintains thermodynamic equilibrium. This term assumes a velocity ratio of 1 between the two gas–liquid phases and no temperature gradient between the gas and liquid, neglecting the high degree of imbalance at the leading edge of the flow, which has less impact on the calculation of the overall flow of the fire extinguishing agent.

The effect of gravity on the flow of the jet is not considered. This assumption ignores the effect of gravitational potential energy on the flow of the extinguishing agent in the line, which has less effect on the calculation of the energy change in the flow of the extinguishing agent, considering the high-pressure constraint in the cylinder.

Only the heat exchange between the extinguishing agent and the pipeline is considered, ignoring the heat exchange between the pipeline and the outside world. This assumption ignores the effect that the environment in which the pipeline is located, such as a high-temperature environment, may have on the flow of the extinguishing agent.

3. Analysis of Simulation Results

Based on the one-dimensional fire-extinguishing pipeline simulation model established in Section 2, the effects of different filling pressures (2, 3, 4, 5, 6 MPa) and different initial temperatures (−50, −40, −30 °C) on the performance of the fire-extinguishing system are studied. Then, under the same filling conditions, the influence of the pipeline conditions of the aircraft fire extinguishing system on the discharge quality of the agent, the pressure of the cylinder outlet, and the flow distribution of the agent at each nozzle are investigated.

3.1. *Effect of Filling Pressure*

According to the change of cylinder outlet pressures over time under different filling pressures in Figure 3, it can be seen that with the start of discharge, the cylinder outlet pressures at the initial stage decrease rapidly over time, and then the pressures drop sharply. Finally, the cylinder outlet pressure fluctuates gently and decreases to the environmental pressure. With the increase in the filling pressures, the pressure drop rate at the outlet of the cylinder accelerates at the initial stage, and then the pressures at the outlet of the cylinder plummet. The phenomenon of pressure drop is caused by the simplified hypothesis of the one-dimensional simulation model established in this paper. The simplified hypothesis (2) proposed in the second chapter of the paper is that only the agent Halon 1301 flows in the pipeline without considering the flow of nitrogen and air in the pipeline. When the mass height of the gas–liquid interface described in the one-dimensional model is reduced to the minimum, the cylinder is emptied, and Halon 1301 enters the pipe completely, nitrogen in the air gap in the model will not continue to be discharged, and the pressures at the cylinder outlet plummet from nitrogen pressure to Halon 1301 vapor pressure. The time between the start of the fire extinguishing agent discharge and the occurrence of a sharp drop in pressure is defined as the emptying time. The emptying time varies from 0.826 to 1.541 s with different filling pressures. Increasing the filling pressures can effectively shorten the discharge time of the agent, but with the increase in pressures, this effect will gradually weaken.

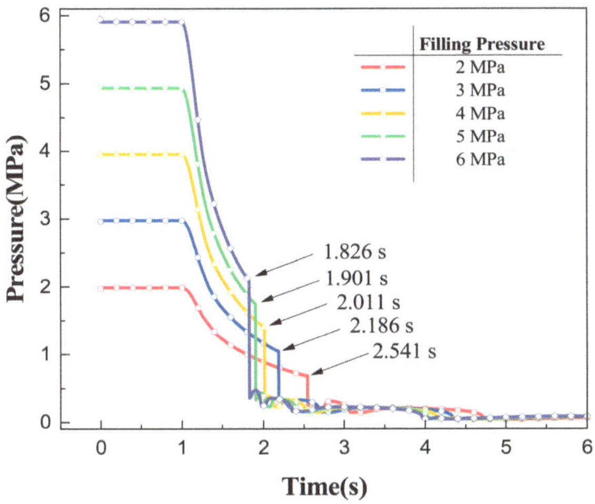

Figure 3. Pressures change curve of the cylinder outlet under different filling pressures.

Figure 4 shows the discharge quality changes of the fire extinguishing system under different filling pressures. It can be seen from the figure that the discharging mass of the fire extinguishing system varies with time under different filling pressures. The mass of the agent discharged is 3.87, 3.98, 4.08, 4.16 and 4.21 kg, respectively, under 2–6 MPa filling pressure 2 s after discharge. Before 6 s, with the increase in discharge pressures, the total discharge quality of the two nozzles in the fire extinguishing system is higher, but the difference decreases with the increase in filling pressures, and the discharge quality curves of the three nozzles with filling pressures of 4, 5 and 6 MPa are close. The results also further verified that in the actual engineering design of the aircraft fire extinguishing system pipe network, the design filling pressure of the cylinder is 4.2 MPa, which can ensure the discharge rate and reduce the structural strength requirements of the cylinder.

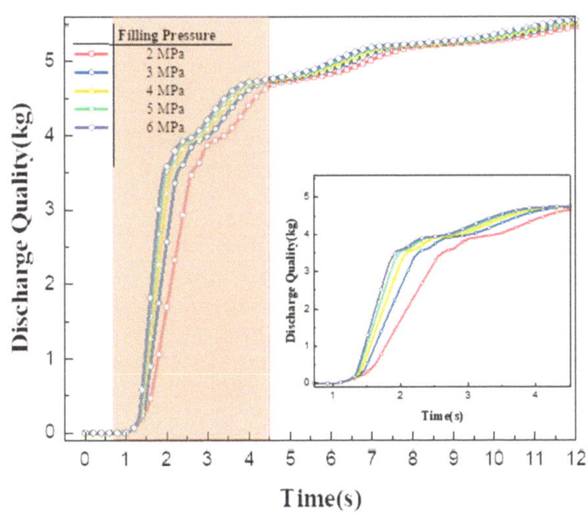

Figure 4. Agent quality from nozzle of fire extinguishing system under different filling pressures.

Figure 5 shows the changes in pressures and mass flow rate at the nozzles of two branch pipes (long branch and short branch). On the whole, the trend of mass flow is similar to that of pressures. With the increase in discharge pressures, the maximum pressures and mass flow at the nozzle of the two tubes increase. The maximum pressures at the nozzle of the long branch pipe increase from 0.42 to 0.68 MPa, and the maximum mass flow rate increases from 0.6 to 1.67 kg/s. The short branch nozzle is more sensitive to changes in initial pressures. The maximum pressures at the short branch nozzle increase from 0.49 to 0.83 MPa with the increase in initial pressures, and the maximum mass flow rate increases from 2.70 to 4.68 kg/s.

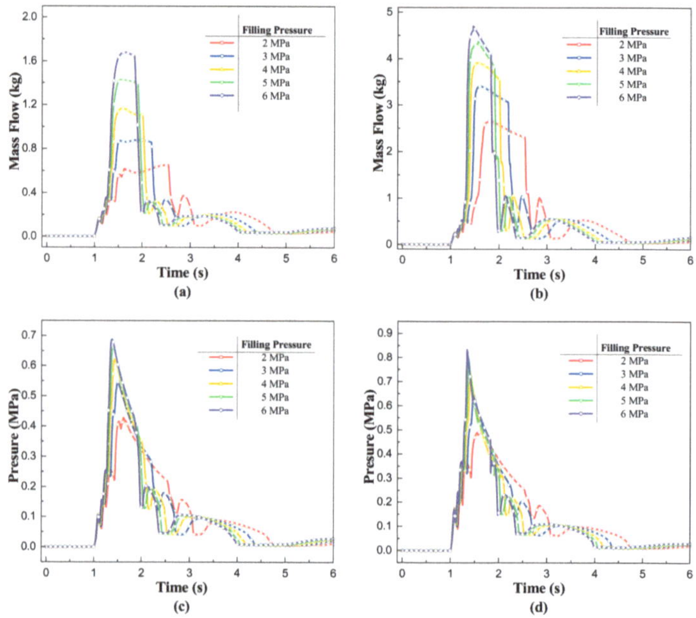

Figure 5. Mass flow rate and pressures change at nozzle under different filling pressures: (**a**) Mass flow in long branch; (**b**) Mass flow in short branch; (**c**) Nozzle pressures for long branch pipes; (**d**) Nozzle pressures for long branch pipes.

3.2. Effect of Initial Temperature

As the flight height of the aircraft increases, the ambient temperature of the aircraft decreases, which will affect the filling state of the cylinder [27,28]. In an airworthiness test, the initial temperature is required to be about −55 °C. According to the ideal gas equation, on the premise that the filling nitrogen mass remains unchanged, the decrease in temperature will lead to a decrease in the pressure in the cylinder, thus affecting the fire extinguishing system. Therefore, compared with the surface, the low-temperature environment in the air is more unfavorable to the discharge of the agent. In this paper, the initial temperature of the cylinder is −50, −40 and −30 °C, respectively, and the discharge behavior of the fire extinguishing pipeline is studied.

The filling pressure P in the cylinder can be expressed as the sum of Halon 1301 saturated vapor pressure P_{1301} and nitrogen partial pressure P_{N_2} in a stable state.

$$P = P_{1301} + P_{N_2} \tag{13}$$

The saturated vapor pressure P_{1301} of Halon 1301 can be calculated using the empirical equation based on the Antoine formula [29]:

$$P_{1301} = A + \frac{B}{T} + C lg T + DT + ET^2 \tag{14}$$

where A, B, C, D and E are constants related to Halon 1301, T is absolute temperature, K.

Assuming that nitrogen follows the ideal gas equation [26], the partial pressure of nitrogen P_{N_2} can be expressed as:

$$P_{N_2} = \frac{nR}{V}T \tag{15}$$

where n is the quantity of nitrogen substance, R is the ideal gas constant, and V is the volume of nitrogen.

Ignoring the volume change of nitrogen in the cylinder caused by temperature, $\frac{nR}{V}$ is a constant set as F, and the pressure P in the cylinder can be expressed as a function of temperature:

$$P = P_{1301} + FT \tag{16}$$

In the formula, F is a constant related to nitrogen quantity, which can be calculated by measuring the temperature and pressure in the actual cylinder in a stable state. It is known that the filling pressure in the cylinder is 4.2 MPa at 22 °C, and when the cylinder is frozen to −55 °C, the pressure in the fire extinguishing cylinder is measured as 2.63 MPa, and F can be calculated as 0.011545 MPa/K. The temperature and pressure in the cylinder and the saturated vapor pressure of the corresponding agent at different temperatures were calculated, respectively, as shown in Table 1.

Table 1. Pressure in fire extinguishing cylinder and saturated vapor pressure of agent at different temperatures.

Initial Temperature/°C	Cylinder Pressure/MPa	Agent Vapor Pressure/barA	Agent Density/kg/m^3
55	2.6300	1.17	2012.3566
−50	2.7169	1.46	1991.5513
−40	2.89236	2.20	1947.7218
−30	3.0725	6.28	1900.6716

Figure 6 shows the change of cylinder outlet pressure at different initial temperatures. Different initial temperatures correspond to different initial pressures in the fire extinguishing cylinder. The lower the initial temperature is, the lower the initial pressure in the fire extinguishing cylinder is and the slower the pressure drop rate. This result is consistent with the analysis in Section 3.1. Under different initial temperatures, the discharging time of the agent in the cylinder is close, about 1.22 s, but the discharging pressure of the cylinder outlet is inconsistent, which is caused by the inconsistent temperature of nitrogen in the cylinder when the mass block height drops to the lowest. When the cylinder is emptied, according to Equation (14), the steam pressure of Halon 1301 depends on the temperature; the higher the initial temperature, the higher the pressures at the outlet of the cylinder after plummeting.

Figure 7 shows the change in discharge quality of the agent at −50, −40 and −30 °C. When the agent is not drained from the cylinder before 2.22 s, the mass of the agent discharged from the fire extinguishing system at the three initial temperatures is similar, about 3.19 kg. With the increase in time, the mass flow rate of the fire extinguishing system at the three initial temperatures begins to decrease. The lower the initial temperature is, the more obvious the effect of mass flow reduction is. The reason for this effect is that the difference in steam pressure of the agent caused by different temperatures leads to the inconsistency of the corresponding time when the final system reaches the maximum discharge mass. The higher the initial temperature is, the shorter the time the system can discharge all the agents, so the elevation of the aircraft's flight height is not conducive to the work of the aircraft fire extinguishing system.

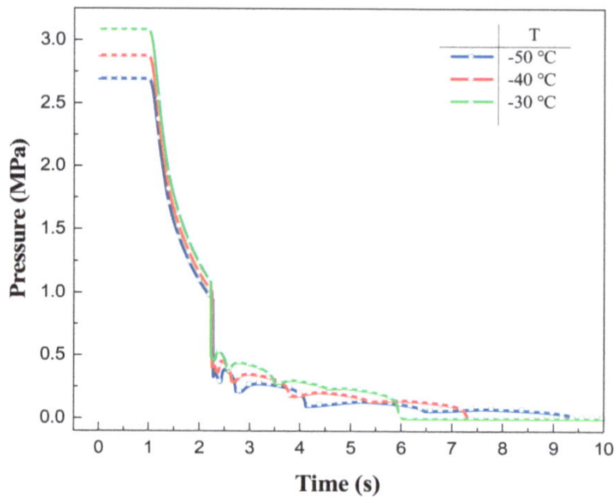

Figure 6. Change of cylinder outlet pressures at different initial temperatures.

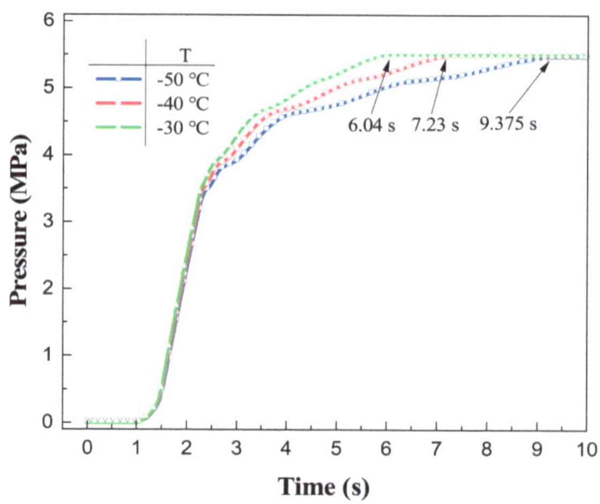

Figure 7. Mass of agent at different initial temperatures.

Figure 8 shows the variation of pressure and mass flow rate at the nozzles of the two branch pipes for different initial temperatures. The pressure at the nozzle rises to a maximum after the start of discharge, at which point it enters the high mass flow discharge phase, with the mass flow rate remaining in a high range of fluctuations, and as the discharge continues until about 2.2 s, the mass flow rate begins to decrease, and the high mass discharge phase ends. As the initial temperature increases, the pressure at the nozzles of both branch pipes increases. The mass flow rate changes in a different pattern; as the initial temperature increases, the maximum mass flow rate at the long branch nozzle increases, and the maximum mass flow rate at the short branch nozzle decreases. At different initial temperatures, the duration of the high mass flow discharge phase is similar for both branch pipes.

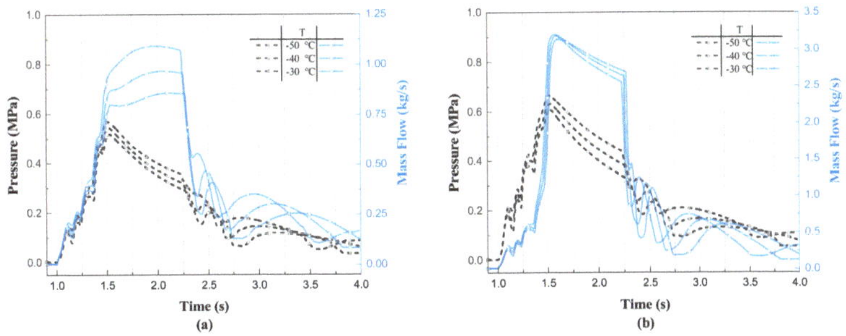

Figure 8. Variation of pressure and mass flow rate at the nozzles of the two branch pipes at different initial temperatures: (**a**) Long branch nozzle; (**b**) Short branch nozzle.

The mass flow rate at the long branch nozzle is more sensitive to changes in temperature, and in the high mass flow discharge stage, the two branch nozzle mass flow rate maximum shows an opposite change pattern, which will inevitably bring about differences in flow distribution, with the initial temperature increase, the long branch nozzle discharge mass increases while the short branch discharge mass decreases, and the total discharge remains the same.

3.3. Effect of Pipeline Roughness

According to Equation (6), it can be seen that the absolute roughness of pipeline ε affects the flow of the agent pipeline by influencing the friction coefficient f. Under the filling condition of a 5.55 kg agent and 4.2 MPa filling pressure, the flow of Halon 1301 under different kinds of roughness is simulated by modifying the absolute roughness of the pipeline to 0 mm (ideal smooth round pipe), 0.13, 0.2, 0.3 and 0.5 mm respectively.

The influence of different absolute roughness on the pressure of the cylinder outlet is shown in Figure 9. It can be seen that the absolute roughness of the pipeline affects the outlet pressure change of the upstream cylinder; when the absolute roughness is smaller, the smoother the pipeline is, the faster the pressure change of the cylinder outlet under the ideal smooth round pipe (ε = 0 mm), the emptying time of cylinder is 0.72 s, while when ε = 0.5 mm, the emptying time is 1.15 s, emptying time increased by 59.72%. The pressure at the outlet of the cylinder before the pressure drop is defined as the evacuation pressure. When the cylinder is emptied, the volume change of nitrogen in the nitrogen piston is consistent, and the pressure change is consistent, so the emptying pressure is close under different roughness.

According to Figure 10, the absolute roughness of the pipeline also has an impact on the discharge mass of the fire extinguishing system; in the ideal smooth round pipe, the agent only needs 7.43 s of all of the pipeline to discharge. When the absolute roughness of ε = 0.5 mm, it takes nearly 16 s to discharge the agent all out. When the absolute roughness increases from 0.13 to 0.5 mm, the time of all agent discharges increases from 12 to 16 s, and the total discharge mass remains the same. Therefore, in engineering design, in order to improve the fire extinguishing system discharge effect, we should try to reduce the absolute roughness of the pipeline.

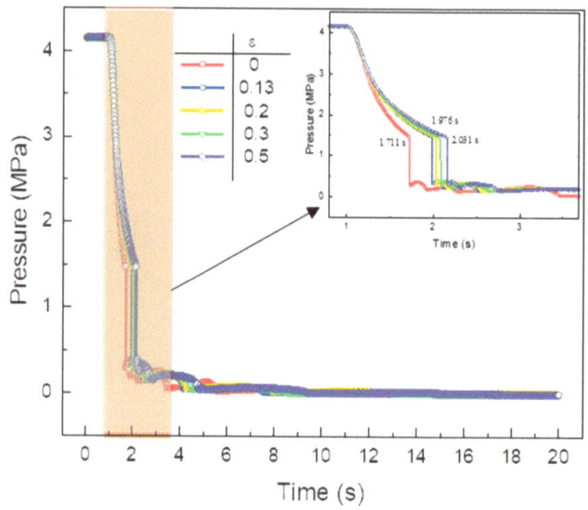

Figure 9. Cylinder outlet pressure under different absolute roughness of pipeline.

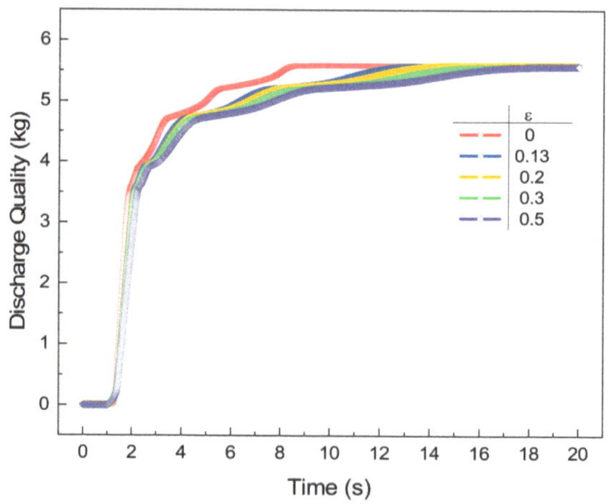

Figure 10. Different absolute roughness of pipeline under the fire extinguishing system discharge mass changes.

3.4. Effect of Main Pipe Diameter

Compared with the branch pipe, the main pipe diameter and longer length is the main component of the fire suppression system piping, this section explores the impact of increasing or decreasing the diameter of the main pipe for the flow of agent.

The change of pressure of cylinder outlet with different main pipe diameters is shown in Figure 11; increasing or decreasing the diameter of the main pipe has no obvious effect on the emptying pressure inside the cylinder, and the emptying time decreases with the increase in the diameter of the main pipe, and the shortest emptying time is 0.89 s when the diameter of the main pipe is 20 mm. Increasing the pipe diameter can improve the rate of agent discharge, but it also needs to take into account the increased volume and weight of the extinguishing system.

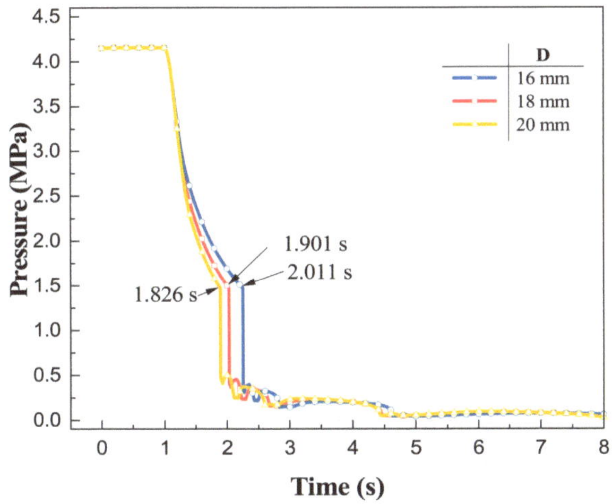

Figure 11. Effect of different main pipe diameters on the pressure of cylinder outlet.

The pressure and mass flow rate at the two branch nozzles under different main pipe diameters are shown in Figure 12, and the main pipe diameter has a small effect on the pressure at the two branch nozzles. The maximum pressure at the long branch nozzle increases with the increase in the diameter of the main pipe, and the maximum pressure at the short branch nozzle changes in the opposite direction. The maximum mass flow rate at both nozzles increases with the increase in the diameter of the main pipe, the maximum mass flow rate at the long branch nozzle increases from 1.50 to 1.94 kg/s, and the maximum mass flow rate at the short branch nozzle increases from 2.59 to 3.05 kg/s. With the increase in the diameter of the main pipe, the duration of the high mass flow discharge phase decreases, and you can achieve an increase in the diameter of the main pipe. The conclusion is that it is conducive to improving the efficiency of agent discharge.

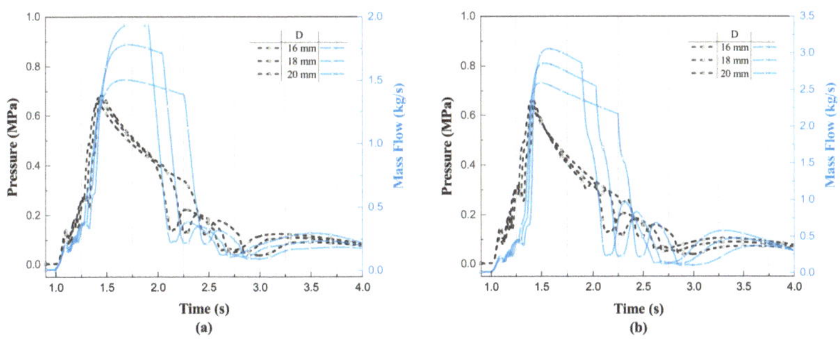

Figure 12. Pressure and mass flow rate at two branch nozzles with different main pipe diameters: (**a**) long branch nozzle; (**b**) short branch nozzle.

According to Figure 13, as the diameter of the main pipe increases, the maximum value of mass flow rate at the nozzles of the two branch pipes increases, the duration of the high mass flow discharge phase decreases, and the discharge mass of both branch pipes increases with the diameter of the main pipe in 1–8 s. The size of the main pipe diameter also affects the flow distribution between the two branch pipes; in the case of the main pipe diameter of 16 mm, the total discharge mass of the short branch increases, and the total discharge mass of the long branch decreases, and the total discharge mass remains unchanged.

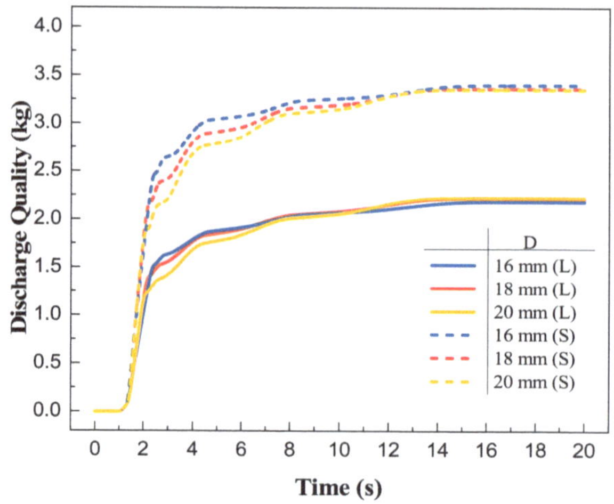

Figure 13. Discharge mass of two branch pipes with different main pipe diameters.

3.5. Branch Pipe Diameter Analysis

According to the pipeline structure model shown in Figure 2, the length of the two branch pipes is different; according to the results of the two branch pipes' nozzle flow in Section 3.2, it can be seen that the difference between the two nozzles at the discharged mass is large, in 4.2 MPa filling pressure, the short branch nozzle discharge mass is more than the long branch discharge mass of about 2.468 kg.

In order to make the concentration of agents in the target protection nacelle more balanced, it is necessary to adjust the flow resistance of the two branch pipes to influence the distribution of agents and thus achieve the purpose of balancing the flow rate. In the case of a compact and fixed aircraft, it is easier to adjust the branch pipe diameter rather than the branch pipe length [11]. Therefore, the results of different branch pipe diameters are investigated for the flow distribution.

Route 1: increase the long branch pipe diameter to reduce the long branch pipe flow resistance; Route 2: reduce the short branch pipe diameter to increase the short branch pipe flow resistance. According to these ideas, set up a total of five simulation scenarios, different scenarios in the long and short branch pipe diameter, as shown in Table 2.

Table 2. Discharge mass of long and short branch pipes at different pipe diameters.

Scene	Long Branch Diameter/mm	Short Branch Diameter/mm	Long Branch Discharge Mass/kg	Short Branch Discharge Mass/kg	Ratio of Discharge Mass of Long Branch to Short Branch
A	14	14	1.5586	4.0268	0.387
B	16	14	1.8621	3.7233	0.500
C	18	14	2.1005	3.4850	0.603
D	14	12	2.0621	3.5233	0.585
E	14	10	2.7636	2.8230	0.979

The values of the long and short branch diameters and the discharge masses and their mass ratios for the corresponding pipe diameters are given in Table 2 for a total of five scenarios. Scene A is the initial scenario corresponding to the branch diameter; from the table, it can be found that increasing the long branch diameter has less effect on the equilibrium flow than decreasing the short branch diameter. When the diameter of the short branch pipe is reduced to 10 mm, the mass ratio of long and short branch pipes reaches 0.979, and the flow rate is close to the same, which can achieve the purpose of balancing the flow rate while reducing the system weight.

As shown in Figure 14, the influence of different branch pipe diameters on the pressure at the cylinder outlet is small, and the emptying time grows from 0.976 s in scene C to 1.066 s in scene E. After the cylinder is emptied, the pressure of the cylinder outlet fluctuates due to the heat exchange between the pipe and the agent. The larger the diameter of the long branch pipe, the faster the pressure change at the cylinder outlet, and the smaller the diameter of the short branch pipe, the slower the pressure change of the cylinder outlet. Scene E to achieve balanced flow and reduce the weight and volume of the fire suppression system while the impact on the discharge time is small. The results of the section for the fire extinguishing system branch pipe diameter design are of guidance.

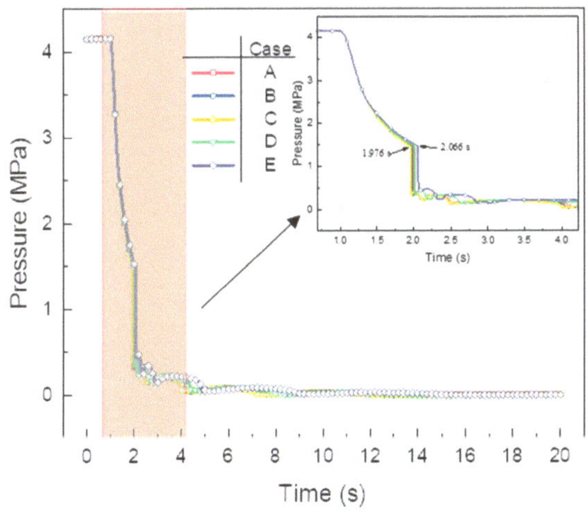

Figure 14. Effect of different pipe diameters on the outlet pressure of cylinders.

4. Conclusions

Amesim software was used to establish a simulation model of an aircraft fire extinguishing system piping, to study the influence of fire extinguishing cylinder filling conditions and piping conditions on the performance of the fire extinguishing system under 5.55 kg of agent filling, and the following conclusions were obtained.

(1) The change in the filling conditions of the fire extinguishing cylinder to the change in the pressure of the cylinder itself, which in turn affects the pressure and mass flow rate of the agent at the nozzle. The higher the filling pressure, the faster the pressure drop; 4, 5, 6 MPa filling pressure under the agent discharged mass curve is similar; with the initial temperature decreases, the pressure at the cylinder outlet decreases, and the emptying time is similar to 1.22 s, but the difference between the discharge mass to reach the peak point is larger.

(2) Under the same filling conditions, roughness affects the flow of the agent by affecting the friction coefficient; the lower the roughness, the faster the discharged agent, but the total mass discharged from the fire extinguishing system under the pipeline thermal boundary conditions is close; when the short branch pipe diameter is 10 mm, and the long branch pipe diameter is 14 mm, the two branch pipe discharge mass ratio is close to 1, which makes the concentration of the agent in the target protection nacelle more balanced. Changing the branch pipe diameter has less effect on the emptying time; the larger the main pipe diameter, the larger the diameter of the main pipe, and the higher the discharge rate, but the actual design requires comprehensive consideration of the energy costs associated with the increase in the diameter of the main pipe.

Although this research was conducted with Halon 1301, the main value of this paper is to optimize the system design and reduce the potential use of Halon 1301 during aircraft design and in actual aircraft firefighting and to provide a research methodology and comparative benchmark for subsequent pipe flow studies and applicability studies of a Halon substitute.

Author Contributions: Conceptualization, R.L.; Data curation, R.L. and W.M.; Formal analysis, S.L. (Song Lu) and H.Z.; Methodology, R.L. and C.Y.; Project administration, H.Z.; Writing—original draft, S.L. (Shaonan Liu) and J.G.; Writing—review and editing, R.L. and C.Y. All authors have read and agreed to the published version of the manuscript.

Funding: This study was supported by the National Natural Science Foundation of China (No.51974284), the Fundamental Research Funds for the Central Universities under Grant No. WK2320000046, WK2320000049.

Institutional Review Board Statement: Not applicable.

Informed Consent Statement: Not applicable.

Data Availability Statement: Not applicable.

Acknowledgments: The numerical calculations in this paper were supported and assisted by the computational support from the Supercomputing Center of the University of Science and Technology of China.

Conflicts of Interest: The authors declare no conflict of interest.

References

1. Niu, X.; Xie, Y.; Hasemi, Y. Analysis of fire spread and fire extinguishing agent distribution in nacelle of helicopter under no-ventilation condition. *Procedia Eng.* **2013**, *62*, 1073–1080. [CrossRef]
2. Friedman, R. Fire Safety in the Low-Gravity Spacecraft Environment. 1999. Available online: https://ntrs.nasa.gov/citations/19990063738 (accessed on 15 April 2022).
3. Halons Program. Available online: https://www.epa.gov/ozone-layer-protection/halons-program (accessed on 1 May 2022).
4. European Union Aviation Safety Agency. *Halon Replacement in the Aviation Industry*; European Union Aviation Safety Agency: 2019. Available online: https://www.easa.europa.eu/downloads/106162/en (accessed on 15 April 2022).
5. Tapscott, R.E.; Morehouse, E., Jr. *Next-Generation Fire Extinguishing Agent. Phase 1. Suppression Concepts*; New Mexico Engineering Research Institute: Albuquerque, NM, USA, 1987.
6. Wang, Y.; Zou, G.; Liu, C.; Gao, Y. Comparison of fire extinguishing performance of four halon substitutes and Halon 1301. *J. Fire Sci.* **2021**, *39*, 370–399. [CrossRef]
7. Xing, E.; Jin, J.; Zhang, Z.; Pan, R.; Li, Q.; Zheng, J. Simulation on flow rate characteristics of gas fire extinguishing agent with expansion nozzle based on AMESim. In Proceedings of the CSAA/IET International Conference on Aircraft Utility Systems (AUS 2018), Guiyang, China, 19–22 June 2018.
8. Jin, J.-M.; An, F.-L.; Shou, Y.; Pan, R.-M.; Xuan, Y.; Li, Q.-W. Simulation on Release Characteristics of the Gas Extinguishing Agent in Fire Extinguisher Vessel with Different Filling Conditions Based on AMESim. *Procedia Eng.* **2018**, *211*, 315–324. [CrossRef]
9. Keyser, D.R.; Hewson, J.C. Assessment of Fire-Suppression Simulations Using Full-scale Engine Nacelle Tests. In *Assessment of Fire-Suppression Simulations Using Full-Scale Engine Nacelle Tests*; Citeseer: 2005. Available online: https://www.nist.gov/publications/assessment-fire-suppression-simulations-using-full-scale-engine-nacelle-tests (accessed on 6 May 2022).
10. NFPA 12A. Standard on Halon 1301 Fire Extinguishing Systems. 1992. Available online: https://www.nfpa.org/codes-and-standards/all-codes-and-standards/list-of-codes-and-standards/detail?code=12A (accessed on 6 May 2022).
11. Williamson, H. Halon 1301 flow in pipelines. *Fire Technol.* **1976**, *12*, 18–32. [CrossRef]
12. Elliott, D.; Garrison, P.; Klein, G.; Moran, K.; Zydowicz, M. Flow of Nitrogen-Pressurized Halon 1301 in Fire Extinguishing Systems. 1984. Available online: https://ntrs.nasa.gov/api/citations/19850007392/downloads/19850007392.pdf (accessed on 6 May 2022).
13. Grosshandler, W.L.; Gann, R.G.; Pitts, W.M. Evaluation of Alternative In-Flight Fire Suppressants for Full-Scale Testing in Simulated Aircraft Engine Nacelles and Dry Bays. 1994. Available online: https://www.nist.gov/publications/evaluation-alternative-flight-fire-suppressants-full-scale-testing-simulated-aircraft (accessed on 8 May 2022).
14. Gann, R.G. *Fire Suppression System Performance of Alternative Agents in Aircraft Engine and Dry Bay Laboratory Simulations*; NIST: 1995. Available online: https://tsapps.nist.gov/publication/get_pdf.cfm?pub_id=917025 (accessed on 8 May 2022).
15. Yang, J.C.; Huber, M.L.; Boyer, C.I. A model for calculating alternative agent/nitrogen thermodynamic properties. In Proceedings of the 1995 Halon Options Technical Working Conference, Albuquerque, NM, USA, 9 May 1995.
16. Gann, R.G. *Advanced Technology for Fire Suppression in Aircraft*; NIST: 2007. Available online: https://www.nist.gov/publications/advanced-technology-fire-suppression-aircraft (accessed on 8 May 2022).
17. Lee, J. Simulation method for the fire suppression process inside the engine core and APU compartments. In Proceedings of the Fourth Triennial International Aircraft Fire and Cabin Safety Research Conference, Lisabon, Portugal, 15–18 November 2004; pp. 15–18.

18. Kim, J.; Baek, B.; Lee, J. Numerical analysis of flow characteristics of fire extinguishing agents in aircraft fire extinguishing systems. *J. Mech. Sci. Technol.* **2009**, *23*, 1877–1884. [CrossRef]
19. Cleary, T.G.; Yang, J.C.; King, M.D.; Boyer, C.I.; Grosshandler, W.L. Pipe flow characteristics of alternative agents for engine nacelle fire protection. In *Proceedings on Halon Options Technical Working (Lmferek~May)*; 1995; Volume 9, p. 11. Available online: https://tsapps.nist.gov/publication/get_pdf.cfm?pub_id=909942 (accessed on 8 May 2022).
20. Rafiee, S.E.; Sadeghiazad, M. Experimental and 3D CFD analysis on optimization of geometrical parameters of parallel vortex tube cyclone separator. *Aerosp. Sci. Technol.* **2017**, *63*, 110–122. [CrossRef]
21. Zhang, M.; Zhenbo, F.; Yuzhen, L.; Jibao, L. CFD study of NOx emissions in a model commercial aircraft engine combustor. *Chin. J. Aeronaut.* **2012**, *25*, 854–863. [CrossRef]
22. Shah, M.M. A general correlation for heat transfer during film condensation inside pipes. *Int. J. Heat Mass Transf.* **1979**, *22*, 547–556. [CrossRef]
23. Lopez-Echeverry, J.S.; Reif-Acherman, S.; Araujo-Lopez, E. Peng-Robinson equation of state: 40 years through cubics. *Fluid Phase Equilib.* **2017**, *447*, 39–71. [CrossRef]
24. Churchill, S.W. Friction-factor equation spans all fluid-flow regimes. *Chem. Eng.* **1977**, *84*, 91–92.
25. Müller-Steinhagen, H.; Heck, K. A simple friction pressure drop correlation for two-phase flow in pipes. *Chem. Eng. Process. Process Intensif.* **1986**, *20*, 297–308. [CrossRef]
26. Amatriain, A.; Rubio, G.; Parra, I.; Valero, E.; Andreu, D.; Martín, P.M. Mathematical modeling of nitrogen-pressurized Halon flow in fire extinguishing systems. *Fire Saf. J.* **2021**, *122*, 103356. [CrossRef]
27. Bacmeister, J.T.; Eckermann, S.D.; Newman, P.A.; Lait, L.; Chan, K.R.; Loewenstein, M.; Proffitt, M.H.; Gary, B.L. Stratospheric horizontal wavenumber spectra of winds, potential temperature, and atmospheric tracers observed by high-altitude aircraft. *J. Geophys. Res. Atmos.* **1996**, *101*, 9441–9470. [CrossRef]
28. Chen, X.T.; Wang, C.X.; Sun, Q.; Wang, H.B.; Xie, S.; He, Y.H. Study on thermomechanical coupling characteristics of embedded sensor lithium batteries under low-temperature environment. *Energy Storage* **2020**, *2*, e107. [CrossRef]
29. Braker, W.; Mossman, A.L. *Matheson Gas Data Book*; Matheson: Lyndhurst, NJ, USA, 1980.

fire

Review

Water Mist Fire Suppression Systems for Building and Industrial Applications: Issues and Challenges

Kyle Farrell, Md Kamrul Hassan *, Md Delwar Hossain, Bulbul Ahmed, Payam Rahnamayiezekavat, Grahame Douglas and Swapan Saha

School of Engineering, Design and Built Environment, Western Sydney University, Penrith, NSW 2751, Australia
* Correspondence: k.hassan@westernsydney.edu.au

Abstract: Interest in water mist fire suppression has increased within the fire protection industry due to its ability to control the spread and development of fire without using environmentally damaging agents. Water mist fire suppression has been used for many years in various applications such as machinery spaces, combustion turbine enclosures, and onboard passenger sea vessels. Now there is a demand to use this firefighting method to protect other fire risks such as cooking areas, commercial buildings, residential buildings, electrical equipment, road tunnels, bushfire (wildland fire) protection, and nuclear power generation facilities. To support this industry demand, this review covers the fundamentals of water mist, its suppression mechanisms, areas of application, existing research and development, and the codes and standards related to design. This comprehensive review provides a clear history of water mist suppression. It is able to identify the issues and challenges related to the technology to help pave the way for future research and development that will improve these systems to a level so that they are suitable for these new applications and meet the industry demand for nontoxic fire suppression systems.

Keywords: water mist; fire behaviour; fire suppression; fire protection; spray dynamics; fire compartment; design codes; test protocols

Citation: Farrell, K.; Hassan, M.K.; Hossain, M.D.; Ahmed, B.; Rahnamayiezekavat, P.; Douglas, G.; Saha, S. Water Mist Fire Suppression Systems for Building and Industrial Applications: Issues and Challenges. *Fire* **2023**, *6*, 40. https://doi.org/10.3390/fire6020040

Academic Editors: Song Lu, Changcheng Liu, Guohui Li and Paweł Wolny

Received: 2 December 2022
Revised: 4 January 2023
Accepted: 10 January 2023
Published: 19 January 2023

1. Introduction

In the firefighting and fire safety industries, halon was widely utilised as a fire control agent. However, such substances are not only hazardous to people but also damage the ozone layer [1,2]. As a result, nations all over the world have prohibited their manufacturing and usage since the end of the twentieth century [3]. In order to identify a substitute for halon, effective and environmentally acceptable fire extinguishing technologies have become a research hotspot in the field of fire safety during the last 20 years [3]. There is a growing interest in using CO_2, inert gas, and water sprinklers as replacements. However, despite all efforts to substitute with alternative clean agents, the goal remains challenging due to several constraints. For example, using CO_2 to control fire can displace oxygen in the air and may be hazardous to human health [4]. Sprinklers are the most widely utilised fire suppression system in many applications, such as buildings, owing to their demonstrated efficacy in minimising the intensity of fire and fire spread beyond the fire origin. However, there are worries that sprinkler systems might cause postfire water damage to structures and equipment [5]. Furthermore, this suppression system has disadvantages in some locations where water supplies are restricted and suppression is required in confined spaces, such as ship fires and spaceship fires, among others [3].

Water mist fire suppression technology has been developed in recent years and is regarded as a prospective replacement agent due to its qualities of speedy fire extinguishment, minimal environmental concerns, minimal water usage, and reduced water damage [1,6,7]. In the 1940s, the National Fire Protection Association (NFPA) Technical Committee designed water mist systems for specialised purposes, such as passenger ferries [8]. In the

1980s, the Manchester Air Accident and an International Maritime Organisation (IMO) rule requiring the installation of fire sprinklers rekindled interest in employing water mist systems to extend the tenability of space for the safe evacuation of passengers through heat management and fire suppression [9]. Following the fire that killed 159 passengers on the MS *Scandinavian Star* ferry on a route from Norway to Denmark on 7 April 1990, the water mist fire suppression system gained prominence in the maritime and insurance industries [10]. Since then, the application areas for water mist systems have expanded at an exponential rate. The technology now has a wide range of applications in preventing and dealing with historical building fires [7], tunnel fires [11], ship fires [12], nuclear power plant fires [13], and explosion fires [9,14–17].

Water in the form of a mist has been found to be an efficient extinguishing agent, and NFPA defines water mist as a water spray in which 99% of the water is in droplets with diameters (D_{v99}) smaller than 1000 microns [8]. The development of fine water mist systems offers a new water-based suppression alternative for many potential applications. Figure 1 shows the typical fire water mist suppression system arrangements. Since the fire-extinguishing mechanisms behind the water mist system are different from standard sprinkler systems, researchers have focused their attention on investigating those mechanisms required to suppress fire successfully [18].

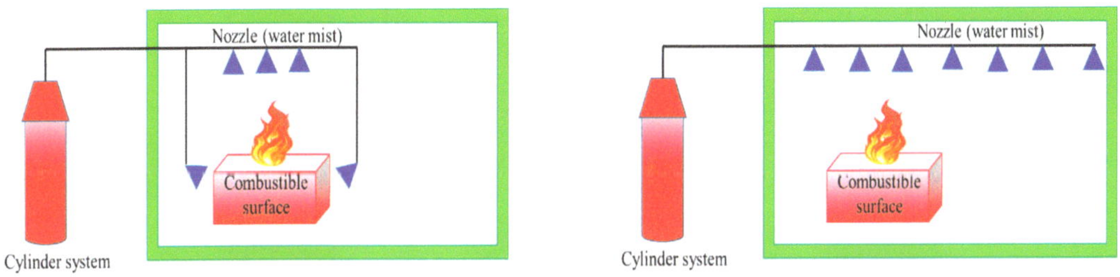

(**a**) Local application (**b**) Total flooding

Figure 1. Water mist fire suppression general arrangement.

Various research studies have been utilised to experimentally and numerically examine the effectiveness of water mist systems [2,3]. In particular, several researchers have examined the dynamics of the water mist system and the effects of different fire compartment factors on the performance of water mist fire suppression [17,19,20]. To identify the future development and direction of water mist fire extinguishing technology, as well as expand its area of application, this paper summarises the latest research progress of extinguishing mechanisms of water mist in fire suppression. It also discusses the influence of the spray dynamics of water mist, such as water mist additives, operating pressure, water droplet size, flow rate, nozzle K-factor, spray momentum, and spray angle on the effectiveness of water mist as a fire extinguishing agent. The effects of ceiling height, ventilation, and different dynamic mixing situations in fire compartments on the fire suppression process of water mist are also discussed in this review. This review discusses how this existing knowledge related to water mist system can potentially be used in new and emerging industries and protect people and property from their associated fire risks. The study also extends to existing design codes of water mist systems for different countries and highlights how legislation and system approvals may be restricting potential applications for this firefighting method. Various challenges are highlighted using current research data, and finally, the paper concludes with various suggestions about the observed challenges and issues.

2. Water Mist Fire Systems

2.1. Methods of Generating Water Mist

Water mist is a water spray for which 90% of the flow-weighted cumulative volumetric distribution of water droplets (DV0.90) is less than 1000 μm. This value is found at the minimum design operating pressure and is measured in a plane 1 m from the nozzle. NFPA 750 [8] defines three mist classifications within this 1000 μm window, which are categorised as Class 1 ($D_{V0.90} \leq 200$ μm), Class 2 (200 μm< $D_{V0.90} \leq 400$ μm), and Class 3 ($D_{V0.90} > 400$ μm). Three different operating pressures are suggested by NFPA [8] to create these droplet sizes of water mist. The pressure ranges are low pressure (pressure ≤ 12.1 bar), intermediate pressure (12.1 bar < pressure < 34.5 bar), and high pressure (pressure ≥ 34.5 bar). Droplet size also depends on the discharge nozzle, which significantly influences the production of the water mist spray. Three different types of nozzles, impingement, pressure jet, and twin-fluid nozzles, are mainly used to create water mist spray, as shown in Figure 2.

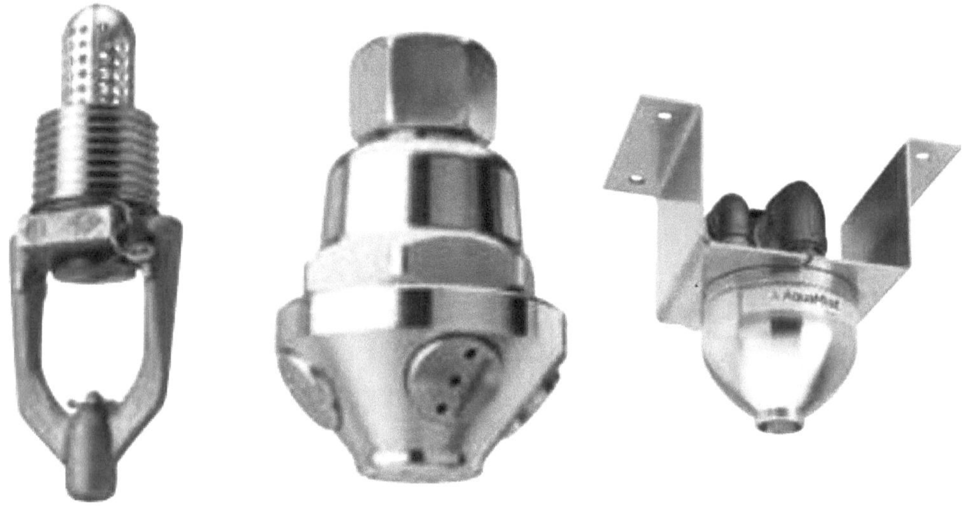

(a) (b) (c)

Figure 2. Types of nozzles used to create water mist in water mist systems. (**a**) Impingement nozzle [21]; (**b**) Pressure jet nozzle [21]; (**c**) Twin-fluid water nozzle [21].

2.1.1. Impingement Nozzles

Impingement nozzles, as shown in Figure 2a, are common in water-based fire suppression, and the operating principle is typical of such a fire sprinkler. A large diameter orifice allows water to flow from the nozzle and impact with a deflector that impinges the flow The deflector is designed to break up the high-velocity flow and create a flow of smaller water droplets. Impingement has been used to generate coarser mists of Class 2 and Class 3 to be used on low- and intermediate-pressure systems.

2.1.2. Pressure Jet Nozzle

Pressure jet nozzles consist of smaller-diameter orifices or swirl chambers, as shown in Figure 2b. A high-velocity jet of water flowing through the nozzle creates a thin jet of water that become unstable with discharge into the atmosphere and disintegrates into fine droplets. Pressure jet nozzles are used across low-, intermediate-, and high-pressure systems and can create Class 1-type mists below 200 μm. Multi-nozzle heads are common for this type of nozzle to increase the spray cone angle.

2.1.3. Twin-fluid Nozzle

Twin-fluid nozzles, as shown in Figure 2c, are designed to operate with water and compressed gas (typically air or nitrogen). This design has two inlets into the nozzle chamber where the high-pressure gas atomises the water before discharging into the atmosphere, with further agitation causing water droplets to break down further. The water and gas inlet lines operate at different pressures; typically, both are below 12.0 bar and within the low-pressure range. These systems can produce a finer water mist of Class 1 and Class 2 sprays. The use of nitrogen in a twin-fluid system can also add additional fire suppression performance by improving the primary suppression mechanisms of water mist systems.

2.2. Fire Control and Suppression Mechanisms

It is widely agreed that fire can take many forms involving chemical reactions between combustible species and oxygen. A common simplification often describes the combustion process of fire using the fire triangle (or tetrahedron), where the inputs are fuel, oxygen, and heat/ignition (and chemical reaction). Most fires involve combustible solids, although liquid and gaseous fuels are found in many industry sectors [22]. When in the appropriate conditions, these fuels will react with the oxygen in the air, generating combustion products and releasing heat. This heat acts as a continuous ignition source for surrounding fuels, which maintains the continued combustion process of the burning fire. Water mist fire suppression mechanisms work to inhibit this burning process using three main mechanisms (heat extraction, the displacement of oxygen, and radiation attenuation) categorised into primary and secondary mechanisms (Figure 3).

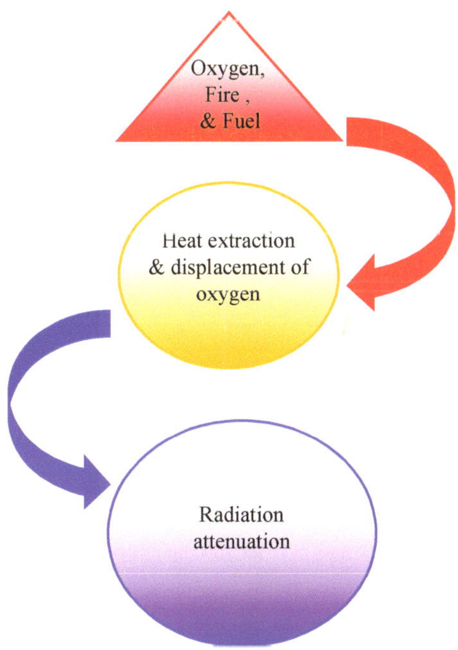

Figure 3. Fire suppression mechanism of water mist.

These suppression mechanisms work by removing two sides of the fire triangle by reducing the oxygen in the atmosphere and reducing the heat output that feeds the continuous ignition of additional fuel sources. In the primary mechanism, heat is extracted to cool the fire plume and wet/cool the fuel surface, and then oxygen is displaced and diluted by blocking it from the fuel sources. In this stage, the water mist expands faster due to the faster vaporisation and cools the hot gases in the fire area. In the secondary mechanism, expanded water mists block the radiant heat transfer to the surrounding fuel surfaces.

2.2.1. Heat Extractions

Water is widely used as an extinguishing medium for fire because of its ability to cool flames, burning fuel, and the surface of surrounding fuels not yet involved in the fire. Water can absorb a significant amount of heat due to a high heat capacity of approximately 4182 J/kg °C and high latent vaporisation heat of 2260 kJ/kg [23]. A flame will be extinguished if its temperature is reduced below its critical temperature. This critical temperature can be determined using Equation (1) [24].

$$T_{f,crit} = T_{f,max} - \left(\frac{(T_{f,max} - T_{ls})m_{og}}{\frac{m_{fs} - \left(\frac{m_{og}}{r}\right)}{1 - m_{fs}} - r} \right) \quad (1)$$

where $T_{f,crit}$ is critical flame temperature (°C) below which the flame will extinguish, $T_{f,max}$ is theoretical flame temperature (°C) assuming no heat losses to the surface, T_{ls} is the surface temperature of the fuel–fire point temperature (°C), m_{og} is the mass fraction of oxygen in the atmosphere, m_{fs} is the mass fraction of fuel vapour immediately above the liquid surface, and r is the stoichiometric ratio (mass of O_2 required to burn a unit mass of fuel).

Water has also the ability to extinguish fire by cooling the burning fuel below its fire point by removing heat from the fuel surface (Figure 4). Fuel cooling by water mist is primarily due to heat removal during water conversion to steam. This heat removal rate per unit area can be estimated using Equation (2) [6].

$$S_h = (H_f - \lambda_f)\dot{m}_b + R_a - R_s \quad (2)$$

where S_h is the heat removal rate per unit area (kJ/m^2.s), H_f is convective heat transferred from flames per unit mass of fuel (kJ/kg), λ_f is the heat required to produce a unit mass of vapour (kJ/kg), $\dot{m}_b$ is the burning rate per unit area (kg/m^2.s), R_a is heat transferred to the fuel surface (kJ/m^2.s), and R_s is heat lost from the surface (kJ/m^2.s) not included in λ_f.

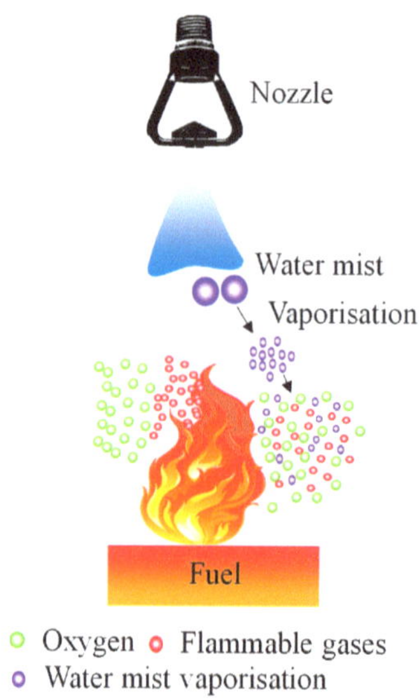

Figure 4. Heat extraction and water displacement of water mist system.

When the water is converted to steam and extracts the heat, it is also contributing to the displacement of oxygen within the combustion zone. The critical extinguishment concentration of water mist surrounding a diffusion flame ranges from 100 to 200 g/m^3.

2.2.2. Oxygen Displacement

Reducing oxygen from a fire has long been used as a reliable fire suppression method (Figure 4). When a water mist system is discharged locally onto a fire or within a fire compartment, the small water droplets are heated and transformed into steam. When this vaporisation of water occurs within the flame zone, the volumetric expansion can disrupt the entrainment of oxygen into the flame and significantly reduces the oxygen concentration in a compartment [25]. Water mist systems achieve oxygen displacement by producing excessive amounts of steam. The water mist droplets tend to be evaporated before reaching the fire flame, and the evaporated steam causes O_2 displacement and contributes to the suppression of the fire [19]. When the water spray penetrates the fire plume, the vaporising water expands to 1600 times its liquid volume [6,26]. This volumetric expansion distributes the entrainment of oxygen into the flame and dilutes the fuel vapour available for combustion. This contributes to the suppression and extinguishment of the flame by diluting the vapour below its fuel–oxygen mixture flammable limit for the given fuel source (removing the chemical reaction required).

The impact of oxygen dilution by water mist is dependent on the properties of the fuel. The minimum amount of oxygen concentration required to support combustion varies depending on the fuel source. Oxygen typically makes up 21% of the air in naturally low atmospheric conditions and is the baseline oxygen concentration for a typical fire. Hurley [25] explained that smouldering and flaming fires need to be below 10% and 16% O_2 concentration, respectively, to be smothered. Xin and Khan [27] investigated the limiting oxygen concentration (LOC) in which a fire will reach extinction. Table 1 shows the LOC percentage of various liquid and solid fuels, which varies from 11% to 15% O_2 concentration. Mawhinney et al. [28] and Liu et al. [29] explained that the hot layer of gases in a fire compartment is rapidly cooled by its first contact with water mist and forced down by the spray to the seat of the fire. This is known as turbulent mixing and contributes to extinguishment through oxygen depletion. This increase in water mist particles surrounding the flame also inhibits the spread of fire through the attenuation of radiation.

Table 1. Limiting oxygen concentration at fire extinction [27].

Material	Extinction (O_2) (% vol.)	Material	Extinction (O_2) (% vol.)
Liquid Fuel		Solid Fuel	
Unleaded gasoline (dish)	12.48	Polyethylene (low density)	11.39
Methanol (dish)	11.64	Polystyrene (high density)	11.21
Methanol (wick)	12.33	PMMA	10.48
Ethanol (dish)	12.40	Red oak	12.26
Ethanol (wick)	13.35	White maple	14.66
Corn oil (dish)	12.29	Corrugated paper (tri-wall)	12.86
		Corrugated paper (single layer)	12.94
		Kraft linerboard	12.33

2.2.3. Radiation Attenuation

Water mist can prevent fire spread by acting as a thermal barrier to inhibit heating by radiation from the burning fuel to nonburning surfaces surrounding the fire (Figure 5). A substantial amount of heat released from flames is transmitted by radiation during typical fire scenarios. This is a form of energy travelling as electromagnetic waves through a material or space [22]. During a fire, surrounding materials absorb, reflect, or transmit this energy, which increases the temperature and initiates pyrolysis. Kashiwagi [30] described the pyrolysis mechanism of a solid material that causes fuel vapour to be released. The fire spreads as these volatile vapours are ignited by travelling embers or from direct flam contact. Balner [31] described how water droplets prevent the transmission of radiant hea due to the absorption and scattering of thermal radiation. This is known as attenuatio (Planck mean absorption coefficient) and is described as the blocking effect [22]. Thi absorption coefficient of materials has a direct relation to its ability to absorb radiation an heat energy. Figure 6 shows the high absorption coefficient of water (H_2O) compared wit other gases at various temperatures [25].

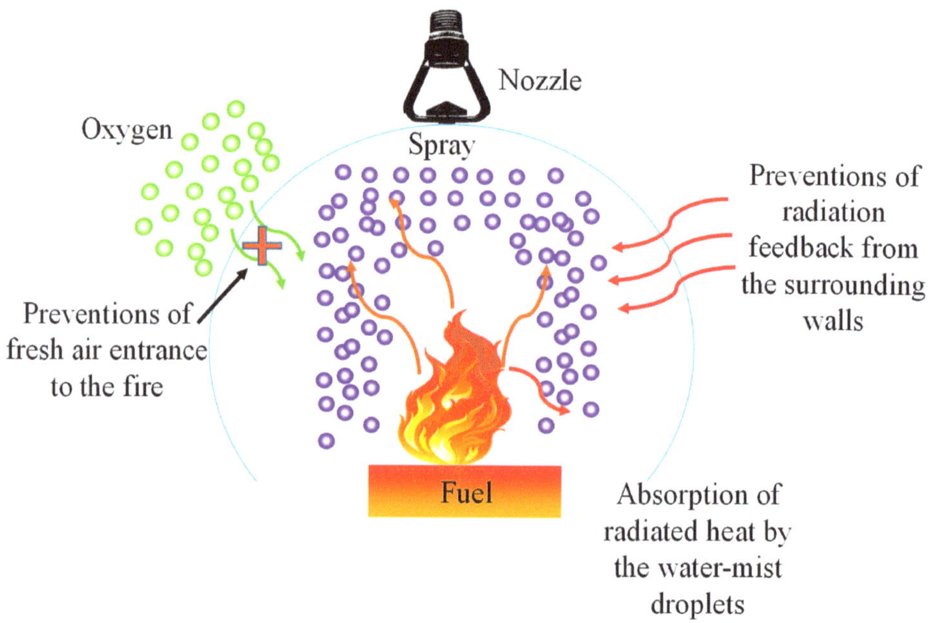

Figure 5. Attenuation of thermal radiation by a water mist system.

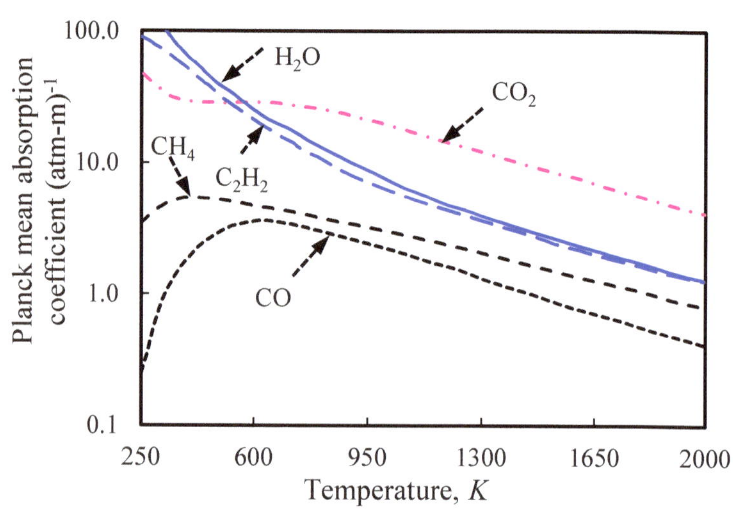

Figure 6. Planck mean absorption coefficient for various gases [25].

To prevent the heating and ignition of a nonburning surface fuel by radiation, water droplets also wet the surface. The minimum water flow required to prevent the ignition of these surfaces can be calculated using Equation (3) [25].

$$\frac{F_m}{A_s} = \frac{\varepsilon \times \sigma \times \phi \times \left(T_r^4 - T_s^4\right) - I_c}{H_{vap}} \quad (3)$$

where F_m is the minimum water flow rate (m^3/s), A_s is the fuel surface area (m^2), ε is emissivity of the radiator, σ is the Stefan–Boltzmann constant (5.67×10^{-8} W/m^2K^4), ϕ is the view factor of the fuel bed, T_r is the mean absolute temperature of the radiation source (K), T_s is the mean absolute temperature of the surface (K), I_c is the critical radiation intensity required for piloted ignition (kJ/kg), and H_{vap} is the heat vaporisation of water (kJ/kg).

The performance of a water mist system in a fire depends on its ability to achieve these three suppression mechanisms: (1) removing heat from the gases, (2) the displacement of oxygen by water vapour, and (3) the attenuation of radiation by water droplets. Understanding these mechanisms has led to various water mist system applications for common and unique fire hazard types across multiple industries.

3. Applications of Water Mist Systems

Water mist systems have been proven to perform in many applications, such as enclosed compartments; open air; and large, well-ventilated spaces [25]. In addition, water mist systems are also used as a fire protection method in many applications for industrial facilities, commercial buildings, electrical equipment, residential buildings (including bushfire protection), transportation, tunnels, and explosion hazard mitigation (Figure 7).

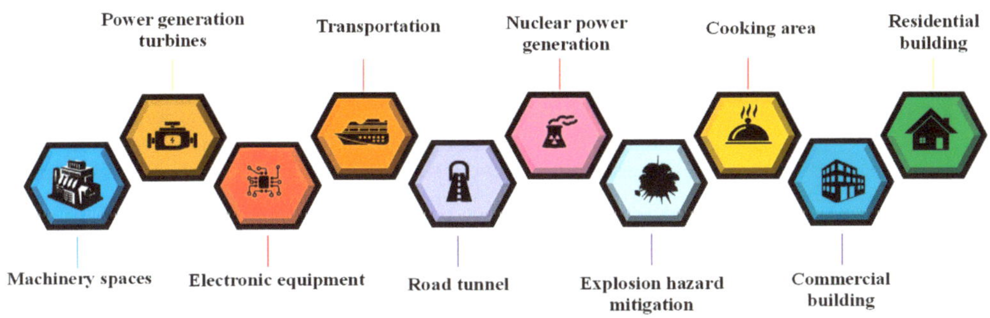

Figure 7. Common applications of water mist systems.

3.1. Machinery Spaces

In recent years, water mist has emerged as an alternative suppression medium over CO_2 and halocarbon gaseous agents in machinery spaces for its ability to remove heat from hot surfaces without environmental impact. Machinery spaces such as large engine rooms on a ship or within a manufacturing process facility create flammable liquid hazards in the form of liquid fuel and lubricating and hydraulic oils. A ruptured hydraulic line can create fuel spray fires or liquid pool fires if the hot engine components ignite. There are many different test protocols established by IMO, Underwriters Laboratories (UL), and Factory Mutual (FM) related to machinery spaces for various fire scenarios that may occur in machinery spaces. These test fire scenarios are for various fire types, fire sizes, compartment obstructions, fuel types, ignition sources, compartment sizes, and ventilation conditions. The recent studies completed by Jeong et al. [28] and Choi et al. [32] applied water mist to engine diesel generators, shipboard engine rooms, and machinery spaces following IMO test protocols and found that water mist can effectively extinguish various fires, including spray fires, pool fires, and cascade fires that may occur in machinery spaces. A larger fire within the machinery space relative to the compartment size is easier to extinguish due to the larger amount of steam generated to displace the oxygen. According to these studies, using water mist to protect machinery spaces from fire is a preferable option. The quantity of water mist test protocols that are written specifically for machine space protection supports this use. The industrial environment is constantly changing as a result of new technology, so it is crucial that fire protection techniques such as water mist adapt as well to account for these new risks.

3.2. *Power Generation Turbines*

Turbine enclosures are seen as another type of machinery space, except the potential fire sources differ. The fire source in a turbine enclosure is mainly limited to the turbine itself and the associated fuel line. Water mist suppression is well suited for protecting turbine enclosures for its ability to extinguish fire rapidly and efficiently while producing minimal thermal shock and corrosion damage, post-fire clean-up requirements, and negative environmental impacts. A requirement of the FM Global Test Protocol 5560 approval for turbine protection is that the cooling of the turbine casing will not result in damage to the turbine blades. To overcome this, manufacturers have incorporated the cycling discharge method, which sprays momentarily on–off pulses to reduce the rapid cooling of the casing and still extinguish the fire within the required time [25]. Liu et al. [33] and Liu and Kim [6,34] established that cycling mist discharges into an enclosure resulted in a faster extinguishment and used less total water when compared with continuous applications. The increase in the compartment temperature during the off cycle of the water mist allowed for the evaporation of suspended droplets and the growing fire to further consume oxygen in the compartment. Once the water mist on cycle was reintroduced, it produced further cooling and turbulent mixing. The cycle is repeated until extinguishment is successfully achieved. This requirement makes turbines one of the more difficult fire hazards to protect effectively. Cycling discharge methods for turbines should be tested for other applications and fire risks to determine if the benefits can be applied to other fire risks. Because power generation is an expensive process with numerous potential safety hazards, fire protection methods are critical to turbine operators and insurers. Water mist has proven to be an effective method of fire suppression, but it will require ongoing research and development as new methods of power generation and storage are established. Water mist must be suitable for electrical equipment in order to be used in the power generation industry.

3.3. *Electrical Equipment*

Halon and CO_2 gas agents are known methods for protecting electrical equipment from fires. There have been concerns regarding the use of water mist suppression as a replacement for halon and CO_2 because of the electrical conductivity of the water causing damage or creating a risk of shock to personnel. Through various studies, water mist has since been proven to be a feasible protection method for electrical equipment in computer rooms, datacentres, and electrical switch gear cabinets [35–37]. Fires within these environments are typically slow-growing, where burning plastic-coated cables and circuit boards spread corrosive smoke. These studies showed that water mist could effectively extinguish fires without introducing any major electrical leakage across circuits or creating any critical damage to the equipment. The operation of the equipment remained unaffected during and after the studies. Water also proved to be superior to gas suppression for its ability to cool hot cable fires and avoid the reignition of plastics [38]. These findings show that water mist can be used safely in various applications, including control rooms for vital machinery or data storage facilities. Water mist suppression can do this as an environmentally safer alternative to previously used gases, which contribute to ozone depletion. Many industrial facilities place a high priority on the protection of this equipment to ensure fire safety in their normal operations. The ability to control fire with minimal water damage to the protected area also makes water mist an attractive alternative to fire sprinklers.

3.4. Transportation

Water mist is commonly used across many forms of transportation systems because it requires minimal water storage and is lightweight compared with other alternatives. Water mist fire suppression has been tested and approved for use on passenger ships, cargo ships, freight aircraft, military aircraft, road vehicles, and heavy machinery vehicles. In the 1980s, many lives were lost onboard during passenger ferry fires, leading to the International Maritime Organization (IMO) regulating sprinklers on all ships capable of carrying 35 or more passengers [25]. Due to the complexity of adding sprinklers to existing ships not designed for the weight and additional space required for the system, there was a large push in research and development to find a system equivalent to sprinklers that could be fitted. Testing showed that the water mist system was more effective at extinguishing shielded fire when compared with conventional sprinklers and provided equivalent or better protection for small cabins [39]. The system should be designed specifically for each fire compartment and risk. Most maritime authorities worldwide will accept a cabin water mist system that meets the IMO test protocol.

The cargo area of ships presents a large fire hazard due to the dense fuel loads within the compartment. Large-scale tests conducted by the Technical Research Institute of Sweden found that roll-on–roll-off (ro-ro) cargo decks on ships could be protected using a water mist system, which provided an increase in flow rate when compared with traditional water spray systems [40]. The system was effective for direct applications, though the system could not extinguish shielded fires when covered by shelving.

Water mist has also been tested for cargo fires on large aircraft. Full-scale testing was conducted by the Federal Aviation Administration [41,42] to determine if water mist is an alternative to halon on aircraft cargo fires. These fires are typically deep-seated, involving a wide variety of fuels. The tests showed that for water mist to be effective it required a high-pressure discharge where it could control a fire for up to 90 min. The performance objective for aircraft cargo fires is to control the fire for 180 min to allow for a safe landing before evacuation [38]. Water mist has not yet been proven to be an equivalent alternative to halon fire suppression on cargo or military aircraft, and, therefore, there will be an exception for the use of this ozone-depleting gas in the aviation industry until an effective substitute is found that meets the lightweight suppression requirements for aircraft safety.

The performance requirements for ground transportation are less onerous, and water mist has been used as a protection method for transport trucks, military combat vehicles, and heavy industrial machinery such as mining trucks and equipment [38]. Water mist is used to suppress diesel vehicle engine fires in a similar way as it is to suppress large industrial engine room fires. This research has significance because every type of transportation vehicle requires low-profile fire protection equipment. The equipment cannot significantly reduce the amount of space that can be used for cargo or passengers and must be installed within aircraft, sea vessels, and road vehicles without adding extra weight that reduces travel distances and increases fuel consumption. Comparing water mist with other fire suppression techniques, such as sprinklers, it can be said that water mist belongs to the category of low-profile equipment. The world is moving towards more freight due to import trade and online shopping, which requires support from the fire protection sector in the form of ongoing research and development. In big cities with a growing population and commuter problems, transportation safety is a crucial issue.

Burn-over protection for firefighting vehicles used externally for bushfire (wildland fire) purposes in Australia has recently incorporated cabin and vehicle spray systems for the protection of occupants and could have wider applications using mist systems (AFAC, 2021). The role of misting systems would need to be assessed in the context of wind conditions before using finer mists than those applied at present.

3.5. Road and Rail Tunnels

Water-based fire protection was originally discredited for use in tunnels until a series of serious highway and rail tunnel fires occurred between 1995 and 2005 [25]. This led to the development of water mist systems for road tunnels being one of the most active research areas. Fire tests carried out in Switzerland [43] involved burning passenger cars arranged to simulate an accident involving three (03) vehicles in a tunnel. The activation of the water mist system was able to control the burning of the three (03) vehicles and keep the fire from spreading to the adjacent vehicle until the fire consumed all the fuels available within the three (03) controlled cars. As fuel tanks were ruptured, the water mist was able to easily extinguish the burning liquid fuel and cool it as it spread across the roadways Tunnel safety researchers claim that water mist also improves visibility within tunnels during fire scenarios and assists in safe evacuation [15,43]. The research is in furtherance of using water mist in road tunnels. As tunnel construction varies, it is crucial that it be used in each individual case. To prove its effectiveness in conjunction with other protection systems, such as extraction exhaust fans or fire detection techniques, research and testing need to be conducted. Current research on fire protection in tunnels is intense, and more studies on the use of water mist should be conducted.

3.6. Nuclear Power Generation

Fire protection systems for nuclear powerplants are designed with the concept of defence-in-depth. This involves fire prevention, rapid detection and suppression, and safe shutdown capability [40]. Ha et al. explained that water mist fire extinguishing systems are a very good candidate for protecting safety shutdown equipment in fire compartments because the application of water spray needs to be restricted [44]. Several recent studies were conducted to apply water mist systems to nuclear powerplants, typically in emergency diesel generator rooms or the electrical rooms of a reactor [9,13,45]. The findings were similar to existing research on water mist suppression in power generation and industrial occupancies. Nuclear power is often considered when discussing alternative energy generation methods. It is critical that fire suppression is performed safely if nuclear power is to be considered for use. Water mist fire protection should be investigated if it can improve nuclear powerplant safety. It may be useful for other aspects of nuclear power safety.

3.7. Explosion Hazard Mitigation

Hazardous explosion areas created by airborne gases can become extremely dangerous if ignited [17]. Water mist discharge into an unconfined environment with high momentum can entrain air into a gas cloud and dilute it below its lower flammable limit, reducing the potential for ignition and explosion. If ignited, during the explosive ignition of a gas cloud with a high flame speed and a shockwave, the spray/mist can reduce deflagration pressure and possibly extinguish the flame. The shockwave breaks up the droplets into a micro-mist that can quickly evaporate to absorb a significant fraction of the combustion energy released during the deflagration. To be effective, water mist must be produced at a smaller droplet size of around 1 μm, which is smaller than what is seen to be achievable for the existing water mist products for fire suppression that have been reviewed [46,47] During this literature review, there was little research found on this topic. It appears to be an intriguing application that should be further developed.

3.8. Cooking Areas

Commercial and industrial cooking oil or fat fryers are difficult to protect from fire due to their high temperatures and ability to reignite easily [45]. In order to extinguish the fire, the entire surface of the oil must be suppressed simultaneously with the rapid cooling of the oil to prevent reignition [45]. Due to the chemical additives consistent in most fire suppression systems not being appropriate for the food production industry, there are limited options for protection. A common method is carbon dioxide (CO_2), which can extinguish flames across the oil surface but is not able to cool the oil sufficiently and prevent reignition. Water mist has demonstrated good fire extinguishing and oil cooling capabilities for these fires [6,46–48]. Specifically, research by Liu et el. in 2006 [45] found that water mist effectively extinguished a large industrial oil fire and cooled the hot oil to prevent it from reigniting regardless of the oil depth. According to our literature review, water mist for oil fat fryers appears to have a significant advantage over alternative methods (such as foams). It is clearly a strong option because there are minimal methods for suppression that are acceptable for food processing. More food processing facilities are being built in metropolitan areas to allow for quick delivery to consumers, as the popularity of mass production of "ready-to-eat meals" has grown in recent years. Large fryer converters require fire protection in these facilities. Water mist should be researched further to see if it can be used on these conveyor-type fryers to protect the equipment and the people working around them. The primary motivation for fire protection is to protect people, so it is crucial to investigate the use of water mist in commercial structures as well.

3.9. Commercial Buildings

Water mist systems have been developed and used in commercial buildings, with FM Global and Underwriters Laboratories (UL) having test protocols for Light Hazard and Ordinary Hazard I and II, which are commonly protected by sprinkler systems. A water mist suppression system's ability to suppress a fire and maintain minimal water damage to the protected area makes it an attractive option for protecting areas such as libraries, archives, heritage buildings, and museums [21,46,49,50]. Milke and Gerschefski [51] completed fire testing for library settings with shelving-type storage and book vaults for sensitive archives. A total compartment application was able to control the fire damage to the documents and extinguish shielded fires using 30% of the amount of water compared with a traditional sprinkler system. No damage occurred due to water, and only a minor amount of discolouration of documents was observed from smoke if stored closest to the shelf flue near the fire. It was noted that the water mist could not control the fire without discharging within 100 s of ignition. Sprinklers have been proven to be a reliable fire protection strategy in commercial structures. Before the industry switches from using standard sprinklers, the advantages of water mist systems over sprinklers would need to be further proved. Although more testing needs to be performed, water mist suppression has potential as an alternative.

3.10. Residential Buildings

Water mist can be suitable for residential occupancies in areas where water supplies are limited [52]. Fire testing on water mist in bedroom and kitchen fuel fires has found that water mist can control a fire with up to half the water required for a typical residential sprinkler [50,53,54]. The use of water mist as an alternative to sprinklers is limited by its reliability and cost [38]. Water mist systems, in combination with other life safety systems such as smoke detection, fire alarms, and passive systems, have been used in residential buildings [7]. These systems are used to mitigate the risk of life loss, protect areas where occupants are housed, increase tenable conditions where extended emergency exit travel distances are present, and protect high-value items or historic features of the building [54]. The use of these systems in residential buildings is limited by the information available for experts and approval authorities to apply and review. Aspects such as reliability, real fire historical data, and long-term maintenance requirements are largely unknown. There is a need for more research and development to enhance the understanding of these systems and how they can be used in all applications.

A further consideration is the protection of external building elements arising from bushfire events. For this application, consideration will need to be given to the "design bushfire" conditions, including ambient wind, when considering droplet sizes. The location and design of nozzles may need to be further researched in the use of misting systems for "real-world" bushfire protection purposes.

4. Research and Development on Water Mist Systems

Water mist suppression research can be completed using various methods including real fire testing experiments and computational numerical modelling. The literature on water mist systems consists of various experiment methodologies, including small-scale and large-scale fire testing. Small-scale experiments are typically conducted using minimal resources. They can be run in a timely fashion and repeated easily to obtain fast results compared with large-scale experiments. A typical example relating to fire water mist experiments would a laboratory-controlled test using one spray nozzle to suppresses a small pool fire. Large-scale experiments can be very costly, require copious amounts of planning and coordination, and are typically difficult to repeat. For example, completing a fire suppression test inside an aircraft hangar or warehouse building using multiple water mist nozzles discharging at the same time. These are enclosure fire tests, which are also recommended by different standards [50,55] and can be used for commercial and residential buildings. The studies reviewed here have been a combination of small-scale and large-scale experiments. The studies found in the literature have identified that the primary design parameters of a water mist system are the spray dynamics of the discharge nozzle and the compartment in which the system is installed. The research findings in the reviewed literature have been summarised in Table 2 by categorising the key system parameters and are further discussed in this section.

Table 2. Experimental and numerical research on water mist suppression.

Key Parameter	Findings	Methodology
Suppression additives	• Extinguishing performance improves with the addition of 0.2 multi-component (MC) additive. [18] • Pure water failed to extinguish the fire in 37.5% of test scenarios, whereas 0.2 MC additive only failed in 30%. [18] • When firefighting additives are introduced, the effectiveness of water mist is significantly improved. [7]	E E E
Operating Pressure	• An increase in discharge pressure has a positive impact on the extinguishing performance of the finer mist system and a negative effect on the courser mist system. [45] • In oil cooker testing, the faster extinguishment time was at a discharge pressure of 689 kPa. [45] • Extinguishing performance was determined by the type of water system, discharge pressure, and hood position for oil cooker fire testing. [45] • High water mist operating pressure results in better mixing of the air and fuel vapour. [52] • A high discharge pressure of 7000 kPa introduces a peak in HRR and CO_2 concentration before extinguishing the fire faster than the operating pressure of 4000 kPa [56]. • Fire suppression time decreased by increasing gas pressure to 8 bar for twin-fluid systems. [57] • An increase in operating pressure could increase the suppression performance for the lower discharge densities. [40,45,52,58]. • The high-pressure mist was better at reducing the convective heat release rate compared with water spray for the shielded fire [40]. • Nozzles with the highest operating pressure provided the fastest fire control [40,52]. • There is a limit in which fire cannot be extinguished if pressure is too low. [59]	E E E E E E E,E,E E E E

Table 2. *Cont.*

Key Parameter	Findings	Methodology
Water Droplet Diameter	• An inflow zone is created where the air is sucked into the spray's low-pressure centre, transporting large droplets at low velocity to the centre of the spray [19,25]. • A transition zone occurs where spray collapses and a large spread in droplet sizes is seen [19,25]. • A turbulent zone occurs with a uniform droplet and velocity distribution across the spray [19,25]. • Critical cooling is achieved at two optimum droplet ranges of 100–250 micrometres and again at 550–1000 micrometres. The range between 250–550 micrometres is not small enough to rapidly evaporate and displace oxygen nor are the droplets big enough to penetrate and cool the flame [19]. • The presence of a solid wall in the vicinity of the spray influences the distribution of water droplets [29]. • Fire extinguishment time was reduced by decreasing the droplet size to 50 micrometres for a large warehouse simulation [60]. • Systems with a smaller droplet size of 100 micrometres had better performance regarding extinguishment time [9,61]. • Water droplet diameter decreases with a decreasing water flow rate and airflow rate for the twin-fluid system. [62]	N,E N,E N,E N E/N N E/N,E E
Flow Rate	• Suppression time decreased as the flow rate was increased for ceiling injection [40,52,63]. • Suppression time increased as the flow rate increased for side wall injection [52,63]. • The liquid gas mass flow rate is increased with increasing gas operation pressure, while the liquid mass flow rate is exponentially decreased with the increasing gas–liquid flow for twin-fluid systems [56,62,64]. • Fire suppression performance is best at an optimal flow rate between 1–1.7 L/min. In the same way, there is an optimal flow rate that will not suppress the fire around 0.6–1 litre/min [25,52,62]. • Fire suppression can be improved by decreasing the flow rate and the spray angle and increasing the water discharge duration [5,52]. • The restraining effect of the flashover and the fire improved with the increase in the water mist pressure and flow rate [63]. • The dominant factor affecting the decrease in heat flux is the flow rate [63]. • Fire suppression time decreases as a result of increasing water mist flow rate injection for some experiments. • Water droplet diameter decreases with decreasing water flow rate and increased airflow rate [62]. • The velocity of the spray decreased with a decrease in the supplied air flow rate for twin-fluid systems [62].	E,E,E E,E E,E,E E,E,E E,E E E E E
Spray Angle	• Water mist with a large spray angle and finer droplets had better performance for oil hood cooker fires [45]. • Fire suppression can be improved by decreasing the flow rate and the spray angle and increasing the water discharge duration [5,25,45]. • Too small spray angles result in large flux intensity. Too large spray angles, such as 120 degrees, can be inefficient in fire suppression [45]. • The large orifice/droplet size typically resulted in a large water spray angle, contributing to the radiation attenuation performance [25].	E E,E,E E E

Table 2. *Cont.*

Key Parameter	Findings	Methodology
Nozzle Height	• Suitable configurations for nozzles lead to greater suppression performance [65]. • The water droplet diameter increased at a greater distance from the vertical axis of the nozzle [6]. • Greater aerodynamic drag exists at a greater distance from the vertical axis of the nozzle. Velocity increase is minimal in this area [65]. • The position of the flame as droplets hit the flame played a role in the extinction behaviour [66]. • Extinction time was higher for a higher vertical distance of nozzles [12,64,67]. • Extinction time was increased with the increase in the pool fire size and was unsuccessful for the largest pool fire tested [64].	E E E E E,E,E E
Compartment Size and Ventilation	• Suitable configurations for nozzles lead to greater suppression performance [64] • Changing the aspect ratio and the opening ratio of the door opening affected the maximum value and reduction rate of the heat release rate due to the flame displacement effect [9].	E E/N

Note: E—Experimental research involving fire testing; N—Numerical research involving a Fire Dynamic Simulator (FDS).

4.1. Spray Characteristics

The ability of a water mist system to extinguish specific fire types is affected by water mist additives, nozzle discharge pressure, water droplet size, flow rate, nozzle K-factor spray momentum, and spray angle.

4.1.1. Effect of Different Water Mist Additives

Additives alter the viscosity of the suppression agent and, ultimately, the spray pattern and suppression performance. Different additives have been tested with water mist to achieve different objectives [17]. Additives, such as aqueous film-forming foam (AFFF), showed improved effectiveness for hydrocarbon pool fires [48]. Additives can be used to avoid the freezing of storage water in colder climates and have been found to improve the effectiveness of systems [68–70]. Adding nitrogen and other inert gases to the water mist can increase firefighting capabilities, leading to the development of twin-fluid systems [71,72] Cong and Liao [18] studied the effects of suppressing liquid pool fires (Figure 8) with three separate firefighting additives: sodium bicarbonate ($NaHCO_3$), multi-component agent (MC), and aqueous film-forming foam (AFFF).

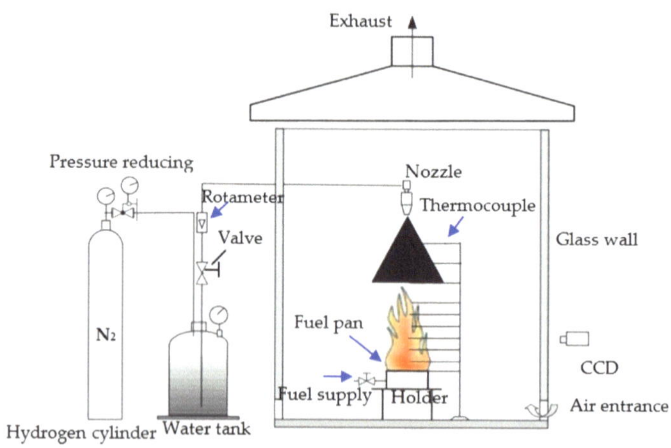

Figure 8. Schematic of liquid pool fire with water mist system additives experimental apparatus [18].

Each additive proved a slight increase in the suppression time compared to pure water. Figure 9 (a-b) shows the greatest improvement due to liquid additives was the AFFF when suppressing hydrocarbon fires such as diesel from the foam, forming a thin layer of fluorinated surfactant on the liquid pool surface. Fluorinated firefighting foams, such as PFAS, being harmful to human health has led to their ban in most of the world. Foam and other additives are not commonly found in water mist suppression systems as they typically rely on water as the suppression medium, and its performance is controlled by other parameters such as the discharge pressure.

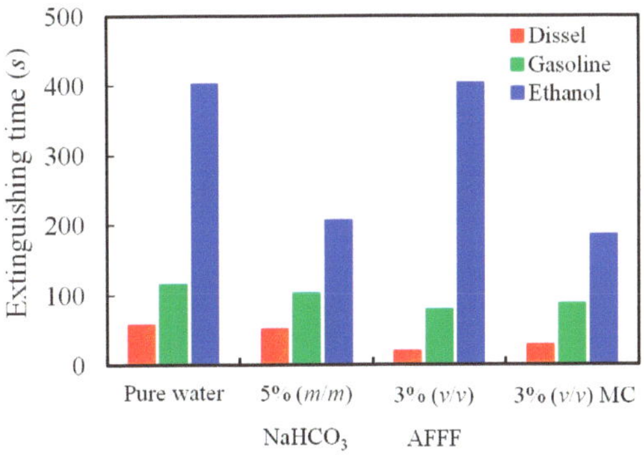

(**a**) Effect on extinguishing time

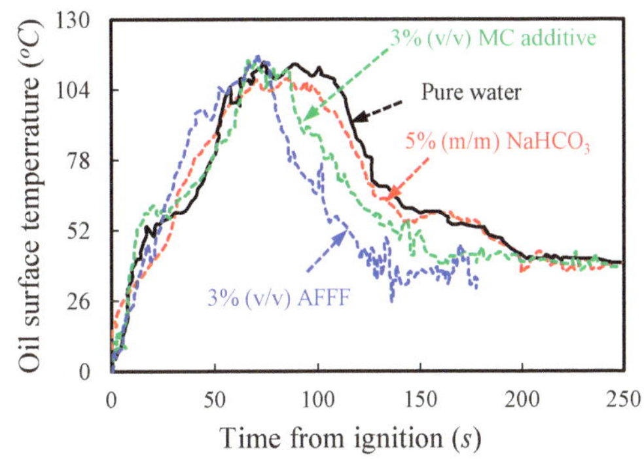

(**b**) Effect on oil surface temperature

Figure 9. Liquid pool fires suppressed by water mist with additives [18].

4.1.2. Effect of Different Operating Pressure

Yinshui et al. [20] and Lui et al. [29] each suggested that an increase in the operating discharge pressure of water mist nozzles contributes to the reduction in the fire extinguishment time. One key reason is that the higher operating pressure of the water mist system results in a stronger penetration into the flame zone. This contributes to the displacement of flammable mixtures in the flaming region for better suppression and reduces the risk of backdraft. The increase in pressure also results in a decrease in the water droplet size, as shown in Figure 10. It is known that a smaller water droplet size is one of the main drivers of water mist suppression performance.

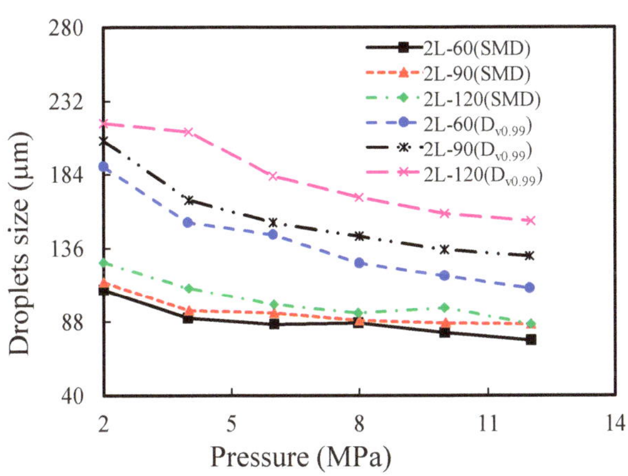

Figure 10. Discharge pressure effect on water droplet size [20].

4.1.3. Effect of Different Water Droplet Sizes

The effect of water droplet sizes has been investigated by different researchers [19]. It has been noted that water droplet size is the main contributor to water mist system performance [19,45]. As droplet size is reduced below 1000 µm in diameter, it significantly improves extinguishment time, as shown in Figure 11.

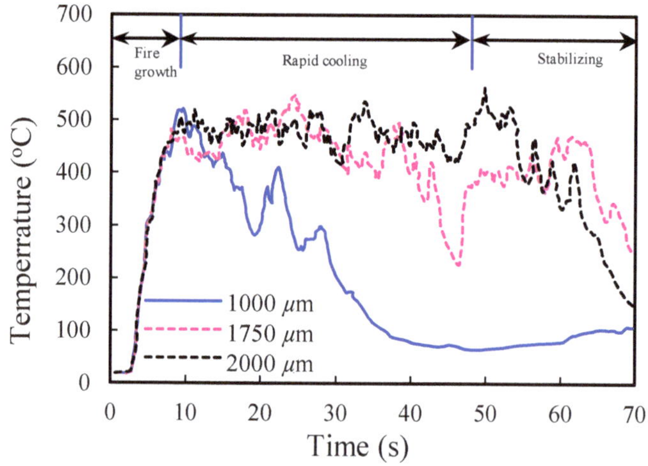

Figure 11. Time histories of temperatures for various spray nozzle droplet sizes [29].

Rapid cooling is the result of a large amount of evaporation generating steam. This steam contributes to the O_2 displacement and radiation attenuation, which are water mist suppression mechanisms. Figure 12a highlights the O_2 concentration reduction when comparing water droplet diameters of 100 µm and 1000 µm. Ferng [19] observed the critical cooling time for different water mist droplet sizes between 100 µm and 1000 µm and found that performance was better at either end of the scale, as shown in Figure 12c. The effect of droplet size on the temperature reduction and extinguishment time is shown in Figures 12b and 13 to provide examples from different experiments. For a smaller water droplet size closer to 100 µm, the fire is extinguished by rapid evaporation, causing O_2 displacement and surrounding gas to cool. On the other end of the scale, closer to 1000 µm in regions II and III, the increased droplet size has greater mass and momentum to directly penetrate the flame region before evaporating. This phenomenon directly cools the burning fire surface and can provide a similar critical cooling time compared with other suppression mechanisms. This finding highlights that it is important for a water mist system to achieve an effective droplet diameter close to 100 µm. The cooling from the flame zone penetration achieved in regions II and III from the momentum can also be achieved for finer droplets if the spray velocity is increased, in other words, the nozzle's flow rate.

The disadvantage of smaller droplet sizes is related to the effects of wind conditions, which can be an important consideration in the external environment or during turbulent, fire-generated wind conditions. The location and application rates may be related to specific components of a building, for example, outside of enclosed spaces.

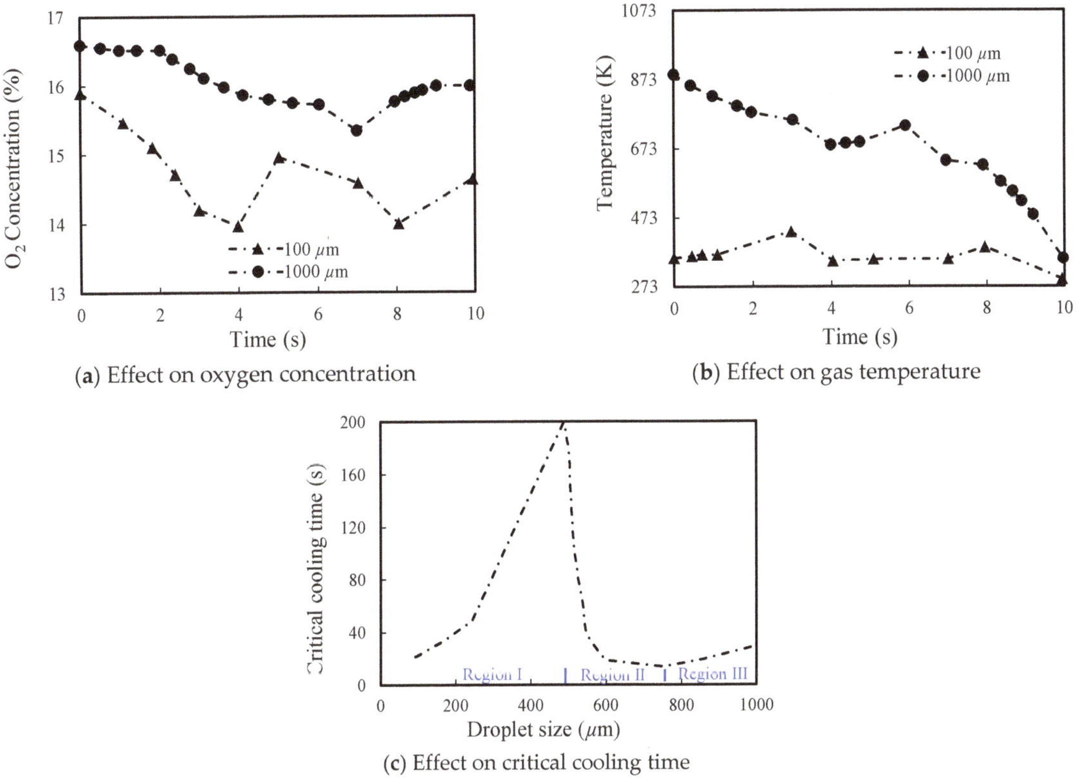

Figure 12. Effect of various water mist droplet sizes [19].

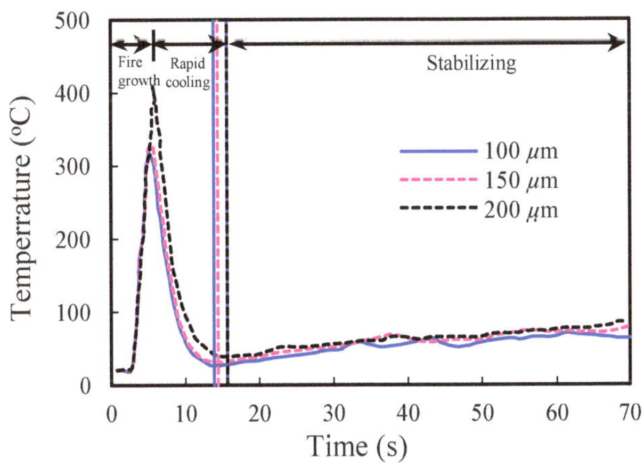

Figure 13. Time histories of temperatures for various water mist droplet sizes [29].

4.1.4. Effect of Different Flow Rates

The water mist flow rate affects the discharging ability to suppress a fire. Yinshui et al. [20] and Jenft et al. [49] conducted separate experiments and modelling. It was reported that the fire suppression time decreases because of an increasing water mist flow rate. These observations, found during water mist testing, support earlier findings from Nam [73], who completed a numerical simulation of the penetration capability of sprinkler sprays. It was observed that the penetration ratio of droplets increased as the flow rate increased, as illustrated in Figures 14 and 15.

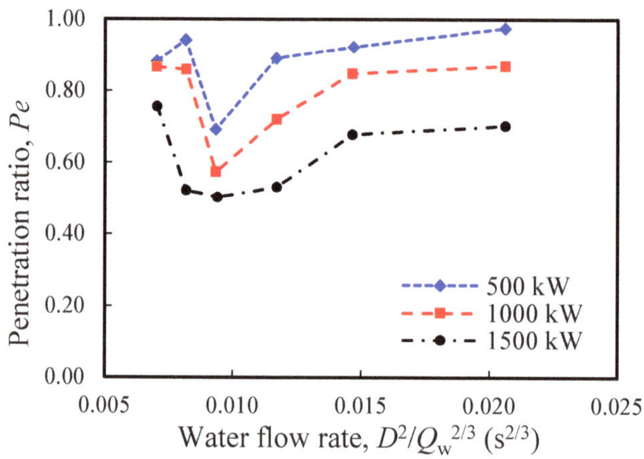

Figure 14. Penetration ratio vs. water flow rates under 6.1 m ceiling [72].

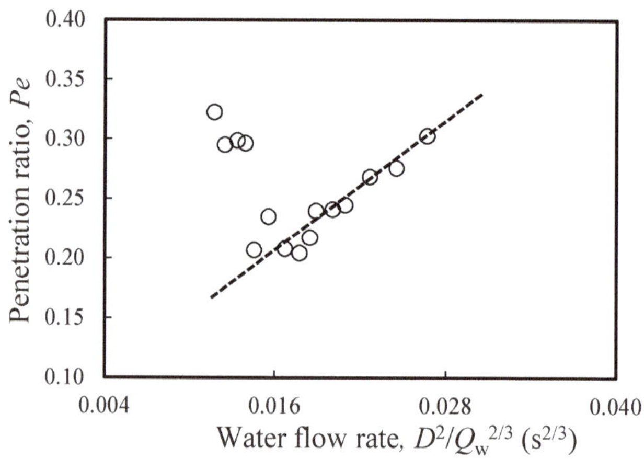

Figure 15. Penetration ratio (Pe) vs. water flow rates [72].

Nam [73] investigated sprinkler penetration capabilities with multiple factors by conducting steady-state simulations, which suggested that an optimal flow rate exists from any given ceiling height to provide the highest penetration ratio in a practical flow range It is expected that an increase in flow rate will result in great suppression performance, as the flow rate is linked to the operating pressure, which has been shown to improve suppression. These results highlight the importance of designing water mist suppression systems with appropriate flow rates. Water quantity contributes to flame cooling, which has been established as one of the extinguishing mechanisms of water mist. The relationship between nozzle operating pressure and the discharge flow rate of a water mist nozzle can be described by the K-factor of the nozzle [5,12,40,74,75].

4.1.5. Effect of Different Nozzle K-Factors

The K-factor of a spray nozzle is defined as the discharge coefficient. It relates to the amount of water permitted through the nozzle at a given pressure. Equation (4) is used to determine the nozzle pressure from the set K-factor in relation to the flow rate [76].

$$P = \left(\frac{Q}{k}\right)^2 \quad (4)$$

where P is the pressure at the inlet of the nozzle (bar), Q is the nozzle flow rate (L/min), and k is the discharge coefficient K-factor $\left(\frac{\frac{1}{min}}{\sqrt{bar}}\right)$.

It has been previously found that the leading suppression mechanism of a water mist system is the O_2 displacement that comes from a smaller droplet size, achieved with a higher operating nozzle pressure. Equation (4) shows that the pressure and discharge coefficient are inversely proportional; the spray momentum provides a K-factor for any given nozzle, highlighting the importance of nozzle selection when designing a water mist system.

4.1.6. Effect of Different Spray Momentum

The spray momentum relates to the mass, velocity, and spray direction relative to the fire [73]. The suppression system's ability to penetrate the fire plume with water mist droplets depends on the spray momentum. Additionally, it determines the entrainment rate of the surrounding air into the fire plume. High momentum effectively cools the flame and fuels and causes faster steam generation, which results in oxygen displacement, whereas low momentum can achieve water droplet suspension in the fire compartment, contributing to radiation attenuation [40]. Due to the water vapour mist in the combustion zone, the spray mass is a combination of both the mass of the liquid-phase water and the mass of the vapour-phase water and air entrained by water mist. Lui and Kim [6] developed an equation for spray momentum, which can be expressed in Equation (5).

$$M_w = (m_{wl} + m_{wv} + m_{wa}) \times V_w \quad (5)$$

where M_w is the spray momentum; m_{wl} is the mass of the liquid phase water; m_{wv} is the mass of the vapour-phase water; m_{wa} is the mass of the air entrained by the mist; and V_w is the velocity vector of the water mist.

Nam [73], Santangelo and Tartarini [77], Pokorný et al. [78], and Sikanen et al. [79] have focused their research on water mist spray momentum. Experiments show that water mist spray momentum varies depending on the distance from the spray nozzle, as the accumulated mass is shown to reduce as distance increases, as shown in Figure 16. These findings agree with Nam's [73] earlier research showing that the spray momentum also decreases further away from the spray nozzle, as shown in Figure 17. It was also found (as shown in Figure 18) that, most of the time, water mist's ability to penetrate the flame increases as the spray momentum increases. The water mist's ability to penetrate the flame helps the water cool the flame and extract heat. Heat extraction has been shown to be one of the main suppression mechanisms of water mist and explains why spray momentum is a key design parameter in ensuring the system can penetrate the flame and aid heat extraction. For all three cases (500 kW, 1000 kW, 1500 kW), the penetration ratio clearly increases when the spray momentum ratio is increased. The spray momentum and flame penetration are directly affected by the spray angle of the discharge nozzle [20].

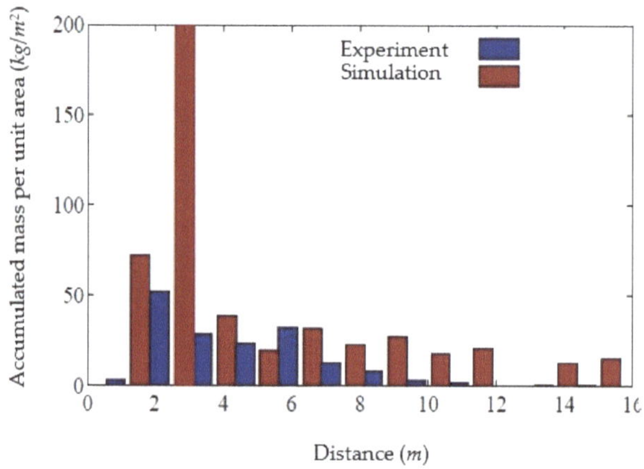

Figure 16. Accumulated mass relative to distance from the nozzle in a large-scale hall experiment [79].

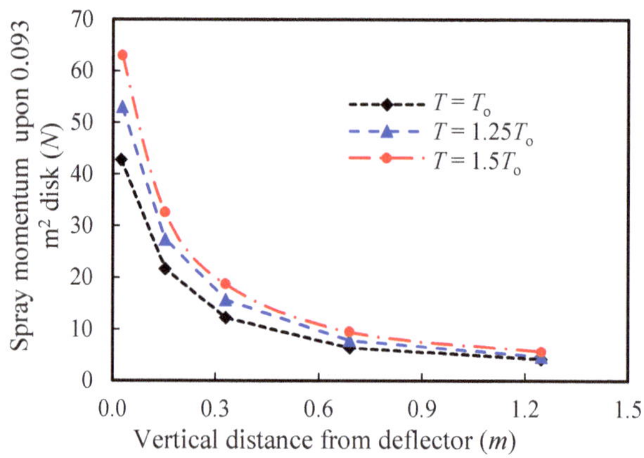

Figure 17. Spray momentum effect at 4.42 L/s [73].

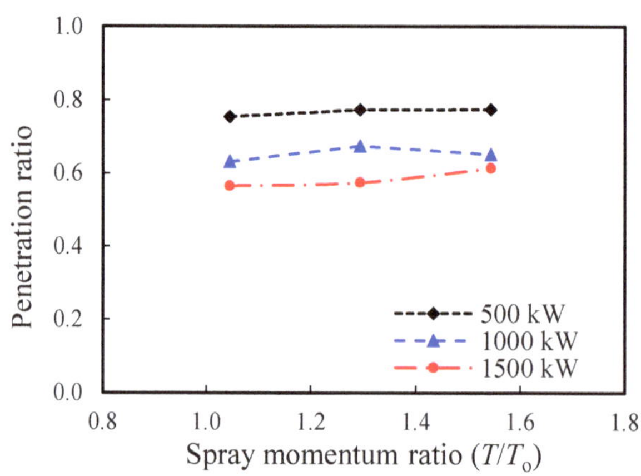

Figure 18. Spray moment effect at 3.161 L/s [73].

4.1.7. Effect of Different Spray Angles

The spray angle of a nozzle (θ) is measured from the central axis of the nozzle, as shown in Figure 19, and is important to water mist system performance. Studies focused on the effect of different spray angles have determined that a small nozzle spray angle provides a faster extinguishment time, as shown in Figure 20, in experiments completed at three different angles [20].

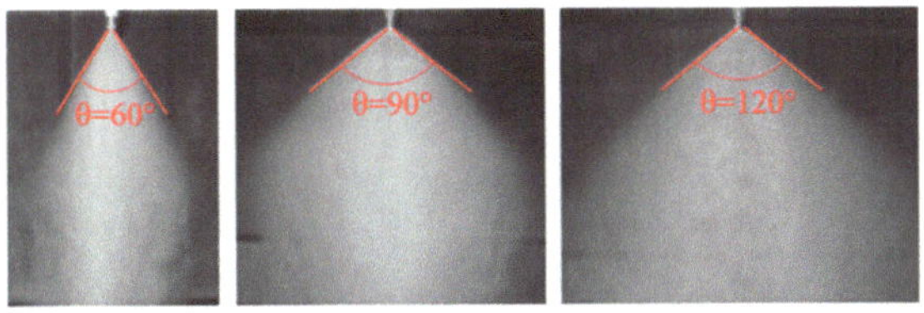

Figure 19. Nozzle spray angles for fire testing [20].

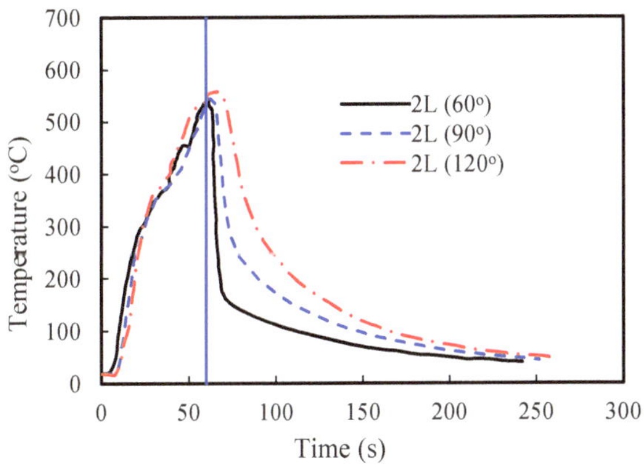

Figure 20. Time vs. temperature for different water mist nozzle spray angles [20].

It can be observed that the smaller spray angle performs better because the buoyancy of the hot smoke layer is reduced and suppressed by the injection of water mist within a very short time [7]. Without the direct injection from the smaller spray angle, the water is unable to effectively penetrate the buoyant hot layer, explaining why the large spray angle is less effective. To overcome this, it has been suggested that when the spray angle is increased, the flow rate needs to be adjusted simultaneously to maintain suppression performance [20]. Given what can be understood from the reviewed literature, this inability to effectively penetrate the buoyant hot layer with a large spray angle would also be expected if a nozzle's proximity to the flame were extended by increasing the nozzle height and other compartment variables.

4.2. Fire Compartments

Water mist suppression performance is not only affected by the design parameters of the system but also by the arrangement of the fire compartment it is installed in. The system's ability to cool the flame and fuel surfaces, displace the oxygen in the room, and maintain radiation attenuation depends heavily on the compartment's ceiling height, enclosure size and ventilation, and the dynamic mixing of gases, which are discussed in the following subsections.

4.2.1. Effect of Different Ceiling Heights

The literature available on the effects of different ceiling and nozzle heights shows that water mist is able to control fires, though its performance decreased with an increase in ceiling height [12,67,75]. It has been suggested that this performance reduction is due to the water mist penetration capabilities being reduced, which cannot sufficiently counter the fire plume. Testing has shown that deluge-type water mist is effective up to a height of 9.1 m and with a closed head system of only 6.1 m [67]. The testing of nozzles located close to the fuel source resulted in cooling contributing to the reduction in flame size, and the spray from the water mist was able to block the movement of the fuel between the fire sources [75]. An increase in ceiling height introduces an increase in the enclosure volume which has been shown to alter suppression performance [80]. Since sprinkler systems have similar properties, early research on sprinkler systems completed by Nam [73] also suggested that penetration capabilities are reduced as the ceiling height increases. This research shows how ceiling height is an important factor in water mist fire suppression performance and why it is an important design parameter that must be considered.

4.2.2. Effect of Different Enclosure Volumes and Ventilation

Enclosure volume effects on water mist suppression are similar to ceiling height. It is understood that the increase in a fire compartment's volume reduces the effectiveness of water mist in fire suppression [7,80,81], though there is little effort or research dedicated to finding ways of overcoming this. The testing completed by Bill et al. [8,12] showed that water mist was unable to suppress certain fire designs when there were no boundary walls, regardless of the number of nozzles added to the system. The maximum size of the compartment in which the system could extinguish fire was 940 m^3. Figure 21 shows the test results when the fire was extinguished.

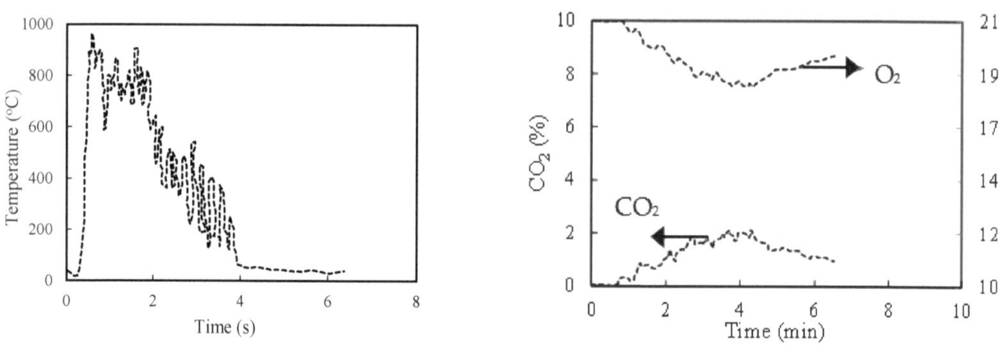

(**a**) Effect on flame temperature

(**b**) Effect on gas concentration

Figure 21. Effect of water mist on a 6 MW diesel-spray fire inside a 940 m^3 compartment [12].

The O_2 concentration reached a low of only 18.5% at the time of extinguishment, indicating that the main suppression mechanism contributing to the fire extinguishment was the cooling effect, not the oxygen displacement. This was confirmed in subsequent tests [8,12] when the fire was shielded from the cooling effect of the water, and extinguishment could not be achieved with only the displacement of oxygen. An enclosed compartment with minimum ventilation allows for a continuous reduction in oxygen, resulting in the timely extinguishment of fires even if they are shielded from the water spray. Inside an enclosed compartment, hot gases from the fire accumulate near the ceiling, and when water mist is discharged downwards into the hot gases, maximum water vaporisation is achieved This also pushes the combustion products down towards the floor, which is known as dynamic mixing.

4.2.3. Effect of Different Dynamic Mixing Situations

The dynamic mixing of gases in a fire compartment due to water mist discharge aids fire suppression through the reduction in the oxygen concentration in the lower portion of the compartment. It also increases the convective mixing of mist, water vapour, and combustion gases near the fire, resulting in the enhancement of the mist's extinguishing capacity [82]. Studies have shown that suppression is less effective when nozzles are located lower within the compartment, below the hot layer of gases [34,83]. Nozzles closer to the ceiling have the ability to push and mix the gases back down into the combustion zone. Knowledge of dynamic mixing has led to the concept of cycling discharge, where water mist is cycled on and off to achieve a faster extinguishment time in some cases. Combustion gases rise to the top of the compartment quickly when mist is cycled off and then are mixed back into the compartment when cycled on. This force continues the dynamic mixing of water vapour and combustion gases near the fire [57]. Figure 22a,b, from fire tests conducted by Lui et al. [57], shows that discharge cycling reduced extinguishment time, which was attributed to the higher depletion and dilution rate of oxygen caused by dynamic mixing. As previously discussed, the dilution of oxygen is one of the leading mechanisms in fire extinguishment from a water mist system. This research shows that dynamic mixing created from discharge cycling may significantly improve the performance of a water mist system. This could be applied to existing water mist products available for use.

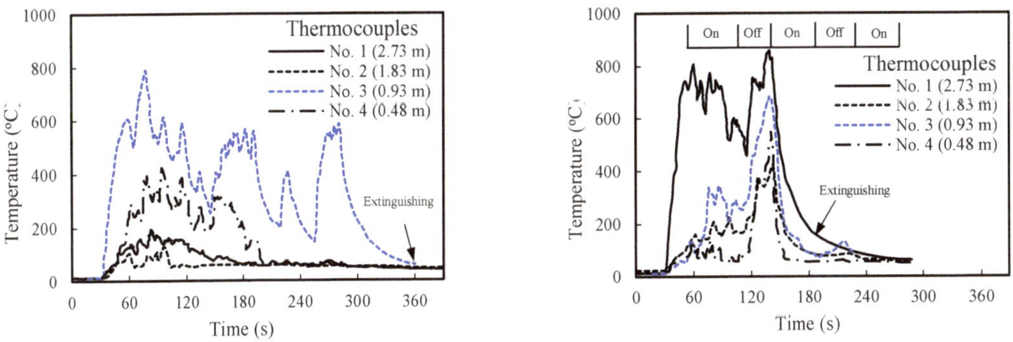

(a) Continues water mist discharge (b) Cyclic water mist discharge

Figure 22. Temperature over a round-pan fire with different water mist discharge rates [57].

4.2.4. Effect of External Environment and wind conditions

In the external environment, ambient conditions such as local wind may influence the dispersal and effectiveness of misting systems. Studies of bushfire–wind interactions arising from "modelled" bushfire events provide a basis for the consideration of droplet size and, hence, the effectiveness of misting systems for external conditions [84]. These investigations found that turbulent winds can be enhanced by the effects of a fire event, increasing pulsating wind speeds. These conditions dictate that roof-mounted nozzles will be less effective than those nozzles placed closer to the ground and can be blown onto the surfaces of critical building elements. Existing standards for bushfire spray systems [85,86], whilst providing a starting point, do not address the limitations of nozzle location and design in the development of bushfire protection systems. The volume of water storage, usage, and application are crucial when considering enhanced bushfire protection, potentially giving misting systems an advantage over traditional drenching systems.

4.3. Current Products for Water Mist Systems

A number of water mist systems available on the market have been reviewed to understand their designed applications, limitations, and operating parameters. Water mist systems can be categorised by their operating pressure as either low-pressure (<12.1 bar), intermediate-pressure (12.1 to 34.5 bar), or high-pressure (>34.5 bar) [8]. The key parameters of the existing water mist products on the market are summarised in Table 3 and compared with the parameters found in previous research. Most of the existing water mist products have either low operating pressure or high operating pressure, which are discussed in the below subsections. The purpose of this section is to highlight the limitations of different water mist systems available on the market. This is important for future research that could lead to the improvement of these systems or the development of new types for different applications.

Table 3. Summary of various water mist parameters from existing research and available products.

Author/ Manufacturer	Operating pressure (bar)	Flow Rate (lpm)	K-Factor (lpm/bar$^{0.5}$)	Spray Angle (°)	Droplet Diameter $D_{v0.9}$ (µm)	Nozzle Height (m)
Published Research System Parameters						
Kim et al. 2003 [81]	13	6	1.66	70	121	1.8
Yang et al. 2010 [87]	100	11.9	1.19	48	46	2.5
Vaari 2012 [88]	70	6.45	0.77	30	200	4
Jenft et al. 2014 [49]	10.13	6.3	1.97	65	112	2.45
Yinshui et al. 2014 [20]	2–12	2–5	-	60–120	80–130	2
Lee et al. 2019 [13]	10	22.45	7.1	76	124.6	2
Ha et al. 2021 [44]	10	22.8	7.25	90	124.6	12.5
Industry-Used System Design Parameters						
Aqua Mist ULF [21]	12.8	12.5	3.5	-	100–200	8
VID Fire Kill APS [89]	10	32	10.3	-	<300	Unlimited
Aqua Mist FOG [90]	100	4.4	0.44	-	50	5
Marioff HI-FOG [46]	40	7.3	1.15	45–68	50	5
SEM-SAFE [47]	100	39	3.9	-	10–50	5.5

4.3.1. Low-Pressure and Intermate-Pressure Systems

Tyco AquaMist ULF

The AquaMist Ultra Low Flow (ULF) is marketed as a low-pressure water mist system and total flooding system. It is designed to operate at working pressures of 7 to 16 bar and to produce droplets of 100 to 200 µm in diameter [21]. This water mist system has various configurations and available nozzle types for different applications and can either be installed with a pumped water supply or with stored water and pressure cylinders, as shown in Figure 23a,b. It is UL- and FM Global-listed for applications in machinery spaces and on machinery components, oil cookers, and combustion or steam turbines. Depending on the nozzles used, the system's application is restricted at various ceiling heights, the largest possible height being 10 m. Tyco also offers a high-pressure water mist system.

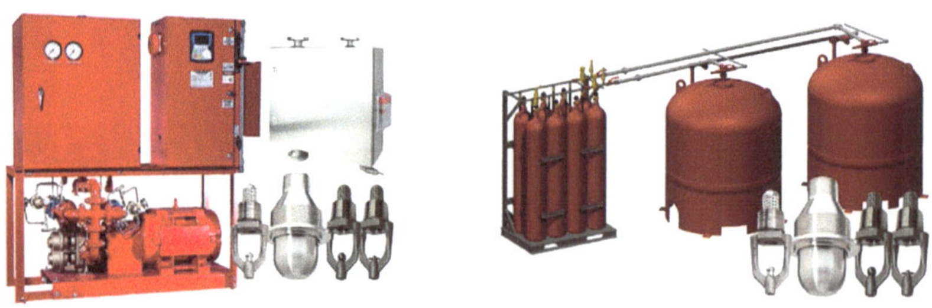

(a) (b)

Figure 23. Typical components of an AquaMist ULF's pumped water supply and stored pressure configuration. (**a**) ULF pumped water supply configuration [91]; (**b**) AquaMist ULF stored pressure system [21].

VID FireKill APS Atrium System

FireKill has designed a product for use in atriums and other large rooms [89]. The open nozzle system is designed to operate with a pressure of 10 bar and to be discharged horizontally from the wall. A droplet size of less than 300 μm is discharged from either side of the fire compartment, as shown in Figure 24. This system differs from many others due to the fact that there is no limit on the ceiling height of the atrium, though its listing is still limited by the width of the walls at up to 26 m.

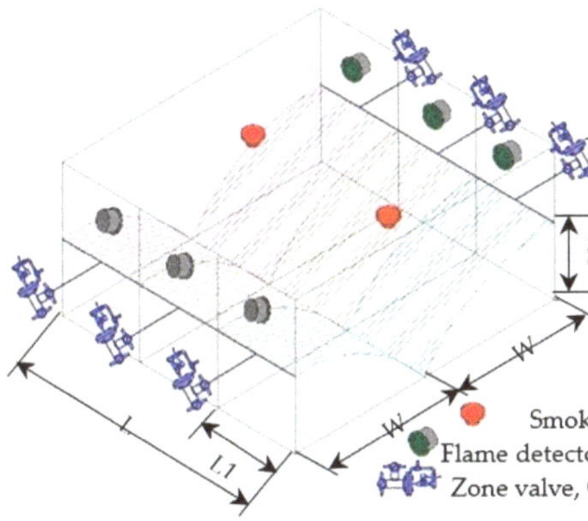

Figure 24. VID FireKill APS Atrium System general arrangement schematic [89].

4.3.2. High-Pressure Systems

Tyco AquaMist FOG

The AquaMist FOG is a high-pressure water mist system that operates at 50 to 200 bar to produce droplets of an approximate diameter of 50 μm [21,92]. This water mist system also has multiple nozzle types suitable for each application and can be either pumped or used with stored pressure cylinders, as shown in Figure 25. This water mist system is FM Global-listed for use in machinery and combustion turbine enclosures, and it is restricted to a volume not exceeding 260 m^3 and a maximum ceiling height of 5 m. This system is similar to Kidde's high-pressure Marioff HI-FOG system.

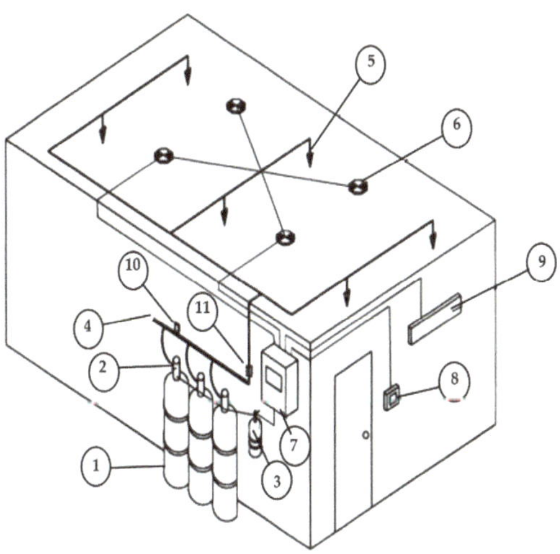

Figure 25. Typical components of an AquaMist FOG stored pressure general arrangement [92].

Marioff HI-FOG

Marioff HI-FOG is a high-pressure water mist system designed for total compartment protection, with a minimum operating pressure of 40 bar, and to produce water droplets of approximately 50 μm [90,93]. This water mist system is FM Global-listed for use on combustion turbines, machinery paces, and special hazard machinery spaces. Figure 26 shows the layout of this water system for turbine enclosure protection.

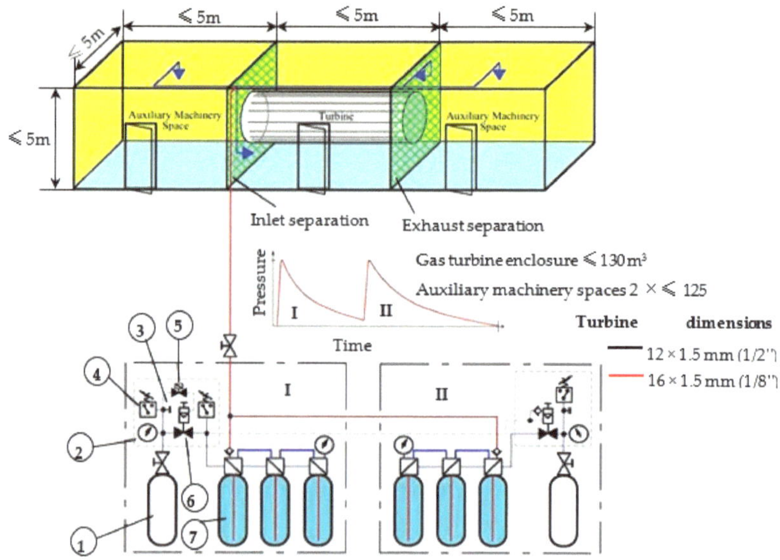

Figure 26. Marifoff HI-FOG installation example for a turbine enclosure [93].

Danfoss SEM-SAFE System

The SEM-SAFE system is a high-pressure water mist system designed to operate at approximately 100 bar of pressure and produce water droplets ranging from 10 to 50 μm in diameter [47]. The general arrangement of this water mist system is shown in Figure 27 FM approval allows this water mist system to be used in datacentres, machinery spaces, corridors, and offices with 5 m high ceilings or less. This water mist system is also suggested for offshore applications.

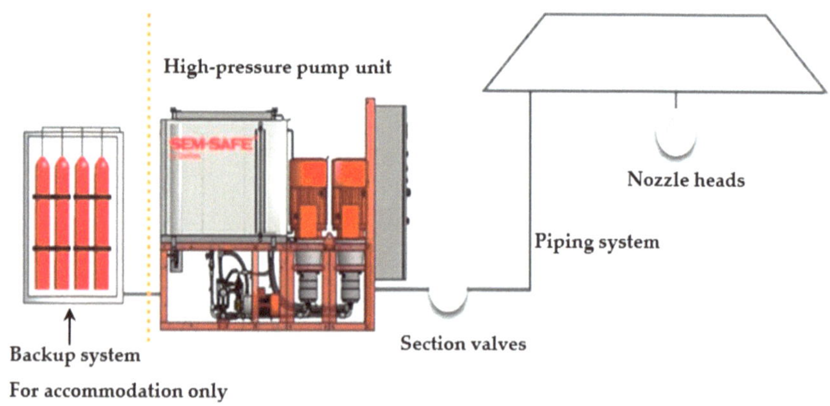

Figure 27. SEM-SAFE water mist system main components [47].

5. Design Codes for Water Mist Systems

Water mist is typically designed under a performance-based approach, as it is not one-size-fits-all. There are some standards that govern the design, installation, and maintenance of water mist fire suppression systems. The relevant codes and standards will depend on the authority with jurisdiction applicable to the geographical location of the system. The main standards currently available for water mist suppression designs are National Fire Protection Association Standard NFPA 750 [8], Australian Standard AS 4587 [94], British Standard BS 8458 [50], and European Standard EN 14972 [95]. In addition, AS5414 [85] has been developed for bushfire water spray systems. The objective of these standards is to provide users of water mist fire protection systems with minimum design, installation, testing, and commissioning requirements. The design sections of the standards direct the users to follow the system's listed requirements. This lack of information and guidelines in the standards can be interpreted as stemming from research gaps related to water mist fire suppression systems. The basic design criteria of the NFPA 750 [8] and AS 4587 [94] water mist design codes are summarised in Table 4. These two standards both specify similar requirements for components such as pipework and fitting material, corrosion protection, the use of strainers, pipe supports, and the use of isolation and check valves, all of which focus on the maintenance and durability of the system to extend the operational life.

Table 4. Water mist design code comparison.

Design Criteria	NFPA 750—2022 [8]	AS 4587—2020 [94]
Pipework and fittings	Maximum pressure developed by the system at 54 °C or higher temperatures as specified in the manufacturer's listing.	Shall be rated or selected for at least the maximum system design pressure at 50 °C.
Nozzle operating pressure	In accordance with the nozzle's listing.	In accordance with the nozzle's listing.
Occupancies	1. Machinery spaces; 2. Combustion turbines; 3. Wet benches and similar processing equipment; 4. Local applications; 5. Industrial oil cookers; 6. Computer room raised floors; 7. Chemical fume hoods; 8. Continuous wood board presses.	1. Machinery spaces; 2. Combustion turbines; 3. Industrial oil cookers; 4. Pool and spray fires—localapplications.
Extinguishing criteria	Approved test plan from the authority with jurisdiction.	Extinguishment in half the prescribed discharge duration.
Discharge duration		Various for open nozzle systems; 60 min for sealed.
Hangers/ Supports	In accordance with NFPA 13 Standard for the installation of sprinkler systems, and shall be listed.	Hanger shall conform to the requirements of AS 2118.1 automatic fire sprinklers.
Nozzle spray angle	The spray angle is set by the listed nozzle used.	The spray angle is set by the listed nozzle used.
Flow rate		Selected as appropriate for the hazard.
Maximum height	The minimum and maximum heights shall be in accordance with the manufacturer's listing.	In accordance with the nozzle's listing.

Table 4. *Cont.*

Design Criteria	NFPA 750—2022 [8]	AS 4587—2020 [94]
Maximum coverage per nozzle	Listing.	In total, 18 m^2 for sealed nozzle systems unless greater in listing.
Maximum spacing of nozzles	In accordance with the nozzle's listing.	In accordance with the nozzle's listing.
Maximum nozzle spacing from walls	In accordance with the nozzle's listing.	In accordance with the nozzle's listing.
Activation method	Detection system, automatic, or manual.	Detection system, automatic, or manual.
Application methods	1. Local application; 2. Total compartment application system; 3. Zoned application system; 4. Occupancy protection system.	1. Local application; 2. Total flooding; 3. Zoned total flooding; 4. Assumed area of operation.
System limitations	Light hazards = 4831 m^2 Ordinary hazards = 4831 m^2	Sealed nozzle system = 9000 m^2 Open nozzle system = as per listing
Temperature rating of nozzle	Temperature rating chosen depending on the maximum ambient temperature. Table for guidance is provided.	Heat rating to be 30 °C above highest ambient temperature.
Water supply	At least one automatic water supply.	Public water, automatic pumps, or containers (pressure cylinders).
Connecting to public water supply	Light hazards—minimum 378 lpm Ordinary Hazards—minimum 946 lpm	NIL
Water droplet size	<1000 μm within the nozzle operating range	<1000 μm at minimum operating pressure.
Acceptable listing authority (test protocol)	JAS-ANS, CSIRO Actifire, FM, LPCB, SP Technical Research Institute of Sweden, UL, VdS	Organisation that is acceptable to the authority with jurisdiction.

Note: Listing is described as equipment in a list published by an organisation (the listing authority) that is acceptable to the relevant authority that states that the equipment meets appropriate designated standards or has been tested and found suitable for a specified purpose against test protocols.

5.1. American Standards

The NFPA Technical Committee on Water Mist Fire Suppression Systems was organised in 1993 by research and engineering communities, water mist system manufacturers, the insurance industry, and enforcement authorities [8]. The NFPA 750 standard was issued and is regularly updated to contain the minimum requirements for the design, installation, maintenance, and testing of water mist protection systems. This standard lists the specific applications of water mist for machinery spaces, combustion turbines, processing equipment, local applications, industrial oil cookers, computer rooms, chemical fume hoods, and continuous wood board presses. The standard includes occupancy protection systems for residential building applications. This standard does not provide definitive fire performance criteria, nor does it offer specific guidance on system design to specifically control, suppress, or extinguish fire [8].

5.2. Australian Standards

The Standards Australia Committee FP-011 for Special Hazard Fire Protection Systems issued their first standard for water mist fire protection systems in 1999 [94]. The AS 4587 standard [94] covers the minimum requirements and recommendations for the design, installation, and commissioning of water mist fire protection systems and describes the characteristics of the various types of systems. It lists the specific applications of water mist for machinery spaces, combustion turbines, industrial oil cookers, and local pool and spray fire applications. The standard does not include residential occupancies or other building applications. The standard does not provide a performance criterion on an application rate or density basis [94]. The standard does not cover water mist systems in residential occupancies or explosion suppression.

AS5414 [85] was developed for ember protection under bushfire conditions and does not adequately address radiant heat exposure or provide design parameters for higher bushfire attack exposures on the critical elements of buildings; rather, it addresses the overall coverage of water with nozzle spacing and locations (including roof ridges).

5.3. British Standards

The British Standards Institution published the BS 8458 standard for fixed fire protection systems for residential and domestic water mist systems in 2015 [50]. The BS 8458 standard covers the design, installation, water supply, commissioning, maintenance, and testing of water mist systems with automatic nozzles installed in both domestic and residential occupancies [50]. This standard does not cover water mist systems used in industrial and commercial buildings. Industrial and commercial water mist systems are covered by BS 8489-1 [96] and BS 8489-7 [55].

5.4. European Standard

The European Committee for Standardization published the European standard for fixed water mist firefighting systems. The EN 14972 standard specifies the requirements and provides recommendations for the design, installation, inspection, and maintenance of all types of fixed, land-based water mist systems [95]. The scope does not cover explosion protection or use within vehicles.

5.5. Test protocols

Water mist systems are performance-based, and they are designed and tested against established fire test protocols that are acceptable to the relevant authority with jurisdiction. Fire test protocols are designed by testing authorities to match or simulate a specific hazard through established performance criteria. Table 5 includes formalised fire test protocols produced by approval agencies that are widely accepted globally [25].

Table 5. Fire test protocol for water mist fire protection systems [25].

Agency	Water Mist Fire Test Protocol
International Maritime Organization (IMO)	IMO Res. A800 (19): Revised Guidelines for Approval of Sprinkler Systems Equivalent to that referred to in SOLAS Regulations II-2, Chap. 12 Appendix 1, "Component Manufacturing Standards for Water Mist Nozzles" Appendix 2, "Fire Test Procedures for Equivalent Sprinkler Systems in Accommodation, Public Space and Service Areas on Passenger Ships", December 1995 IMO MSC/Circular 668: Alternative Arrangements for Halon Fire Extinguishing Systems in Machinery Spaces and Pump Rooms Appendix A: "Component Manufacturing Standards of Equivalent Water-Based Fire Extinguishing Systems", 1994 Appendix B: "Interim Test Method for Fire Testing Equivalent Water-Based Fire Extinguishing Systems for Machinery Spaces of Category A and Cargo Pump Rooms", 1994 As amended in MSC/Circ. 728: "Amendments to the Test Method for Equivalent Water-Based Fire-Extinguishing Systems for Machinery Spaces of Category A and Cargo Pump-Rooms contained in MSC/Circ. 668, Annex B", June 1996 MSC/Circ. 913: "Guidelines for the Approval of Fixed Water-Based Local Application Fire-Fighting Systems for use in Category A Machinery Spaces", 4 June 1999 MSC/Circ. 1165, "Revised Guidelines for the Approval of Equivalent Water-Based Fire-Extinguishing Systems for Machinery Spaces and Cargo Pump-Rooms", 10 June 2005 FM Global, Approval Standard for Water Mist Systems, Class Number 5560, 2009

Table 5. *Cont.*

Agency	Water Mist Fire Test Protocol
FM Global Research Corporation	(a) Appendix A, B, C: Fire Tests for Water Mist Systems for the Protection of Machinery Spaces, Special Hazard Machinery Spaces, Combustion Turbines with Volumes up to, and including, 2825 ft3 (80 m^3) (respectively) (b) Appendices D, E, and F: Fire Tests for Water Mist Systems for the Protection of Machinery Spaces, Special Hazard Machinery Spaces, Combustion Turbines with Volumes up to and including 9175 ft3 (260 m^3) (respectively) (c) Appendix G: Fire Tests for Water Mist Systems for the Protection of Machinery Spaces and Special Hazard Machinery Spaces with Volumes Exceeding 9175 ft3 (260 m^3) (d) Appendix H: Fire Tests for Water Mist Systems for the Protection of Combustion Turbines with Volumes Exceeding 9175 ft3 (260 m^3) (e) Appendix I: Fire Tests for Water Mist Systems for the Protection of Light Hazard Occupancies (f) Appendix J: Fire Tests for Water Mist Systems for the Protection of Wet Benches and Other Similar Processing Equipment (g) Appendix K: Fires Tests for Water Mist Systems for the Protection of Local Applications (h) Appendix L: Fire Tests for Water Mist Systems for the Protection of Industrial Oil Cookers (i) Appendix M: Fire Tests for Water Mist Systems for the Protection of Computer Room Subfloors (j) Appendix N: Other Occupancies Which FM Global Has an Interest in Protecting with Water Mist Systems
Underwriters Laboratories Inc. (UL)	ANSI/UL 2167, Proposed First Edition of the Standard for Water Mist Nozzles for Fire Protection Service, June 1998 Machinery Spaces; Passenger Cabin Fire Tests; Passenger Cabins Greater than 12 m^2; Public Space Fire Tests; Residential Area Fire Tests; Light Hazard Area Fire Tests; Ordinary Hazard I and II Tests; Nozzle Construction Design, Marking, and Performance Requirements.

Note: All appendix contents are available in Ref. [25].

6. Challenges and Issues

The current progress in water mist fire suppression technology and industry use has been established by reviewing the methods of generating water mist suppression, the suppression mechanisms, applications, and key performance parameters; and the applicable design standards. The existing research has shown the current challenges and issues associated with using water mist as a fire suppression solution. The main challenges and issues that were observed based on the literature review are discussed in the below subsections.

6.1. Operation and Maintenance

The main challenge of water mist fire suppression systems is the methods of generating small water droplets using high-pressure discharge and small nozzle orifices, which contributes to technical concerns [29]. These factors combined create the potential for plugging and blockages caused by corrosive products inside the pipework. The alternative for small droplet sizes to be maintained is using a lower pressure when using twin-fluid systems, except that the dual supply lines and storage of water and compressed gas significantly increase the operation and maintenance cost of the water mist fire suppression system [7]. Systems could require more frequent maintenance than other fire suppression systems and the design of the system could require more effort to suit the environment within which it will be installed. It has been noted in current research that there are several maintenance and inspection questions that need to be examined, such as how often should piping be flushed, strainers be cleaned or changed, and pumps be tested [64]. Unfortunately, long-term maintenance history data are largely unknown or unproven for water mist systems [54]. The literature recommends utilizing existing automatic water sprinkler maintenance guidelines for water mist systems. More appropriate guidelines should be written specifically for individual types of water mist systems, as they vary depending on the application. This should be written by the manufacturers or the authority with jurisdiction in the form of a legislated standard.

6.2. Design and Standardisation

It has been established that the physical phenomenon of fire suppression by water mist is complex. Extensive research has shown that there are a lot of factors contributing to the suppression of fire, including water droplet size [9,19,29,52,56,60–62], spray momentum [73], nozzle angle [5,25,94], nozzle height [5,9,65], compartment size [6,9], ventilation [6,9], burning fuel load [18,57,95], and the spray nozzle activation method [97,98]. The varying effects of these parameters make it difficult to design a water mist suppression system by applying first principles. There is currently no theoretical basis for considering all these parameters in the design process, and a standardised approach to system design is yet to be fully developed. Designers must apply a performance-based approach if the application does not fall under an approved solution. When completing a performance-based solution, the complexities of water mist fire suppression systems are difficult to model accurately using computational fire dynamic simulation software due to the low amount of research data available to extract input parameters to model accurate water mist fire suppression scenarios. For a performance-based assessment to be verified and supported, fire testing will sometimes be required to validate the model. Design parameters are normally established from large-scale fire test data on a case-by-case basis for specific applications. This method of design can make a water mist system less cost-effective, which restricts the expansion of water mist fire suppression applications.

6.3. Application Challenges

Though water mist has proven to be an effective fire suppression solution for a variety of fire scenarios, its use is restricted in many locations around the world to industrial applications by product listings, and it cannot be used in commercial or residential buildings. We found that water mist is becoming more commonplace for residential use in areas that follow the NFPA and British Standards. Before it is accepted in Australia and other parts of the world, an increase in performance data and historical evidence of its ability to control fires need to be established [99,100]. This will provide engineers and approval authorities with jurisdiction with the ability to assess if a water mist system is a suitable fire protection solution for the specific application. When considering the approval of water mist systems for residential or commercial applications regarding life safety, the information available is limited [99,100]. To avoid the complex design requirements of large-scale fire testing to prove water mist works in specific applications, it is only used in applications where test data have already been established, such as machinery spaces, turbine enclosures, marine accommodation spaces, and industrial cookers. There is potential for water mist to suit other applications such as light-hazard occupancies, electronic environments, storage facilities, and residential and commercial buildings, as well as bushfire protection [86]. More efforts in developing test protocols for these applications are required.

6.4. Economic Challenges

The economic issues resulting from complex design and system components have been mentioned in different studies [4,44,77]. There are multiple methods of generating water mist, though they each come with expenses that can make water mist a costly option [77] when compared with alternative suppression methods, such as standard sprinklers or gaseous systems [44]. As long as gaseous agents are available, affordable, and permitted for use, water mist may remain an expensive option for asset protection. Further research can be performed to minimise the total cost of water mist fire suppression systems.

7. Conclusions and Recommendations

This literature review analysed previous research and industry data to summarise the suppression mechanisms, applications, products, design standards, test protocols and limitations of water mist fire suppression systems. The ability of water mist systems to suppress and control a fire by cooling, displacing oxygen, and attenuating radiation has made it an effective fire suppression solution for many applications, although there are some limitations. Various products have been designed for specific applications, typically in industrial and marine environments. Challenges and issues still exist around using water mist in new and emerging fire risk applications. It is recommended that further research be conducted that focuses on the performance of water mist in various fire scenarios to better understand and improve water mist systems to make it a suitable and cost-effective fire protection solution in building and industrial applications. The following suggestions can be considered to enhance the potential use of water mist fire suppression systems in building and industrial applications.

(a) Further research can be conducted to develop corrosion-resistant materials and other methods of reducing the potential to block small orifice nozzles.
(b) More research on the performance of water mist suppression at various nozzle heights, fire compartment configurations, and various fuel types and fire scenarios can be conducted.
(c) A theoretical basis can be developed for the design process that can be applied to a wide range of hazards and can be used for a prescriptive design and installation standard.
(d) New test protocols should be developed and tested for water mist in light-hazard occupancies and residential building spaces as an alternative to fire sprinklers.
(e) Environmental benefits associated with water mist can be investigated for an alternative to gaseous suppression systems across the life cycle of products.
(f) Research should be conducted to look at the potential water mist may have in assisting in fire protection in new and emerging fire hazards that the industry is trying to adapt to, such as bushfire protection, combustible cladding, electric battery storage facilities, electric cars, automated warehouse facilities, switchboards, datacentres, and compact residential buildings.

Author Contributions: Conceptualization, M.K.H. and K.F.; Formal analysis, K.F., M.K.H. and M.D.H.; Investigation, K.F.; Methodology, M.K.H., M.D.H. and K.F.; Resources, M.K.H. and S.S.; Supervision, M.K.H. and S.S.; Visualization, M.K.H. and M.D.H.; Writing—original draft, K.F., M.D.H. and M.K.H.; Writing—review and editing, P.R., G.D., S.S. and B.A. All authors have read and agreed to the published version of the manuscript.

Funding: This research received no external funding.

Acknowledgments: The authors would like to thank Western Sydney University for providing access to academic literature and standards.

Conflicts of Interest: There are no conflict of interest.

Nomenclature:

μm	Micrometre or micron	I_c	Critical radiation intensity required for piloted ignition
O_2	Oxygen	H_{vap}	Heat vaporisation of water
K	Kelvin	P	Pressure inlet at nozzle
$T_{f,crit}$	Flame extinguishing temperature	Q	Nozzle flow rate
$T_{f,max}$	Theoretical flame temperature	k	Discharge coefficient K-factor
T_{ls}	Surface temperature of fuel	M_w	Spray momentum
m_{og}	Mass fraction of oxygen	m_{wl}	Mass of liquid-phase water
m_{fs}	Mass fraction of fuel vapour	m_{wv}	Mass of vapour-phase water
r	Stoichiometric ratio	m_{wa}	Mass of air entrained by mist
S_h	Heat removed per unit area	V_w	Velocity vector of water mist

H_f	Convective heat transfer per unit area	°C	Degree Celsius
λ_f	Heat required to produce a unit mass of vapour	kg	Kilograms
$\dot{m}_b$	Burning rate per unit area	kJ	Kilojoules
R_a	Heat transfer to the fuel surface	H_2O	Dihydrogen monoxide
R_s	Heat lost from the surface	C_2H_2	Acetylene
F_m	Minimum water flow rate	CH_2	Methylene
A_s	Fuel surface area	CO	Carbon monoxide
ε	Emissivity of the radiator	m	Meter
σ	Stefan–Boltzmann constant	CO_2	Carbon dioxide
ϕ	View factor of the fuel load		
T_r	Mean absolute temperature of radiation source		
T_s	Mean absolute temperature of the surface		

Abbreviations:

ULF	Ultra-low flow
FM	Factory Mutual
HRR	Heat release rate
NFPA	National Fire Protection Association
AS	Australian Standard
FDS	Fire Dynamics Simulator
LOC	Limiting oxygen concentration
PMMA	Polymethyl methacrylate
Lpm	Litres per minute
MC	Multi-component

References

1. Cui, Y.; Liu, J. Research progress of water mist fire extinguishing technology and its application in battery fires. *Process Saf. Environ. Prot.* **2021**, *149*, 559–574. [CrossRef]
2. Adiga, K.; Hatcher, R.F., Jr.; Sheinson, R.S.; Williams, F.W.; Ayers, S. A computational and experimental study of ultra fine water mist as a total flooding agent. *Fire Saf. J.* **2007**, *42*, 150–160. [CrossRef]
3. Mahmud, H. Simulation of the Suppression of Fires Using Water Mists. Ph.D. Thesis, Victoria University, Melbourne, VIC, Australia, 2016.
4. Yazir, D. Application of IF-TOPSIS method on fixed fire fighting systems for cargo hold fires on the dry/bulk cargo ships. *Ocean Eng.* **2022**, *260*, 111891. [CrossRef]
5. Ko, Y.J.; Elsagan, N.; Gibbs, E. *Water Mist Systems for Protection of Mass Timber Structures—Phase 2 Residential Fire Suppression Tests*; National Research Council Canada: Ottawa, ON, Canada, 2020.
6. Liu, Z.; Kim, A.K. A review of water mist fire suppression systems—Fundamental studies. *J. Fire Prot. Eng.* **1999**, *10*, 32–50.
7. Liu, J.; Liao, G.; Li, P.; Fan, W.; Lu, Q. Progress in research and application of water mist fire suppression technology. *Chin. Sci. Bull.* **2003**, *48*, 718–725. [CrossRef]
8. *NFPA 750*; Standard on Water Mist Fire Protection Systems. 2022.
9. Mawhinney, J.R.; Back, G.G. Water mist fire suppression systems. In *SFPE Handbook of Fire Protection Engineering*; Springer: New York, NY, USA, 2016; pp. 1587–1645.
10. Chang, W.Y.; Fu, P.K.; Chen, C.H.; Shu, Y.L. Performance evaluation of a water mist system in semiconductor wet bench fires. *Process Saf. Environ. Prot.* **2008**, *86*, 213–218. [CrossRef]
11. Nmira, F.; Consalvi, J.L.; Kaiss, A.; Fernandez-Pello, A.; Porterie, B. A numerical study of water mist mitigation of tunnel fires. *Fire Saf. J.* **2009**, *44*, 198–211. [CrossRef]
12. Bill, R., Jr.; Hansen, R.L.; Richards, K. Fine-spray (water mist) protection of shipboard engine rooms. *Fire Saf. J.* **1997**, *29*, 317–336. [CrossRef]
13. Lee, J. Numerical analysis on the rapid fire suppression using a water mist nozzle in a fire compartment with a door opening. *Nucl. Eng. Technol.* **2019**, *51*, 410–423. [CrossRef]
14. Log, T.; Cannon-Brookes, P. Water mist for fire protection of historic buildings and museums. *Mus. Manag. Curatorship* **1995**, *14*, 283–298.
15. Sturm, P.; Pucher, K.; Rodler, J.; Mühlenbruch, G.; Krezschmar, P. Test of Water Mist Equipment to increase Escape Chances in Cases of Fire in Tunnels. In Proceedings of the International Water Mist Association Conference, Vienna, Austria, 4–6 April 2001.
16. Lee, J. Numerical analysis of how ventilation conditions impact compartment fire suppression by water mist. *Ann. Nucl. Energy* **2020**, *136*, 107021. [CrossRef]

17. Cao, X.Y.; Bi, M.S.; Ren, J.S.; Chen, B. Experimental research on explosion suppression affected by ultrafine water mist containing different additives. *J. Hazard. Mater.* **2019**, *368*, 613–620. [CrossRef] [PubMed]
18. Cong, B.; Liao, G. Experimental studies on water mist suppression of liquid fires with and without additives. *J. Fire Sci.* **2009**, *27*, 101–123. [CrossRef]
19. Ferng, Y.M.; Liu, C.H. Numerically investigating fire suppression mechanisms for the water mist with various droplet size through FDS code. *Nucl. Eng. Des.* **2011**, *241*, 3142–3148. [CrossRef]
20. Yinshui, L.; Zhuo, J.; Dan, W.; Xiaohui, L. Experimental research on the water mist fire suppression performance in an enclosed space by changing the characteristics of nozzles. *Exp. Therm. Fluid Sci.* **2014**, *52*, 174–181. [CrossRef]
21. TycoAquaMist. Tyco Fire Protection Products. Available online: https://tycoaquamist.com/uploads/media/AquaMist-Machinery_Space_EMEA_0317_LoRes.pdf (accessed on 3 January 2023).
22. Drysdale, D. *An Introduction to Fire Dynamics*; John Wiley & Sons: New York, NY, USA, 2011.
23. Yang, D.; Converse, M.C.; Mahvi, D.M.; Webster, J.G. Expanding the bioheat equation to include tissue internal water evaporation during heating. *IEEE Trans. Biomed. Eng.* **2007**, *54*, 1382–1388. [CrossRef]
24. Drysdale, D. Ignition of liquids. In *SFPE Handbook of Fire Protection Engineerings*; Springer: New York, NY, USA, 2016; pp. 554–580.
25. Hurley, M.J.; Gottuk, D.T.; Hall, J.R., Jr.; Harada, K.; Kuligowski, E.D.; Puchovsky, M.; Watts, J.M., Jr.; Wieczorek, C.J. *SFPE Handbook of Fire Protection Engineering*; Springer: New York, NY, USA, 2015.
26. Majdalani, A.H.; Carvel, R. *Effectiveness of the Gas Cooling Technique in Larger Compartment Fires*; Final Report; 2016.
27. Xin, Y.; Khan, M.M. Flammability of combustible materials in reduced oxygen environment. *Fire Saf. J.* **2007**, *42*, 536–547. [CrossRef]
28. Mawhinney, J. Water mist fire suppression systems for marine applications: A case study. In *The TRIS and ITRD Database*; Institute of Marine Engineers IMAS 94: Hong Kong, 1994.
29. Liu, H.; Wang, C.; Cordeiro, I.M.D.C.; Yuen, A.C.Y.; Chen, Q.; Chan, Q.N.; Kook, S.; Yeoh, G.H. Critical assessment on operating water droplet sizes for fire sprinkler and water mist systems. *J. Build. Eng.* **2020**, *28*, 100999. [CrossRef]
30. Kashiwagi, T.; Ohlemiller, T.J. A study of oxygen effects on nonflaming transient gasification of PMMA and PE during thermal irradiation. In *Symposium (International) on Combustion*; Elsevier: Amsterdam, The Netherlands, 1982; pp. 815–823.
31. Balner, D.; Barcova, K. Attenuation of thermal radiation through water mist. *Process Saf. Prog.* **2018**, *37*, 18–24. [CrossRef]
32. Choi, B.-I.; Han, Y.-S.; Oh, C.-B.; Kim, M.-B.; Kim, C. The assessment of fire suppression capability of water-mist system for machinery engine room. *Fire Sci. Eng.* **2007**, *21*, 111–117.
33. Liu, Z.; Kim, A.; Su, J. Improvement of efficacy of water mist in fire suppression by cycling discharges. In Proceedings of the Second International Conference on Fire Research and Engineering, Gaithersburg, MD, USA, 10–15 August 1997; pp. 3–8.
34. Liu, Z.; Kim, A.; Su, J. The effect of air convection on the performance of water mist fire suppression systems. *Asme-Publ.-Htd* **1998**, *357*, 227–236.
35. Yudong, F.; Jianghong, L.; Guangxuan, L.; Qiang, X.; Xiaomeng, Z. Experimental studies on electrical breakdown field strength of electrode with water mist containing MC additives. *Chin. Sci. Bull.* **2005**, *50*, 2783–2788. [CrossRef]
36. Lu, J.; Chen, B.-H.; Fang, Z.; Hu, J.; Wang, B.; Wu, C.; Huang, S. Electrical safety of suppressing wildfires near high-voltage transmission lines using water mist. *J. Fire Sci.* **2018**, *36*, 295–314. [CrossRef]
37. Hills, A.T.; Simpson, T.; Smith, D.P. Water mist fire protection systems for telecommunication switch gear and other electronic facilities. In Proceedings of the Water Mist Fire Suppression Workshop, Gaithersburg, MD, USA, 1–2 March 1993; pp. 123–144.
38. Liu, Z.; Kim, A.K. A review of water mist fire suppression technology: Part II—Application studies. *J. Fire Prot. Eng.* **2001**, *11*, 16–42. [CrossRef]
39. Zhou, Y.; Bu, R.; Gong, J.; Zhang, X.; Fan, C.; Wang, X. Assessment of a clean and efficient fire-extinguishing technique: Continuous and cycling discharge water mist system. *J. Clean. Prod.* **2018**, *182*, 682–693. [CrossRef]
40. Arvidson, M. Large-scale water spray and water mist fire suppression system tests for the protection of Ro–Ro cargo decks on ships. *Fire Technol.* **2014**, *50*, 589–610. [CrossRef]
41. Marker, T.R.; Sarkos, C.P.; Hill, R.G. Water Spray System Development and Evaluation for Enhanced Postcrash Fire Survivability and In-Flight Protection in Cargo Compartments. In *Federal Aviation Administration, Aircraft Fire Safety*; 1997; Volume 1997, pp. 1–3.
42. Sarkos, C.P. Development of improved fire safety standards adopted by the Federal Aviation Administration. In *Agard, Aircraft Fire Safety 14 (See N 90-17581 10-03)*; 1989.
43. Mawhinney, J.R. *Approval Testing of a Hi-Fog Water Mist System for Protection of the A86 Passenger Vehicle Highway Tunnel*; Marioff Corporation Oy: Bern, Switzerland, 2005.
44. Ha, G.; Shin, W.G.; Lee, J. Numerical analysis to determine fire suppression time for multiple water mist nozzles in a large fire test compartment. *Nucl. Eng. Technol.* **2021**, *53*, 1157–1166. [CrossRef]
45. Liu, Z.; Carpenter, D.; Kim, A.K. Characteristics of large cooking oil pool fires and their extinguishment by water mist. *J. Loss Prev. Process Ind.* **2006**, *19*, 516–526. [CrossRef]
46. Qin, J.; Yao, B.; Chow, W.K. Experimental study of suppressing cooking oil fire with water mist using a cone calorimeter. *Int. J. Hosp. Manag.* **2004**, *23*, 545–556. [CrossRef]
47. Systems S-SFS. SEM-SAFE by Danfoss High-Pressure Water Mist Fire Fighting System. 2022. Available online: https://www.danfoss.com/en/about-danfoss/our-businesses/fire-safety/ (accessed on 3 January 2023).

48. Xu, J.; Duan, Q.; Zhang, L.; Liu, Y.; Sun, J.; Wang, Q. The enhanced cooling effect of water mist with additives on inhibiting lithium ion battery thermal runaway. *J. Loss Prev. Process Ind.* **2022**, *77*, 104784. [CrossRef]
49. Jenft, A.; Collin, A.; Boulet, P.; Pianet, G.; Breton, A.; Muller, A. Experimental and numerical study of pool fire suppression using water mist. *Fire Saf. J.* **2014**, *67*, 1–12. [CrossRef]
50. British Standards BS 8458:2015: Fixed Fire Protection Systems—Residential and Domestic Water Mist Systems—Code of Practice for Design and Installation. NBS Database: 2015.
51. Milke, J.; Gerschefski, C. Overview of Water Research for Library Applications. In Proceedings of the International Conference on Fire Research and Engineering, Orlando, FL, USA, 10–15 September 1995; p. 133.
52. Wang, Z.; Wang, X.; Huang, Y.; Tao, C.; Zhang, H. Experimental study on fire smoke control using water mist curtain in channel. *J. Hazard. Mater.* **2018**, *342*, 231–241. [CrossRef] [PubMed]
53. Budnick, E. *USFA Residential Water Mist Test Program (Initial Feasibility)*; National Institute of Standards and Technology: Gaithersburg, MD, USA, 1996.
54. Arvidson, M. An evaluation of residential sprinklers and water mist nozzles in a residential area fire scenario, Research Institutions of Sweden. *RISE Rep.* **2017**, *4*, 2017.
55. British Standards BS 8489-7 2016: Fixed Fire Protection Systems—Industrial and Commercial Watermist Systems-Part: Fire performance tests and requirements for watermist systems for the protection of low hazard occupancies, British Standards Institution.
56. Huang, X.; Wang, X.; Liao, G. Characterization of an effervescent atomization water mist nozzle and its fire suppression tests. *Proc. Combust. Inst.* **2011**, *33*, 2573–2579. [CrossRef]
57. Liu, Z.; Kim, A.K.; Su, J.Z. Examination of the extinguishment performance of a water mist system using continuous and cycling discharges. *Fire Technol.* **1999**, *35*, 336–361. [CrossRef]
58. Qin, J.; Weng, W. Preliminary study of water mist suppressing ghee flame in historical building in the northwest China. *J. Cult. Herit.* **2006**, *7*, 329–333. [CrossRef]
59. Xiaomeng, Z.; Biao, Z.; Xiang, J. Study of fire-extinguishing performance of portable water-mist fire extinguisher in historical buildings. *J. Cult. Herit.* **2010**, *11*, 392–397. [CrossRef]
60. Prasad, K.; Patnaik, G.; Kailasanath, K. A numerical study of water-mist suppression of large scale compartment fires. *Fire Saf. J.* **2002**, *37*, 569–589. [CrossRef]
61. Mahmud, H.I.; Moinuddin, K.A.; Thorpe, G.R. Experimental and numerical study of high-pressure water-mist nozzle sprays. *Fire Saf. J.* **2016**, *81*, 109–117. [CrossRef]
62. Jeong, C.S.; Lee, C.Y. Experimental investigation on spray characteristics of twin-fluid nozzle for water mist and its heptane pool fire extinguishing performance. *Process Saf. Environ. Prot.* **2021**, *148*, 724–736. [CrossRef]
63. Zhang, P.; Tang, X.; Tian, X.; Liu, C.; Zhong, M. Experimental study on the interaction between fire and water mist in long and narrow spaces. *Appl. Therm. Eng.* **2016**, *94*, 706–714. [CrossRef]
64. Zalosh, R. Water Mist for Deflagration Control. In *Water Mist for Deflagration Control*; National Fire Protection Association: Quincy, MA, USA, 1994.
65. Shrigondekar, H.; Chowdhury, A.; Prabhu, S. Characterization of a simplex water mist nozzle and its performance in extinguishing liquid pool fire. *Exp. Therm. Fluid Sci.* **2018**, *93*, 441–455. [CrossRef]
66. Yu, H.Z. The ceiling height limits for water mist protection of two solid combustible fire hazards in open environment. *Fire Technol.* **2021**, *57*, 163–187. [CrossRef]
67. Chow, W.K.; Yao, B. The potential application of water mist systems for fire protection in atria. *ASHRAE Trans.* **2001**, *107*, 171.
68. Moore, T.A.; Weitz, C.; McCormick, S.; Clauson, M. Laboratory Optimization and Mediumscale Screening of Iodide Salts and Water Mixtures. In Proceedings of Halon Option Technical Working Conference, Albuquerque, NM, USA, 7–9 May 1996; pp. 477–498.
69. DeSipio, P. Evaluation of Fine Water Mist for Aircraft Dry Bay Fire Suppression. 1996. p. 27. Available online: https://www.nist.gov/system/files/documents/el/fire_research/R0301051.pdf (accessed on 10 July 2022).
70. Finnerty, A.E. Water-Based Fire-Extinguishing Agents. 1995. p. 461. Available online: https://www.nist.gov/system/files/documents/el/fire_research/R0000244.pdf (accessed on 15 August 2022).
71. Yu, H.-Z.; Kasiski, R.; Daelhousen, M. Characterization of twin-fluid (water mist and inert gas) fire extinguishing systems by testing and modeling. *Fire Technol.* **2015**, *51*, 923–950. [CrossRef]
72. Pei, B.; Li, J.; Wang, Y.; Wen, X.; Yu, M.; Jing, G. Synergistic inhibition effect on methane/air explosions by N2-twin-fluid water mist containing sodium chloride additive. *Fuel* **2019**, *253*, 361–368. [CrossRef]
73. Nam, S. Numerical simulation of the penetration capability of sprinkler sprays. *Fire Saf. J.* **1999**, *32*, 307–329. [CrossRef]
74. Yu, H.-Z.; Zhou, X.; Carpenter, J. Physical scaling of water mist fire extinguishment in industrial machinery enclosures. *Fire Saf. J.* **2017**, *91*, 596–605. [CrossRef]
75. Pepi, J.S. Advances in the technology of intermediate pressure water mist systems for the protection of flammable liquid hazards. In Proceedings of the Halon Options Technical Working Conference, Albuquerque, NM, USA, 12–14 May 1998.
76. Tanner, G.; Knasiak, K.F. Spray characterization of typical fire suppression nozzles. In Proceedings of the Third International Water Mist Conference, Madrid, Spain, 22–24 September 2003.
77. Santangelo, P.E.; Tartarini, P. Fire control and suppression by water-mist systems. *Open Thermodyn. J.* **2010**, *4*, 167–184. [CrossRef]

78. Pokorný, M.; Eliáš, M.; Kregl, F. Fire Protection of Steel Structures by Low Pressure Water Mist in Large-Scale Fire Test. In *Conference Series: Earth Environmental Science*; IOP Publishing: Bristol, UK, 2019; p. 012028.
79. Sikanen, T.; Vaari, J.; Hostikka, S. Large scale simulation of high pressure water mist systems. *WIT Trans. Eng. Sci.* **2013**, *79*, 77–88.
80. Vaari, J. A transient one-zone computer model for total flooding water mist fire suppression in ventilated enclosures. *Fire Saf. J.* **2002**, *37*, 229–257. [CrossRef]
81. Kim, S.C.; Ryou, H.S. An experimental and numerical study on fire suppression using a water mist in an enclosure. *Build. Environ.* **2003**, *38*, 1309–1316. [CrossRef]
82. Lui, Z.; Kim, A. A Review of water mist fire suppression technology. Part II—Fundamental Studies. *J. Fire Prot. Eng.* **2001**, *10*, 16–42.
83. Hansen, R.; Back, G. Fine Water Mist–Fire Control Using Local Application. In Proceedings of the 1997 International CFC and Halon Alternative Conference, Washington, DC, USA; 1997.
84. He, Y.; Kwok, K.; Douglas, G.; Razali, I. Numerical Investigation of Bushfire-Wind Interactions and its Impacts on Building Structure. *Fire Saf. Sci.* **2011**, *11*, 1449–1462. [CrossRef]
85. Australian Standard AS5414-2012: Bushfire Water Spray Systems. *SAI Global*. Accessed 6 November 2020.
86. Green, A. Sprinkler Systems for Protection of Buildings from Wildfire. Ph.D. Thesis, Wollongong University, Hong Kong, 2019.
87. Yang, P.; Liu, T.; Qin, X. Experimental and numerical study on water mist suppression system on room fire. *Build. Environ.* **2010**, *45*, 2309–2316. [CrossRef]
88. Vaari, J.; Hostikka, S.; Sikanen, T.; Paajanen, A. Numerical simulations on the performance of water-based fire suppression systems. *VTT Technol.* **2012**, *54*, 150.
89. FireKill. Atrium Fire Protection System Model APS VID FireKill. Available online: https://vidfirekill.dk/solutions/solutions/] (accessed on 15 June 2022).
90. Maioff Hi-FOG Water Mist System. Available online: https://www.suppression.com/uploads/pdf/suppression-kidde-hifog/accumulator-units/mau/ls01551.pdf (accessed on 15 June 2022).
91. AquaMist. AquaMist Mist Control Center (MCC) Pump Skin Unit, DATASHEET TFP2270.1 OCTOBER 2021. Available online: https://tycoaquamist.com/uploads/media/TFP2270.1_10_2021_v2_.pdf (accessed on 15 August 2022).
92. Fire, L. *Installation, Maintenance and User Manual for Water Mist Cylinder Bank Systems*; 2009.
93. Luciano Nigro. Performance Based Fire Protection With Water Mist. Available online: https://iwma.net/fileadmin/user_upload/Seminar_Dubai_2013/Marioff_Nigro_Dubai_2013.pdf (accessed on 15 August 2022).
94. Australian Standard. AS 4587:2020: Water Mist Fire Protection Systems—System Design, Installation, and Commissioning. SAI Global Database, 2020.
95. European Standards EN 14972-1:2020: Fixed Fire-Fighting Systems—Water Mist Systems—Part 1: Design, Installation, Inspection and Maintenance. *European Standards EN 14972-1:2020*; CEN: Oak Brook, IL, USA, 2020.
96. British Standards BS 8489-1 2016: Fixed Fire Protection Systems. Industrial and Commercial Water Mist Systems—Code of Practice for Design and Installation.
97. Hopkin, C.; Spearpoint, M.; Muhammad, Y.; Makant, W. Estimating the suppression performance of an electronically controlled residential water mist system from BS 8458: 2015 fire test data. *Fire* **2022**, *5*, 144. [CrossRef]
98. Spearpoint, M.; Hopkin, C.; Muhammad, Y.; Makant, W. Replicating the activation time of electronically controlled water mist system nozzles in B-RISK. *Fire Saf. J.* **2022**, *130*, 103592. [CrossRef]
99. Frank, K.; Gravestock, N.; Spearpoint, M.; Fleischmann, C. A review of sprinkler system effectiveness studies. *Fire Sci. Rev.* **2013**, *2*, 6. [CrossRef]
100. Gravestock, N. *Effectiveness of Fire Safety Systems for Use in Quantitative Risk Assessments*; New Zealand Fire Service Commission: Wellington, New Zealand, 2008.

www.ingramcontent.com/pod-product-compliance
Lightning Source LLC
Chambersburg PA
CBHW041409210926
51850CB00013B/3222
* 9 7 8 3 7 2 5 8 2 4 0 1 4 *

MDPI AG
Grosspeteranlage 5
4052 Basel
Switzerland
Tel.: +41 61 683 77 34

Fire Editorial Office
E-mail: fire@mdpi.com
www.mdpi.com/journal/fire